Volvo 120
Amazon
Ultimate Portfolio

Compiled by R M Clarke

ISBN 1 85520 4053

BROOKLANDS BOOKS LTD.
P.O. BOX 146, COBHAM,
SURREY, KT11 1LG. UK

A-VO56UP

Printed in Hong Kong

ACKNOWLEDGEMENTS

Regular readers of the Brooklands Books Road Test series will realise that we have had a book about the Volvo 120s in our lists before. We would like to reassure them that this Ultimate Portfolio is not simply a reprint of that book as a wealth of new material has been added to this larger volume and only the better articles from the earlier edition have been retained.

Our thanks go to Thomas Salomonsson and his colleagues at AB Lafri of Karlskrona, Sweden, who went to great lengths to assist us with this book. They also go to the Svenska Volvo Amazon Klubben for supplying us with a selection of excellent photographs for our covers, the majority of which were professionally taken by Johan Skarner and under the most difficult of conditions. Finally our thanks go to writer James Taylor, who has kindly penned a short introduction to the model.

This volume would not have been possible without the co-operation of those publishers whose copyright material we include here. We are pleased to express our gratitude to the managements of *Australian Motor Sports, Autocar, Autosport, Auto Topics, Car and Driver, Car Life, Car South Africa, Classic Cars, Classic and Sportscar, Foreign Car Guide, Modern Motor, Motor, Motor Life, Motor Sport, Motor Trend, Popular Classics, Popular Imported Cars, Practical Motorist, Road & Track, Sporting Motorist, Sports Car Graphic, Sports Car World, Technicar, Track and Traffic* and *Wheels* for again supporting our reference series.

R.M. Clarke

Volvo built half a million Amazon saloons and estates between 1956 and 1970 - a remarkable testimony to the cars appeal worldwide. Building on their reputation for toughness which the PV444 and PV544 models had established, the Amazons added greater interior room and - perhaps surprising to those more used to the Volvos of the 1970s and 1980s - real good looks.

These Volvos were quick, too. The very first versions produced only 66bhp from their B16 engines, but the last cars had the enlarged B20 which delivered 115bhp - and tuning specialists were able to increase even that figure quite substantially without compromising the basic integrity of the cars. Handling - a Volvo strong point in the Fifties and Sixties - was excellent, and today, a 123 GT or a late 122S remains an enjoyable car to drive, and one which really is capable of keeping up with the demands of modern traffic.

Although the Amazons were used competitively, they never quite emulated the successes of their PV444 and PV544 forebears. As owners can testify, this was not because they had gone soft; it was simply that the opposition was so much stronger than it had been in the heyday of the earlier models.

At long last, people other than Volvo devotees are beginning to appreciate the Amazon's virtues. Fortunately for them, the model's legendary longevity means that there are still plenty of examples around for them to enjoy. But for those Doubting Thomases who remain unconvinced, this book provides a first-class introduction to the range. It is, I think, difficult to come away from reading it without having a pretty clear idea of what made the Amazon so good.

James Taylor

Front Cover and Back Cover Centre - 1967 Volvo Amazon 123 GT owned by Jan-Inge Persson.
Back Cover Top Left - 1959 Volvo Amazon 121 owned by Glenn Boysen.
Back Cover Top Right - Unknown Amazon - Photographed in Penang 1990 by the Editor.
Back Cover Bottom - Volvo Amazon 1969 221 Estate Police Car owned by Ingmar Rydh.

CONTENTS

VOLVOs for Earls Court

Four-door Saloon to be Added to Range for 1957

LATEST creation of the Volvo company, the Amazon model to be shown at Earls Court will go into production alongside the existing saloon and 2-seater next spring. This stylish newcomer from Sweden has a 1.6-litre o.h.v. engine developing 60 b.h.p. in a four-door saloon body.

a power output of 60 b.h.p. at 4,500 r.p.m., with a single carburetter.

Modern styling has not been allowed to detract from the practical virtues of this new Volvo. Ground clearance is generous, the two front seats are individually adjustable, and a central lever continues to be used for control of the three-speed gearbox. Proven features of the 1.4-litre Volvo, such as pushrod-operated overhead valves in the engine, and use of coil springs for both front and rear suspension, are to be continued on this new model.

FOR the first time, the list of exhibitors for the London Motor Show includes this year the name of Volvo. Three types of car will be shown by Sweden's largest car factory, and are illustrated on this page.

At present and for next season, production of private cars at the Gothenburg factory of A B Volvo is of the type P V 444, a 1.4-litre saloon of two-door type which was the subject of a *Motor* Road Test Report on April 4, 1956. At Earls Court this car will appear in the modified form in which it has recently been very successful in America, with the high-compression twin-carburetter sports version of the engine which raises the power output from 51 to 70 b.h.p.

Also on show will be the Volvo sports two-seater which, with its comfortable furnishing and promise of good performance, will undoubtedly arouse considerable interest.

Centre of attraction, however, will be the new 1.6-litre Volvo "Amazon," shown in prototype form, which will go into production early in 1957 as a more expensive alternative to the type 444. Photographs show that this new car is very modern in appearance, the use of four doors giving it an appeal distinct from that of the existing two-door model. Enlargement of the engine by 14% and use of a compression ratio of 7.5/1 provide

CONTINUED in production, the sports 2-seater and type PV444 two-door saloon will both be shown at Earls Court, the latter in American form with twin - carburetter sports 1.4-litre engine.

The divided grille is a distinguishing feature of the Amazon. Two-colour paintwork and whitewall tyres are standard

Volvo Amazon

A HIGH quality family four-seater of modern appearance, giving a comfortable ride and having an 85 b.h.p. engine producing a top speed of 94 m.p.h. with acceleration to match: these are qualities which help to make the Volvo Amazon one of the most interesting cars in the up-to-1,600 c.c. class. Sensitive steering, a fine four-speed gear box with central lever, and good road-holding qualities are further attractions.

The Amazon was first driven by *The Autocar* staff in its native Sweden just before its announcement in 1956, and again in the same country earlier this year. The introduction of the car to production was not hurried, but now the assembly lines in Gothenburg are in full swing, and it has been possible to complete the first comprehensive test of the Export model. The Volvo company arranged for their Dutch agents to provide a car from their stock, run it in, and hand it over in The Hague for test. So Holland and Belgium became the testing grounds after Automobiel Maatschappij "De Nieuwe Haagsche," N.V., 10–11, Koninginnegracht, The Hague, had delivered the car, appropriately on the day which saw the outright victory of a PV444 Volvo in the Tulip Rally.

Evolution of post-war Volvo models was seen in a series of PV444s, culminating in the announcement of the Amazon, each being an improved successor or alternative to the one already familiar. The engine size was originally 1.4-litre giving, when tested by *The Autocar* early in 1954, 44 b.h.p. In 1956 a 444 California saloon having similar body shape and engine size but developing 70 b.h.p. was tested in Sweden. Then, in June of last year, a test was made of the 444, with sports engine giving 85 b.h.p. from the new, larger 1.6-litre unit. This is the engine used in the Amazon. It is available on the Swedish home market in detuned, 60 b.h.p. form, but the Export model has twin S.U. carburettors, and the 85 b.h.p. output is standard. The 444 may still be obtained with the same engine and, partly because the car has been in production with virtually unchanged chassis layout and suspension for many years, it is cheaper.

Among the attractions of the Amazon the performance must rank as the most outstanding, not only because of the impressive data recorded in full on another page but

because, unlike so many of its rivals, the Volvo engine appears prepared to give maximum power for long periods with consummate ease. During the test, which covered several hundred miles of really fast roadwork, including runs over the length of the Brussels-Ostend motorway at or near maximum speed, the oil consumption was so slight as to make topping-up unnecessary. When the car was returned in The Hague after a total mileage of more than 600 no more than a cupful of oil would have replenished the sump.

It is fair to point out that there is no other full four-seater family car of like engine size even to challenge the Amazon's performance. Coupled with the attainment of a 94 m.p.h. mean maximum are a standing quarter-mile in 19.9sec, and from standstill to 30 m.p.h. in 3.7sec, 50 in 9.7, 70 in 20.0 and 80 m.p.h. in 28.9sec. To put such a performance in perspective, it may be recalled that most family cars of well over 2,000 c.c. engine size could be promptly left behind, and that one must aspire to the much larger-engined sports machine to outperform the Amazon with any ease.

The engine proved smooth throughout the speed range; its noise characteristic was soft, sweet and unobtrusive in normal driving, hard and fierce when the throttle was snapped right open. This was attributed in part to the

Overriders are standard. The combined side lights and winking indicators can be seen clearly from the sides as well as from the front

Volvo
Amazon . . .

All the brightwork, including the long horizontal rubbing strip, is of stainless steel. The rear window wraps well round, and there is a swivelling ventilator in each door window. Mud flaps are fitted behind the wheels

will take the speed up to more than 30 m.p.h. if required; second runs up to a maximum of just over 50 while being low enough to cope with the steepest main road hills, and third will get the car crackling up to 76 flat out, or into the 60s without fuss. The only difficulty experienced with the box on the car tested was reluctant engagement of first or second when the car was stationary or moving very slowly. The car concerned had done but some 2,000 miles, however, and this particular fault might be reduced after further use. The remainder of the transmission was silent, and free from any kind of vibration or jerkiness regardless of speed or driving technique.

A comfortable ride is maintained when rough dirt roads are taken quickly, firmness on hard surfaces being approximately the same as that of most other European cars of similar size. The road surface can be felt, which is no bad thing; there is no pitching on indifferent surfaces. On certain stretches of undulating road of the close-pitched washboard type, the Amazon behaved exceptionally well even at high speed. The amount of roll on corners was not excessive, but seemed a little greater than that of the 444 models, and there was a degree of front-tyre squeal on smooth surfaces that amounted to a fault, even though

customary fitting of small, individual air cleaners on the brace of S.U.s, as opposed to the big cleaner-silencer normally used on the single-carburettor installations usually found on family models.

In the fairly warm spell which lasted throughout the test, the engine fired from cold without use of the choke; this is, perhaps, appropriate to a car which has to respond promptly in a Swedish winter. A radiator blind is fitted as standard, but in the weather described the engine warmed quickly without its use.

The degree of flexibility is entirely satisfactory for a family car—in top gear the engine pulls strongly and without a trace of snatch from about 20 m.p.h., but the majority of drivers in a car of this character will prefer to use the delightful gear box. There are three- and four-speed boxes available, the car tested having four forward speeds. The central change lever is placed conveniently for a driver of any height, and the synchromesh mechanism on the upper three ratios is unbeatable. Even in the artificially severe conditions of acceleration testing, there was no protest from the gear box; indeed, the lever could be pulled through with the throttle wide open.

Spacing of the ratios is without fault. Second can be used for starting from rest, though first, the natural choice,

The spring-loaded locker lid swings right out of the way. There is plenty of room for luggage of orthodox shape, little of which need be removed for access to the spare wheel. To the left of the wheel are stored the well-finished tools

A combination of leathercloth and fabric is used for the upholstery, while the flooring is of synthetic rubber. There are armrests and support straps in the rear compartment. Gear lever and handbrake are conveniently placed

adhesion was not affected. Even in town driving, tyre squeal was heard occasionally.

Thoroughly in keeping with the model's virtues is the steering which is very light to the touch, and gives instant, precise response. It seems almost that the car begins to adopt the right line on a corner before the driver has given movement to the wheel. For a turning circle of 32ft there are but 3½ turns of the wheel from lock to lock. There is no appreciable kick-back.

The Wagner brakes are similar to those of the 444, but lining area at the front has been increased. A servo mechanism is available at extra cost, but it was not fitted to the test car. In commenting that the brakes are not so impressive as, for example, the performance, it must be made clear that they cope adequately with it. The maximum efficiency of 75 per cent could be improved, but the 95 lb pedal pressure required to get the best retardation is within the capability of any driver. Additional pressure results in the front wheels locking. Brake pedal travel increased a little during the test—this may have been a result of completing the bedding-in process—but the braking power did not flag at any time; there was neither fade nor unevenness. The hand brake lever is floor-mounted at the driver's side; it is well placed and works powerfully on the rear wheels. To avoid inadvertent release while the driver is getting in or out, there is a small circular shield round the locking button.

The driving position is comfortable, and the speedometer and auxiliary instruments are easily seen through the upper half of the cross-spoke wheel; the top arc of the horn ring is flattened to avoid obstructing vision. Minor controls are mostly on the facia, near enough for them to be reached easily. The driver's left hand (in this left-hand drive car) operates the standard heater-demister controls, choke, and indicators—whose operating lever also flashes the head lamps for signalling. To the right are the ignition and starter switch, lamps, and cigarette lighter. The main pedals are well placed, but the dip switch is much nearer the front of the car than the clutch pedal; if the driving seat is correctly adjusted, the reach to the dip switch is inconveniently long. There is ample adjustment of the seat location; driver's height, one might say, is no obstacle, although rear leg room inevitably is affected when the seat is well back.

Interior dimensions give adequate room for four people, with a child additionally in the rear compartment. The driver has the elbow room he requires, and for people of average height, leg room at the rear is sufficient.

A deceptively large luggage locker is shaped so that big, ordinary suitcases may be stored easily without wasting appreciable space. The spare wheel stands vertically in

the left side of the locker, and the tools are stored in the cavity between wheel and body side. Though the number of tools is limited, their quality of construction and finish is exceptionally high. The locker lid is spring-loaded, and below its lip is its release button and the fully exposed petrol filler cap.

Curiously, the fuel consumption of the Amazon in hard driving is appreciably higher than that of the 444 with similar engine. This is partly explained by the Amazon having been required to cover long distances at very high speeds on Continental motorways, whereas the 444 had England as a testing ground. Even so, the quiet driving figure is also less favourable. Yet, 22.5 m.p.g. is not unreasonable when it represents long journeys covered at speeds consistently around the 90 m.p.h. mark, and for cruising at about 50 to 60 m.p.h., the figure rises to an impressive 36 m.p.g. In normal driving one could reasonably expect not less than 26, probably more. With a tank capacity of nearly 10 gallons the cruising range is, therefore, comfortably in excess of 200 miles.

The head lamps are powerful, and the dip cut-off gives a wide spread. On the car tested there appeared to be a faulty contact in the flashing system operated by hand. A gentle pull on the indicator lever with the finger-tips should cause the lamps to flash automatically while pressure on

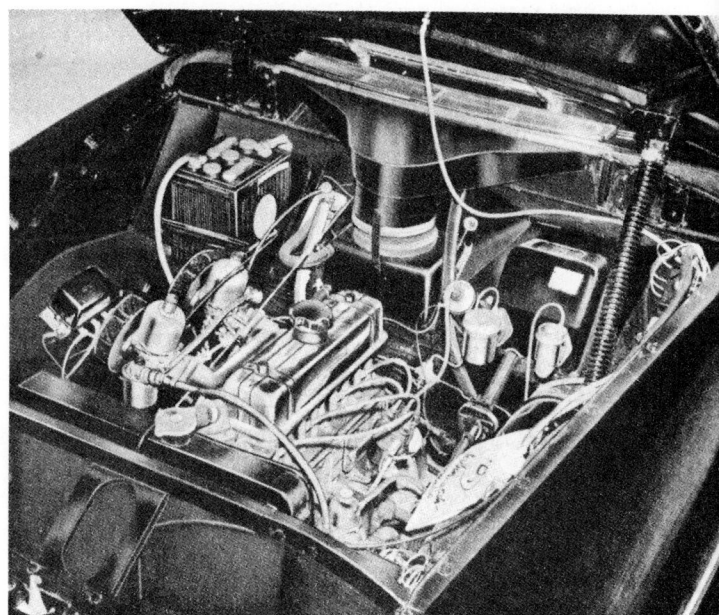

The pushrod-operated o.h.v. Volvo engine has twin S.U. carburettors in its 85 b.h.p. Export form. The electrical system is six-volt

Volvo Amazon . . .

the lever is maintained; in practice they flashed at reduced intensity and then stopped.

Underbonnet accessibility is first rate, both for routine checking and for servicing. This aspect of service and repair has been given a large measure of priority in the 444 models, and it appears that this policy has been continued in the Amazon.

On the car tested two leaks from the exterior into the luggage locker, one of which resulted in the toolbag getting soaked, were noted; quality of the coachwork as a whole, however, is of a very high standard of construction and detail finish. The two-colour paintwork is well executed, and all the external bright metal is of brightly polished

stainless steel. The manufacturers wisely use a lighter colour on the roof in their two-colour schemes. To British eyes, only the use of synthetic rubber flooring in the Continental manner seems to let down a car of this class. The car is completely treated with bitumastic undersealing at the factory. All Volvos are given this protection, but the Amazon has a specially de luxe type of specification, for included as standard are also the heater, two-speed wipers, clock, overriders, cigarette lighter, and tubeless whitewall tyres, and other, more minor, improvements over the 444 specification.

It was mentioned earlier that, in effect, if one were to put into competition all the comfortable family four-seater cars of up to 1,600 c.c. engine size, the Amazon would quickly take itself a clear stride ahead. To that comment no addition or qualification is required.

VOLVO AMAZON

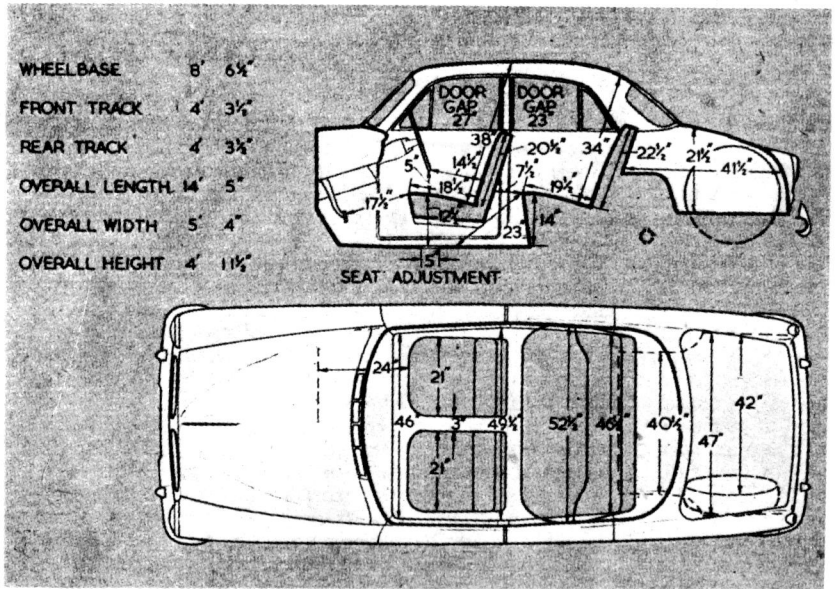

WHEELBASE	8' 6½"
FRONT TRACK	4' 3½"
REAR TRACK	4' 3½"
OVERALL LENGTH	14' 5"
OVERALL WIDTH	5' 4"
OVERALL HEIGHT	4' 11½"

SEAT ADJUSTMENT

Measurements in these ⅛in to 1ft scale body diagrams are taken with the driving seat in the central position of fore and aft adjustment and with the seat cushions uncompressed

PERFORMANCE

ACCELERATION: from constant speeds.
Speed Range, Gear Ratios and Time in sec.

M.P.H.	4.6 to 1	6.0 to 1	9.9 to 1	15.7 to 1
10—30..	—	7.5	4.1	2.9
20—40..	9.5	7.4	4.0	—
30—50..	9.7	6.9	5.2	—
40—60..	11.4	7.9	—	—
50—70..	12.5	9.4	—	—
60—80..	15.5	—	—	—

From rest through gears to:

M.P.H.	sec.
30	3.7
50	9.7
60	14.0
70	20.0
80	28.9
90	41.3

Standing quarter mile, 19.9 sec.

SPEEDS ON GEARS:

Gear		M.P.H. (normal and max.)	K.P.H. (normal and max.)
Top	(mean)	94	151.2
	(best)	94	151.2
3rd		60—76	96.6—122.3
2nd		42—51	67.6—82.1
1st		24—33	38.6—53.1

TRACTIVE EFFORT:

			Pull (lb per ton)	Equivalent Gradient
Top	..	..	190	1 in 11.7
Third	..	..	270	1 in 8.2
Second	..	..	435	1 in 5.0

BRAKES: (at 30 m.p.h. in neutral)

Efficiency	Pedal Pressure (lb)
20 per cent	25
46 per cent	50
59 per cent	75
75 per cent	95

FUEL CONSUMPTION:
26 m.p.g. overall for 669 miles. (10.9 litres per 100 km.)

Approximate normal range 22.5—36.0 m.p.g. (12.5—7.8 litres per 100 km.)

Fuel, Premium.

WEATHER: Sunny, slight cross breeze.
Air temperature 62 deg F.
Acceleration figures are the means of several runs in opposite directions.
Tractive effort and resistance obtained by Tapley meter.

DATA

PRICE ex works (in Sweden) with saloon body, 12,600 Kronor = £868 approx.

ENGINE: Capacity: 1,583 c.c. (96.58 cu in).
Number of cylinders: 4.
Bore and stroke: 79.4 ×80 mm (3.12 × 3.15 in).
Valve gear: o.h.v., pushrods.
Compression ratio: 8.2 to 1.
B.H.P.: 85 at 5,500 r.p.m. (B.H.P. per ton laden 70.8).
Torque: 87 lb ft at 3,500 r.p.m.
M.P.H. per 1,000 r.p.m. on top gear, 16.2.

WEIGHT: (with 5 gals fuel), 21 cwt (2,352 lb). Laden as tested: 24 cwt (2,688 lb).
Lb per c.c. (laden): 1.7.

BRAKES: Type: Wagner.
Method of operation: hydraulic.
Drum dimensions: F, 9in diameter; 2in wide. R, 9in diameter; 2in wide.
Lining area: F, 81.6 sq in.; R, 72.1 sq in (128 sq in per ton laden).

TYRES: 5.90—15in.
Pressures (lb sq in): F, 20; R, 24 (normal).

TANK CAPACITY: 10 Imperial gallons.
Oil sump, 6¼ pints.
Cooling system, 13 pints (plus 2 pints if heater is fitted).

TURNING CIRCLE: 32ft 6in (L and R).
Steering wheel turns (lock to lock): 3¼.

DIMENSIONS: Wheelbase: 8ft 6½in.
Track: F, 4ft 3½in; R, 4ft 3½in.
Length (overall): 14ft 5in.
Height: 4ft 11½in.
Width: 5ft 4in.
Ground clearance: 7¼in.

ELECTRICAL SYSTEM: 6-volt; 85 ampere-hour battery.
Head lights: Double dip; 45—40 watt bulbs.

SUSPENSION: Front, independent, wishbones and coil springs, with anti-roll bar. Rear, live axle controlled by torque arms with coil springs.

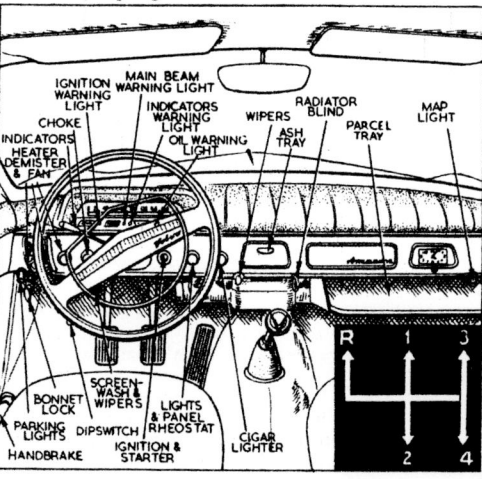

SPEEDOMETER CORRECTION: M.P.H. (converted from k.p.h.)

Car speedometer:	10	20	30	40	50	60	70	80	90	94
True Speed:	8	18	27	38	48	57	68	78	91	94

WINTER MOTORING WITH THE VOLVO 122S

A Good Swedish 1,583 c.c. Saloon, Offering Exceptional Performance and Commendable Detail Arrangements.

IN A WINTER SETTING.—The Volvo 122S saloon, an excellent car of distinctive frontal aspect, is seen here, not in its native Sweden but in the New Forest near Brockenhurst.

THE ugly PV444 Volvo made a great impression as a fast car of modest engine capacity and when the re-styled 122S was introduced at the last London Motor Show I was naturally extremely interested and suggested to Brooklands of Bond Street that an early road-test be arranged. Unfortunately, their Press agents did not lend MOTOR SPORT a car until the middle of January, at which time serious testing was curtailed by the presence of fog and icy roads. However, I put in sufficient driving to form a good opinion of this Swedish saloon, which, with the B 16B 85 b.h.p. engine, provides quite remarkable speed and acceleration for a 1.6-litre vehicle, apart from being sensibly as well as completely equipped and comfortable in which to travel.

MATTERS OF DETAIL

This Volvo 122S is really a four-seater, because separate front seats are used and the corners of the back seat are rounded off so that comfortable accommodation is limited to two persons. The interior of the red test car was rather too brightly finished, with two shades of red, and white, upholstery, black facia sill covering with matching vizors, and red side rails ; upholstery and facia off-set by white roof lining. The facia is not ornate, the seats are of generous dimensions, and all four doors trail. The floor covering of synthetic rubber is rather austere but makes for easy cleaning, the transmission tunnel does not impede leg room, and space under the front seats provides leg room for the back-seat occupants. This Volvo is of conventional appearance, but has a pleasingly distinctive radiator grille ; it is only the provision of cross-wise front-seat safety belts and the purposeful note when the engine is opened up, which suggest the performance capabilities of this 1,583-c.c. four-cylinder saloon.

It was when I prepared to drive away into the foggy night that I appreciated the practical layout of the Volvo's major and minor controls and its satisfactory driving position. A driver of average height cannot quite see the near-side wing and the screen pillars are quite thick, but the high-set driving seat, easily adjustable, and low-placed wheel offer effective visibility. The front seats are somewhat hard and not particularly well shaped, being good rather than outstanding. Gear-lever, right-hand hand-brake and steering wheel are all well located in relation to each other, and the pendant pedals are all at the same height, the foot does not have to be lifted onto the brake pedal, and it is possible to " heel and toe."

The metal facia is set behind a broad, upholstered sill, which is padded and hoods the instruments. The latter consist of four oblong windows, above which is a Vdo 100-m.p.h. ribbon-type speedometer, easy to read but requiring an 8½-in. travel to register 0-100 m.p.h. in graduations every 5 m.p.h. (figures every 10 m.p.h.). The " dials " below are water-temperature gauge, marked C, H ; the trip mileometer with clearly-defined decimal reading ; total mileometer reading, and a very pessimistic petrol gauge ; marked E, ½, F. Between these " dials " or windows are the various warning

lights, green for low oil pressure, red for no dynamo charge, blue (rather too bright) for full headlamps beam, and yellow, augmented by an aural warning, for direction-flashers in use, the last-named indication being sensibly represented by a single warning light.

Below the instruments, on the facia proper, are the minor controls, laid out for maximum convenience of operation. On the extreme right are the vertical heater quadrants, which control demisting, hot and cold air. These are augmented by a two-speed fan. Let it be said that the Volvo heater will supply enormous volumes of heat and that control of it is extremely sensitive, so that, once mastered, interior temperature can be adjusted to a nicety.

Beside the heater-fan knob the Wilmot Breeden ignition key turns to actuate the starter and, turned the other way, retains all electrical services (including the petrol gauge) except ignition, the green warning lamp then remaining on. To the left of the steering column are three more large knobs matching the heater-fan knob, these, from left to right, being a detachable cigarette lighter, lights control and mixture enrichener. The lamps knob pulls out to bring in, firstly, the sidelamps, then the headlamps, while, turned, it controls the excellent rheostat instrument lighting. The steering wheel somewhat blanks the lamps knob and the cigarette lighter is apt to be mistaken for it. The enrichener knob acts as a hand throttle and has a serrated spindle for sensitive setting.

All these knobs are within easy reach of the driver, and just below the facia his left hand falls easily to a tiny turn-switch that controls the really efficient two-speed, self-parking Bosch wipers. Adjacent is a pull-switch which brings in permanently the screen washers—a splendid means of maintaining visibility on dirty days. An exceedingly well placed stalk on the left of the steering wheel, close to the wheel rim, operates the direction-flashers. This excellent control is further enhanced because, flicked up and down, it flashes the headlamps, even when the lamps switch is off, providing an instant warning when overtaking slower traffic without recourse to the horn.

To the left of the essential controls is a drawer-type ash-tray, space for a radio, and, before the passenger, an electric clock. The two-spoke steering wheel, labelled Volvo and with plated spokes which sometimes reflect the sun, carries a good half-horn ring for a surprisingly mediocre horn. The Volvo lacks a facia cubby-hole or door-pockets. There is a big shelf down by the front passenger's knees, mounted flexibly for safety, but the driver could, with advantage, be provided with additional stowage. The radio loud-speaker is on this parcels shelf. Yet another practical point is the provision of a map-light shining onto the aforesaid shelf, brought in by tiny pull-switch under the facia on the extreme left, this lamp in no way dazzling the driver. A matching switch for the driver controls the parking lamps. The driver finds ample room for his left foot away from the clutch pedal and the foot dimmer button is well placed. The centre-roof rear-view mirror gives a rather narrow field of vision ; it swivels to kill dazzle, and twin anti-dazzle vizors made of

REMARKABLE FOR A 1.6-LITRE SALOON, the Volvo 122S will exceed 50 m.p.h. in second gear, show a speedometer reading of more than 80 m.p.h. in third gear, and has a maximum in top of nearly 95 m.p.h., with acceleration and controllability in keeping!

soft material are fitted, both of which swivel sideways. The front doors have quarter-windows with anti-thief catches on their handles, similar windows being provided for the back-seat passengers. The main door windows possess handles with swivelling grips, which require three turns from fully-up to fully-down. The driver's window emitted a squeak when operated, and the near-side back window was too stiff to open properly. Incidentally, the padded facia sill is continued onto the front doors, each of which has a sill interior lock. The front doors have key-locks and all four doors can be locked by pressing down the sill knobs, and thus it is possible for the owner to be locked out if the key is left inside the car. This is no great problem, merely entailing carrying of a spare key on one's person! It *does* enable the Volvo to be locked merely by closing the last door, with no fiddling with the key, which is a decided advantage on a wet day. The doors have good exterior handles incorporating push-buttons, and the door "keeps" are effective. The interior handles have rather sharp edges. The doors shut nicely, in quite "expensive" fashion.

In the back compartment rubber-cord "pulls" are provided on the backs of the front seats, drawer-type ash-trays in the doors, while the seat is comfortable and behind it there is a spacious parcels well. All doors contain good arm-rests, formed as "pulls," and mud-flaps are fitted to front and back wings to comply with Swedish traffic requirements.

It is pleasing to be able to report that the exterior metalwork is of polished stainless steel and that the car is completely treated with bitumastic undersealing. The bumpers have over-riders. The bayonet-type petrol-filler cap has no securing chain. The filler is well placed for refuelling from a can. The back window matches the screen in generosity of area, a combination of Slocherkeltsglas, Duro-glas and Sunex AS-2 safety glass being used about the car.

A roof-light near the mirror gives really useful interior illumination and comes on when the front doors are opened or, using its own switch, can be switched on by the driver.

That concludes a study of the interior details of this interesting Swedish car but before we drive it, let us look under the bonnet and into the boot. The bonnet is released by pulling a toggle down under the scuttle on the off side and pushing back the usual safety catch. The bonnet then springs up and is supported automatically by a spring-loaded strut. The virtually over-square, 79.4 by 80 mm. four-cylinder push-rod o.h.v. engine, with its polished valve cover, looks very ordinary, except that the B 16B version, which develops 19 more S.A.E. horse-power than the B 16A engine, which isn't available in Export markets, has twin horizontal 38-mm. type H4 S.U. carburetters, with A.C. air-cleaners, on the off side. There is a belt-driven two-bladed fan, the dip-stick is accessible, the 6-volt Tudor battery is mounted at the off-side rear of the engine compartment, and the ignition and electrics, reassuringly, are Bosch, with an accessible fuse-box containing separate fuses for reversing lamp, fog- or spot-lamps, if fitted, parking lamps, flashers indicator, heater and, in unit, horn, flashers and petrol gauge. Water and oil fillers are well placed and the fuel pump has a priming lever.

The English Wilmot Breeden door key unlocks the boot press-button, whereupon the lid should rise automatically under the action of torsion-bars. Alas, the lock stuck on the test car and we had to

take a crow-bar to it! The boot is rather shallow, there is a sharp edge round the off-side wheel arch which could damage luggage, the floor is slightly obstructed by the petrol-filler pipe and the spare wheel is strapped vertically on the near side of the boot. The wheels are balanced and tubeless whitewall tyres are fitted.

The cooling system holds 15 pints, the engine sump $6\frac{1}{4}$ pints including the Fram full-flow oil filter, the back axle $2\frac{1}{4}$ pints, the gearbox $1\frac{3}{4}$ pints. The comprehensive instruction book contains details of how to remove blood, lipstick, fruit, chocolate, acid, grease, chewing gum, vomit and urine from the upholstery, which should cover almost every class of owner! Front seats, which will fold down to form beds, are available as an extra.

ON THE ROAD

Having examined the Volvo and liked what I found, I set about driving the car. The gears in the close-ratio box are changed with a long but not unduly whippy central lever, the large knob of which comes conveniently close to the left hand. This is far better than a steering-column lever although rapid changes are slightly baulked by a stiffness of action, while strong spring-loading of the lever towards top and third-gear positions makes for errors when really in a hurry, and the lever wobbles and vibrates. Normally this is not at all a bad gear-change, although the Volvo deserves a remote-control lever. The hydraulically-operated clutch is rather sudden and has to be fully depressed to make effective gear-changes. The gears are quiet. Reverse is away beyond first, safely positioned.

The cam-and-roller steering, low-geared at $3\frac{1}{2}$ turns lock-to-lock, is fairly light, even for parking, delightfully smooth, devoid of sponginess, and has excellent, not too fierce, castor return action. It is sensitive, accurate steering, which transmits only very faint kick and vibration, and as such is an excellent feature of the Volvo. The turning circle is small ($32\frac{1}{2}$ feet). The car corners extremely well, with an understeer tendency, tail slides easily controlled.

The suspension is by coil-springs and wishbones at the front and at the back a rigid axle is used, but this is well tied with radius-arms, has an anti-roll bar, and is sprung on coil-springs, which reduce unsprung weight. The result is firm springing, which transmits considerable road shock and some up-and-down motion, but enables the car to be cornered without roll and which gave notably "sure-footed" motoring over ice and snow. There is a difficult to define "dead" aspect to the ride, which merely accentuates the pleasantly solid feel of the car, which is further reflected in the absence of body rattles and scuttle or bonnet shake. Rather better damping of the suspension would be an improvement.

The brakes are no doubt excellent but on the test car were in need of adjustment, and there was additional lost motion in the pedal linkage, so that considerable pressure was needed to get powerful retardation. The right-hand hand-brake lever couldn't be better placed. It does not impede exit through the driver's door yet is delightfully placed. It holds the car securely and its ratchet button is protected by a wire guard. Vacuum-servo braking is available as an extra. Incidentally, I approve fully of safety-belts, if only to safeguard children who ride in the front seat, and those on the Volvo, if of unusual type, are very easy to put on.

THIS VIEW OF THE VOLVO shows the big rear window area of this outstanding Swedish car, which won the 1958 European Rally Championship.

WELL-BALANCED LINES distinguish the Volvo 122S. Consider-
ing its modest price and its very considerable performance, it is a car
which must be causing much concern to manufacturers of other sports
saloons of similar engine size!

It is when you come to consider the engine that the Volvo excites the enthusiast. The B 16B unit develops its maximum power, probably 79 b.h.p. by our rating, at 5,500 r.p.m., so that the close-ratio gearbox has to be used to get the highest performance. The gear ratios are splendidly plotted and it is possible to obtain progressive acceleration by changing up at peak r.p.m. in each gear, torque being maintained and practically no falling off of power resulting throughout the speed range. The engine emits considerable power-roar when accelerating but does not " pink," is notably smooth, and, in spite of its high-speed, high-output characteristics, is quite docile and tractable, pulling away from less than 20 m.p.h. in top gear. At such speeds the engine runs silently and the Volvo rolls very smoothly and unobtrusively through built-up areas. The engine did " run-on " slightly after performance testing, but its temperature is subject to sensitive control by means of a radiator-blind, adjusted by a long, hanging chain convenient to the driver's left hand. This blind also enables the heater to quickly attain optimum temperature.

The absolute maxima in the gears, allowing for speedometer correction—the strip speedometer was accurate at 30 m.p.h., 1½ m.p.h. fast at 50 m.p.h., 2½ m.p.h. fast at 60 m.p.h.—were : 30 in first, 52 in second and 77 m.p.h. in third gear, impressive speeds from a 1.6-litre saloon! Beyond these speeds savage valve bounce immediately sets in. It pays to change-up at about 45 m.p.h. in second gear to secure maximum acceleration. With some initial wheelspin we clocked a mean time of 20 seconds and a best-time of 19.8 seconds for the s.s. ¼-mile; 0-50 m.p.h. occupied a mean time of 11.7 seconds, with a best time of 11.6 seconds, 0-60 m.p.h. taking 16.7 seconds. In top gear the Volvo will attain a speed of 94 m.p.h. under favourable conditions. It cruises fast without excessive wind-noise. The Robo headlamps give a good but concentrated beam. The sidelamps are Hella.

Starting was a little reluctant after a winter night in the open but the engine warms up quickly with radiator-blind shut. A full tank of petrol took the car 215 miles, mainly over slippery roads calling for light throttle work but offset by cold starts and performance testing. The makers quote a tank capacity of 10 gallons, so this represents less than 22 m.p.g. However, another check, under give-and-take conditions with some cold starts, gave a figure of 27 m.p.g. A check after 300 miles showed that not a drop of oil had been used. The radiator took a quart of water but this may have been because the car was inadvertently driven with the temperature

---

THE VOLVO 122S SALOON (B 16B ENGINE)

Engine : Four cylinders, 79.4 by 80 mm. (1,583 c.c.). Push-rod-operated overhead valves. 8.2-to-1 compression-ratio. 85 b.h.p. (S.A.E.) at 5,500 r.p.m.

Gear ratios : First, 15.7 to 1; second, 9.9 to 1; third, 6.0 to 1; top, 4.56 to 1.

Tyres : 5.90 by 15 Trelleborg Safe Star 4PR whitewall tubeless, on balanced, bolt-on steel disc wheels.

Weight : 1 ton 1 cwt. 0 qtr., ready for the road, without occupants, but with approximately half a gallon of petrol.

Steering ratio : 3½ turns, lock-to-lock.

Fuel capacity : 10 gallons. (Range approximately 215 miles—but see text.)

Wheelbase : 8 ft. 4.4 in.

Track : 4 ft. 3.77 in.

Dimensions : 14 ft. 5 in. by 5 ft. 3½ in. by 4 ft. 11¼ in. (high).

Price : £932 (£1,399 7s., inclusive of purchase tax and import duty).

Concessionaires : Brooklands of Bond Street, Ltd., 103, New Bond Street, London, W.1.

Makers : Aktiebolaget Volvo, Gothenburg, Sweden.

PERFORMANCE DATA

Speeds in gears (after speedometer correction) :

First	...	...	30 m.p.h.
Second	...	...	52 „
Third	...	...	77 „

Acceleration :

0-50 m.p.h., two-way runs ...	...	...	11.7 sec.
0-50 m.p.h., best run	...	...	11.6 „
0-60 m.p.h., best run	...	...	16.7 „
Standing-start ¼-mile (slippery road), two-way runs ...	...	...	...	20.0 „
Standing-start ¼-mile, best run	...	...	19.8 „

above boiling point for some miles—the radiator-blind very quickly affects temperature and a warning light when this gets dangerously high would be useful.

When it is considered that the Volvo has excellent steering, sound road-holding, a passable gear-change and combines in one vehicle the performance of a sports saloon with the comfort of a high-quality family car, it is seen as an outstanding motor car. In spite of its modest capacity of 1,583 c.c. (untidy by our standards but admirable for rally work, in which sphere the Volvo has proved its worth) the car out-performs ordinary cars of considerably greater capacity and even shows better performance figures than some renowned sports cars. The acceleration figures we obtained, two-up and with about five gallons of petrol, show that the Volvo runs away from many larger cars and makes rings round so-called sports saloons. It has road-holding which enables good use to be made of this striking performance, it is pleasant to drive, and the arrangement of the controls is highly commendable, while the quality of the minor controls would not disgrace a car costing twice the price. Altogether, the Volvo 122S is a splendid proposition for discerning drivers! The agents here are Brooklands of Bond Street and the price, with tax, is £1,399 7s.—W. B.

Swedish Amazons invade the U.K.

A new, slick model from an old Swedish firm has made its British debut.
1957 may see sizeable imports to the U.K. of this attractive, ingenious car.

From GORDON WILKINS in London

AFTER having produced their famous PV 444 for ten years, the Swedish Volvo Company launched a new Amazon model at the London show: a larger, five-seater car with four-cylinder 1.6 litre engine, said to do 90 m.p.h. This was the first appearance of these Swedish cars in London, and first appearance of the new model, which was greeted with special interest in view of Volvo's recent achievements in export markets (including, they claim, a current second place in cars imported into California).

The new Volvo is a unit-construction job in steel, fully rust proofed to stand being left out in

Rear view. Body of the Amazon is fully rustproofed, soundproofed. Car does 33 m.p.g. at normal cruising speeds.

The Swedish-built Volvo Amazon sedan. Engine is 4 cylinder o.h.v., giving 60 b.h.p. on 7.5 to 1 compression. Top speed is 90 m.p.h.

The attractive sports two-seater currently being produced by Volvo. Engine is 1.4 litre, body of fibreglass plastic.

ignition coil. Direction indicator switch incorporates a headlamp flasher button, and the dipper is foot operated.

Grouped under a cowl in front of the driver are a large speedometer with trip and total recorders, a thermometer, and fuel gauge. In front of the passenger are an electric clock, ashtray and space for a radio. Instrument lighting is rheostat controlled.

TECHNICAL DETAILS.

Engine.—4-cyl. o.h.v. 79.37 x 80 m.m. 1600 c.c. Compression 7.5 to 1. 60 b.h.p. at 4,500 r.p.m. Torque 81.8 lb. ft. at 2,500 r.p.m.

Transmission.— Three speed gearbox with synchromesh on second and top. Hypoid final drive. Open propeller shaft. Overall ratios; 14.69, 7.37, 4.55 to 1. Central floor gear lever.

Dimensions.—Wheelbase 102 in. Track front and rear 51½ in. Length, 173 in. Width, 63½ in. Kerb weight, 2260 lb. ●

Arctic winters (another Volvo claim is that they were the first car ever manufacturer in the world to instal a Rotodip plant for complete bodies). For additional protection, the undersides of wings are coated with rubber asphalt.

The new car is a sleek modern design, with four doors, curved windscreen and rear window, and ventilating panes on all doors. Interior features include a two-spoke steering wheel of contemporary style, and padding on the facia which extends round into the front doors. Special emphasis is laid on the efficiency of the heater, with fresh air intake on the scuttle, away from exhaust fumes. Upholstery is in textiles and plastics, in single or dual colours.

The engine is similar in design to that of the earlier Volvo, with four cylinders and pushrod operated overhead valves; but the bore has been increased to give 1,600 c.c. On a compression ratio of 7.5 to 1, output is 60 b.h.p. at 4,500 r.p.m., and maximum torque 81.8 lb. ft. at 2,500 r.p.m. Compression rings are chromium plated for longer engine life.

The Amazon's transmission is conventional, through a single plate clutch and three speed gearbox with cast iron casing and an aluminium top cover. Final drive is hypoid. Suspension is by coil springs at the front; and coil springs, torque arms, and Panhard rod at rear. Brakes are self adjusting hydraulic, with two leading shoes in front, and a total friction area of 142 sq. in. 15 inch wheels carry tubeless tyres of 5.90 section.

Electrical system is 6-volt. The ignition key operates the starter and there is an armoured antitheft cable between switch and

NOTES ON THE MANUFACTURERS

Volvo began making cars in Sweden thirty years ago, and today the company manufactures cars, trucks, buses, taxi cabs, and tractors.

The factories are extremely well equipped, although Sweden's automobile industry, as yet, is not sufficiently large to permit economic production of all motor car components, and some are imported from other countries, including the United Kingdom.

Because of this, there is a substantial British content in each Volvo car, and the company's vehicles are subject to a worthwhile concession in import duty when shipped to Britain as part of the current Swedish export drive.

Besides the Amazon, Volvo at present manufacture the Series PV 444 sedan and the plastic-bodied sports convertible (pictured), both of which have 1.4 litre 70 b.h.p. engines. Other products of Volvo, who employ 12,000 workers, include marine engines, jet and reciprocating aero engines, and industrial power units.

The company's products have a reputation of reliability.

Ed.

Pendant pedals, modernistic steering wheel and dash, and hooded instruments add to a tastefully finished interior. Heater is standard and most efficient.

Make: Volvo. **Makers:** AB Volvo, Gothenburg, Sweden. **Type:** 122S.
Concessionnaires: Brooklands Motor Co., Ltd., 103, New Bond Street, London, W.1.

Test Data

CONDITIONS: Weather: Cold and damp, with gusty wind of approx. 15 m.p.h. and exceptionally low barometric pressure. (Temperature 46°—50°F., Barometer 28.5—28.6 in. Hg.) Surface: Wet tarred macadam and concrete. Fuel: Mixed Premium and Super premium pump petrols (approx. 98 Research Method Octane Rating).

INSTRUMENTS
Speedometer at 30 m.p.h.	.. ..	2% fast
Speedometer at 60 m.p.h.	.. ..	4% fast
Speedometer at 90 m.p.h.	.. ..	4% fast
Distance recorder	.. ..	1% fast

WEIGHT
Kerb weight (unladen, but with oil, coolant and fuel for approx. 50 miles) .. 21 cwt.
Front/rear distribution of kerb weight 52/48
Weight laden as tested .. .. .. 24¾ cwt.

MAXIMUM SPEEDS
Mean lap speed on banked circuit .. 89.3 m.p.h.
Best one-way quarter-mile time straight .. .. .. 92.8 m.p.h.

"Maximile" Speed. (Timed quarter-mile after one mile accelerating from rest.)
Mean of four opposite runs .. 86.5 m.p.h.
Best one-way time equals .. .. 90.0 m.p.h.

Speed in Gears
Max. speed in 3rd gear .. .. 80 m.p.h.
Max. speed in 2nd gear .. .. 50 m.p.h.
Max. speed in 1st gear.. .. .. 32 m.p.h.

FUEL CONSUMPTION
38.0 m.p.g. at constant 30 m.p.h. on level.
35.0 m.p.g. at constant 40 m.p.h. on level.
32.0 m.p.g. at constant 50 m.p.h. on level.
28.0 m.p.g. at constant 60 m.p.h. on level.
24.0 m.p.g. at constant 70 m.p.h. on level.
21.0 m.p.g. at constant 80 m.p.h. on level.

Overall Fuel Consumption for 1,225 miles, 50.3 gallons, equals 24.4 m.p.g. (11.6 litres/100 km.).

Touring Fuel Consumption (m.p.g. at steady speed midway between 30 m.p.h. and maximum, less 5% allowance for acceleration) 26.7 m.p.g.
Fuel tank capacity (maker's figure) 9⅞ gallons

STEERING
Turning circle between kerbs:
Left .. .. 31½ feet
Right .. .. 31½ feet
Turns of steering wheel from lock to lock 3½

BRAKES from 30 m.p.h. (damp surface)
0.87g retardation (equivalent to 34½ ft. stopping distance) with 85 lb. pedal pressure.
0.80g retardation (equivalent to 37½ ft. stopping distance) with 75 lb. pedal pressure.
0.49g retardation (equivalent to 61½ ft. stopping distance) with 50 lb. pedal pressure.
0.26g retardation (equivalent to 116 ft. stopping distance) with 25 lb. pedal pressure.

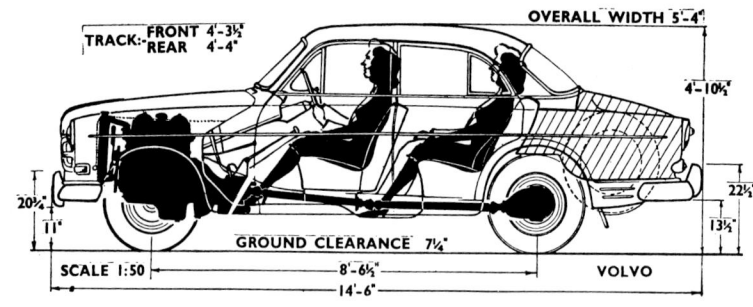

TRACK: FRONT 4'-3½" REAR 4'-4"
OVERALL WIDTH 5'-4"
4'-10½"
20¾"
11"
GROUND CLEARANCE 7¼"
22½"
13½"
SCALE 1:50
8'-6½"
14'-6"
VOLVO

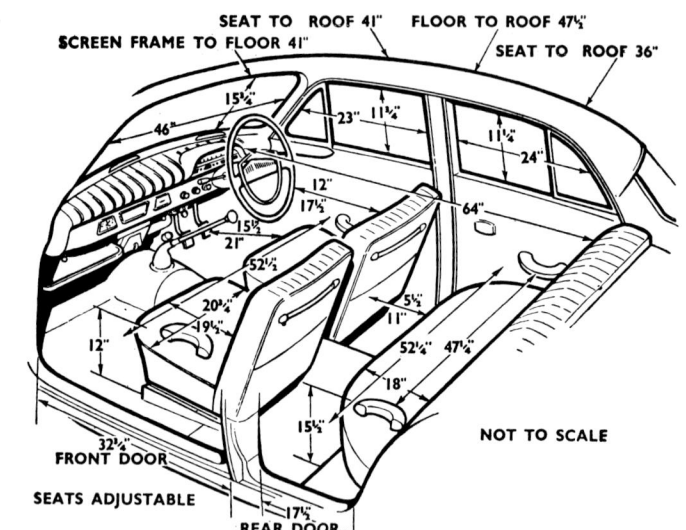

SCREEN FRAME TO FLOOR 41"
SEAT TO ROOF 41" FLOOR TO ROOF 47½"
SEAT TO ROOF 36"
15½" 23" 11½" 11½" 24" 46" 12" 17½" 15½" 21" 52½" 64" 20½" 19½" 5½" 11" 12" 52½" 47½" 18" 32½" 15½" 15½" 17½"
NOT TO SCALE
FRONT DOOR
SEATS ADJUSTABLE
REAR DOOR

ACCELERATION TIMES from standstill
0-30 m.p.h.	.. ..	5.2 sec.
0-40 m.p.h.	.. ..	7.7 sec.
0-50 m.p.h.	.. ..	11.9 sec.
0-60 m.p.h.	.. ..	17.8 sec.
0-70 m.p.h.	.. ..	26.6 sec.
0-80 m.p.h.	.. ..	43.0 sec.
Standing quarter mile	.. ..	21.0 sec.

ACCELERATION TIMES on upper ratios
		Top gear	3rd gear
10-30 m.p.h.	.. ..	12.1 sec.	8.7 sec.
20-40 m.p.h.	.. ..	12.2 sec.	8.0 sec.
30-50 m.p.h.	.. ..	12.4 sec.	8.8 sec.
40-60 m.p.h.	.. ..	14.8 sec.	10.5 sec.
50-70 m.p.h.	.. ..	18.3 sec.	13.6 sec.
60-80 m.p.h.	.. ..	26.0 sec.	—

HILL CLIMBING at sustained steady speeds
Max. gradient on top gear .. 1 in 12.7 Tapley 175 lb./ton)
Max. gradient on 3rd gear .. 1 in 9.3 Tapley 240 lb./ton)
Max. gradient on 2nd gear .. 1 in 5.1 Tapley 430 lb./ton)

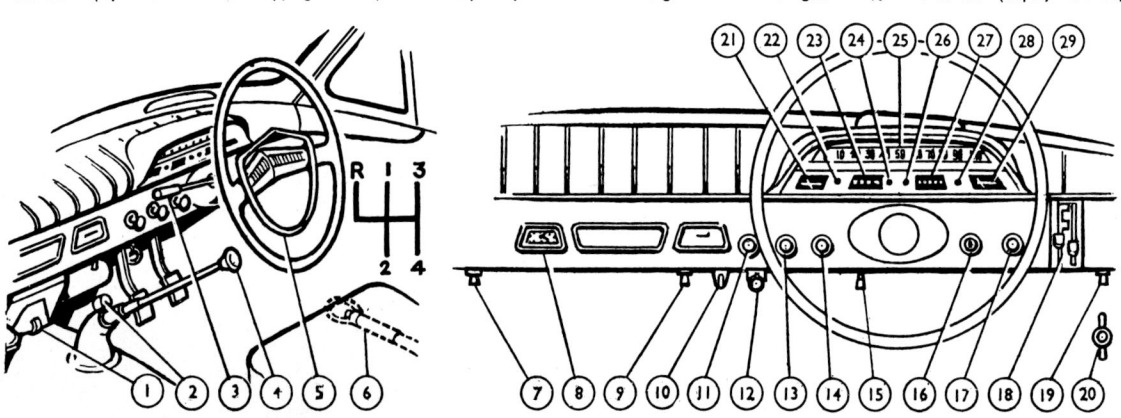

1. Heater air shutter. 2. Dip switch. 3. Trafficator and headlamp flasher. 4. Gear lever. 5. Horn ring. 6. Hand brake. 7. Map light switch. 8. Clock. 9 Screen washer (extra). 10. Windscreen wiper (2-speed). 11. Cigar lighter. 12. Radiator blind. 13. Light switch. 14. Choke. 15. Trip adjuster. 16. Ignition and starter key. 17. Heater fan switch. 18. Heater and demister controls. 19. Parking light switch. 20. Bonnet release. 21. Temperature gauge. 22. Dynamo warning lamp. 23. Trip recorder. 24. Main beam warning lamp. 25. Speedometer. 26. Trafficator warning lamp. 27. Distance recorder. 28. Oil pressure warning lamp. 29. Fuel gauge.

The VOLVO 122S

A Swedish Family Saloon Offering Performance with a Kick in it

WITH quite remarkable success the Swedish-built Volvo, which we have been testing on British roads in rather persistently discouraging weather conditions, manages to combine two potentially conflicting virtues. At one and the same time it is an extremely practical medium-sized 4-door saloon for family or business transport, and also a fast car which the sporting enthusiast finds it fun to drive.

As has already been indicated, our test period was beset by seasonable unpleasantnesses of fog, rain and strong winds which could not prevent us enjoying this car. Persistent bad weather did however force us to record acceleration figures on a wet and consequently somewhat slippery surface, at a time when an exceptionally low barometric pressure was also diminishing engine power. Yet, to find a 2-litre, 4- or 5-seat saloon which would match this car's rest-to-60 m.p.h. time of 17.8 sec. (other than the preceding Volvo 444L model of lesser roominess) it is necessary to go back through our Road Test Reports for almost seven years, to a car which even then cost more than twice today's Volvo price. Very obviously, the first thing which this model offers is performance well beyond expectations, especially in respect of

through-the-gears acceleration and of a top speed which was virtually 90 m.p.h. in thoroughly unhelpful conditions.

But if its speciality is high performance, this model also offers just about everything that is expected of a 1.6-litre saloon in respect of roominess, comfortable furnishing and full equipment. In no sense is it a car of limited usefulness, the only two disadvantages which accompany its high performance being that, when driven hard, it ceases to be quite as quiet or as economical of fuel as are some other, less vivacious cars of comparable size.

Comfortable Seating

Extremely modern in many aspects of its mechanical design and performance, the Volvo 122S is nevertheless in some ways a car of slightly old-fashioned shape. This impression is at its strongest around town and on first taking the wheel, when a driver is aware that he sits behind a longer and higher bonnet than is fashionable, looks ahead through a windscreen with rather thick pillars set close to the normal line of vision, and has a conveniently long central gear lever rising up from the floor. The last-mentioned component is in fact making a welcome return to fashion, and a central gearchange certainly suits this model which, although catering for three-abreast travel in the rear compartment, has two separate and individually adjustable front seats. Upright in shape, the seats proved very comfortable

throughout long days of motoring, their springs firm enough to give support but a soft covering effectively eliminating any impression of hardness. The general style of interior decoration is straightforward, but using painted metal and leather-like plastic fabric in strikingly contrasted colours for the main surfaces.

In relation to the car's sporting manners, some British buyers may criticize the instrument panel as being too sparsely furnished, there being an electric clock, strip-type speedometer with trip and total distance recorders, fuel contents gauge and radiator thermometer. On the other hand, praise can be given to very full equipment, including such items as a radiator blind controlled from the facia, a horn ring which is distorted to improve speedometer visibility but is without any missing sector, a fingertip control which automatically flashes the headlamps at a touch, two-speed wipers which clear almost the whole windscreen very effectively indeed, variable-brilliance instrument lighting, a shaded map-reading lamp as well as the interior light (with courtesy switches), a cigar lighter and three ashtrays. The car would gain in convenience if a glove box and door pockets supplemented the modest parcel shelf.

Under the spring-counterbalanced bonnet, the 1.6-litre 4-cylinder engine is seen to have a pair of S.U. carburetters but otherwise looks as straightforward as it in fact is. In-line overhead valves are operated by pushrods and rockers, the piston

---

In Brief

Price (in Britain) £932 plus purchase tax £467 7s. equals £1,399 7s.

Capacity	1,580 c.c.
Unladen kerb weight ...	21 cwt.
Acceleration:	
20-40 m.p.h. in top gear ...	12.2 sec.
0-50 m.p.h. through gears	11.9 sec.
Maximum direct top gear	
gradient	1 in 12.7
Maximum speed	89.3 m.p.h.
"Maximile" speed ...	86.5 m.p.h.
Touring fuel consumption ...	26.7 m.p.g.

Gearing: 16.2 m.p.h. in top gear at 1,000 r.p.m.; 30.8 m.p.h. at 1,000 ft./min. piston speed.

ORIGIN of exceptionally high performance is this straight-forward and readily accessible four-cylinder engine fitted with British-made S.U. carburetters.

The VOLVO
122S

BRIGHT interior decoration, individual front seats, and a low
floor level which gives front and
rear seat passengers ample legroom, are shown in these photographs of the four-door body.
Instruments face the driver, and a
flexible parcel tray is hung in
front of the passenger seat.

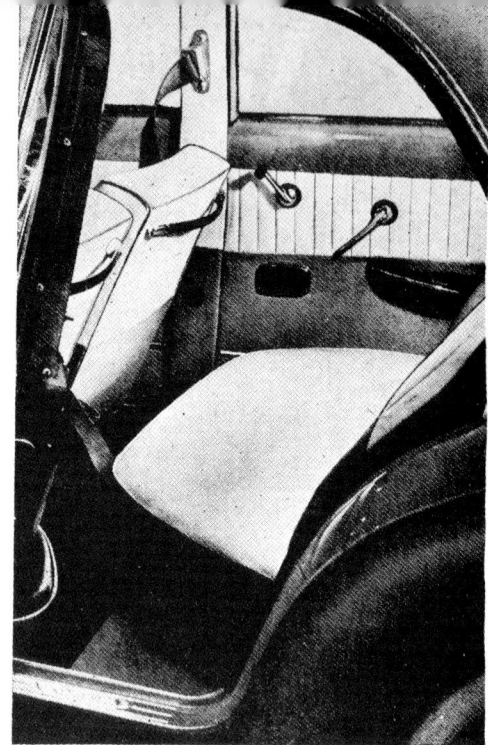

stroke is only fractionally greater than the cylinder bore, and a compression ratio of 8.2/1 permits use of any premium fuel without pinking, although 100-octane petrols give appreciably smoother and quieter full-throttle pulling. But, careful development work and a camshaft giving 66° of valve overlap have allowed this engine to develop remarkable amounts of power at high r.p.m. without temperament.

Easy to start provided that the choke is pulled out for the whole of its exceptionally long travel, this engine warms up from cold rather slowly, even when helped by use of the radiator blind. Once warm, it idles almost as smoothly as do much tamer units, and whilst there is an aural impression that the engine's performance is rather "flat" below 35 m.p.h. in top gear, the stop-watch shows that pulling power remains good as well as surprisingly smooth right down to 10 m.p.h. in this ratio. But it is above 2,000 r.p.m. that the engine really feels to come to life, roaring healthily during hard acceleration but fading to virtual inaudibility when throttled back to sustain a steady cruising speed—even if that cruising speed be in the 75-80 m.p.h. range. This healthy "bark" during acceleration is not found tiring on long runs in the way that more continuous noise can be.

Mated to this engine is a four-speed gearbox which has highly effective synchromesh on *all* ratios—whilst it is possible to start smoothly in 2nd gear, 1st gear is more frequently used by the sort of keen driver who should buy this car, and it is very convenient to be able to push the lever silently into this position before the car has rolled to a standstill at a "Halt" sign. There is a rather wide gap between 2nd and 3rd ratios, but the easy and unprotesting attainment of very high r.p.m. in 2nd gear makes it easy to forgive this, 3rd being a ratio which is really useful up to 70 m.p.h. and which can in extremes be employed up to an astonishing 80 m.p.h. There is very little gear noise, so it is natural to use 3rd almost continuously when driving briskly on congested or winding roads and to drop frequently into 2nd ratio.

Four coil springs which, with the aid of an anti-roll torsion bar, are firm enough to allow very little roll on corners, nevertheless give good insulation against road shocks. The radius arms which locate the rigid rear axle are, however, so well cushioned with sound-insulating rubber that the too-familiar "Hotchkiss drive" complaint of axle tramp can be induced by really fierce braking from a high speed or by a brutally quick start from rest. The brakes respond to quite moderate pedal pressures, and although abnormally hard usage produced a slightly "hot" smell there was no appreciable fade, but unless the car was fully laden premature locking of the rear wheels impaired the ultimate stopping power. A pull-up handbrake to the right of the driving seat is able to hold the car securely on a 1 in 3 hill.

Blessed with an excellently compact turning circle to either left or right, the Volvo has reasonably light steering at all times, and at high speeds there is too little "feel" associated with small divergences from a straight path to please everyone. In wet weather, a stranger is also apt to find it rather easy to provoke wheelspin but familiarity with the car usually silences this criticism, which seems to result from the fact that the Volvo torque curve is still rising as the car accelerates up to speeds at which the torque from most touring car engines is falling off quite rapidly. Whilst it may need "learning" before being fully appreciated by anyone accustomed to less responsive cars, the Volvo soon gets driven at speeds which are consistently above-average for prevailing conditions, without frightening incidents arising.

There has in fact been much work done to make this a safe car, this aspect being spotlighted by the offer of exceptionally well planned safety harness for the front seats at a price of £15. Each safety strap runs from a centre body pillar diagonally across the chest of either the driver or front-seat passenger, then back to an anchorage below the rear seat, and without needing to be adjusted tightly the strap will effectively prevent a head or body moving very far forwards in a sudden stop. Quickly put on or off, the harness is accepted as comfortable in use, once familiarity with it is obtained, only slightly impeding the driver from turning round to see behind him when manoeuvring the car but completely preventing him reaching across to the parcel shelf below the far side of the facia panel. Also as safety features to minimize crash injuries, the parcel shelf is arranged to fold under knee pressure, some vital surfaces are covered by

HARNESS to prevent occupants
of the front seats being thrown
forwards in a collision is an
optional safety feature, the strong
and convenient straps having
anchorages on body pillars and
on the rear floor.

sensibly firm padding, and the sun vizors are flexible—surprisingly, the rear view mirror has a sharp-edged metal frame. At first glance, the long and nearly horizontal steering column extending from near the driver's chest to a point ahead of the front suspension looks potentially dangerous, but a universal joint midway along its length and firm location of the rear half of the column in the scuttle structure should allow this column to "buckle" under impact and so, in conjunction with the harness, eliminate all risk of the driver being impaled on the steering column even in a head-on collision of considerable severity. Fresh air ventilation of the body interior, with the air intake below the windscreen where it does not inhale exhaust fumes from other traffic, is a factor helping safety by keeping the driver alert and the windscreen mist-free, but the standardized heater needs very delicate adjustment of the heat control if it is to provide slightly-warmed (as distinct from hot) air in temperate weather, and

DETAILS of the rear-end layout seen here are a large luggage locker with the spare wheel at one side, torsion-bar springs to support the locker lid, and quadruple tail lamps.

its de-misting effect does not extend to the big but steeply inclined rear window.

Right out of the ordinary in its appearance and in the combination of touring car usefulness with highly sporting verve, the Volvo 122S of 1.6 litre size is obviously worthy of very careful consideration by any keen driver whose motoring budget

might otherwise run to a more orthodox car in the 2½ litre class. The fact that routine chassis lubrication (to 8 grease nipples only) need not be undertaken more frequently than once per 3,000 miles, and the claim that a life of 100,000 miles without need for a major overhaul can be expected, suggest that depreciation should be modest even if the capital outlay required initially is fairly high.

TYPICALLY Swedish is the use of mud-flaps behind the wheels, which may prevent gravel being thrown up but do not keep the car's own coachwork clean in winter weather. Also visible here is the broad wrap-around rear window, and standardized white-wall tyres.

Specification

Engine

Cylinders		4
Bore		79.37 mm.
Stroke		80 mm.
Cubic capacity		1,580 c.c.
Piston area		30.64 sq. in.
Valves		In-line o.h.v. (pushrods)
Compression ratio		8.2/1
Carburetter		Twin S.U. type H4
Fuel pump		AC mechanical with filter bowl
Ignition timing control		Centrifugal and vacuum
Oil filter		Full-flow
Max. power (gross)		85 b.h.p.
at		5,500 r.p.m.
Piston speed at max. b.h.p.		2,890 ft./min.

Transmission

Clutch		Single dry plate (hydraulic actuation)
Top gear (s/m)		4.56
3rd gear (s/m)		5.97
2nd gear (s/m)		9.94
1st gear (s/m)		15.7
Reverse		15.84
Propeller shaft		Divided open
Final drive		Hypoid bevel
Top gear m.p.h. at 1,000 r.p.m.		16.2
Top gear m.p.h. at 1,000 ft./min. piston speed		30.8

Chassis

Brakes		Self-adjusting hydraulic (2 l.s. front)
Brake drum internal dimensions		9 in. x 2 in.
Friction lining area		153 sq. in.
Suspension:		
Front		Independent by coil springs, transverse wishbones and anti-roll torsion bar.
Rear		Coil springs and rigid axle located by radius arms
Shock absorbers		Double-acting telescopic
Steering gear		Cam and roller
Tyres		5.90—15 tubeless whitewall tyres

Coachwork and Equipment

Starting handle	None
Battery mounting	Behind engine
Jack	Lazy-tongs type
Jacking points	External, under sides of body

Standard tool kit: Jack and handle, wheel brace, sparking plug spanner, adjustable spanner, pliers, screwdrivers, tool bag.

Exterior lights: 2 headlamps, 2 parking/flasher lamps, 2 tail/stop/flasher lamps, 2 tail/number plate lamps (4 tail lamps normally illuminated, 2 tail and 2 front lamps only illuminated when parking switch is operated).

Number of electrical fuses: 4 operative, 2 spares for installation of fog and reversing lamps provided in fuse-box.

Direction indicators: Self cancelling flashers combined with side and stop/tail lamps.

Windscreen wipers	Electrical two-blade two-speed, self parking
Windscreen washers	Optional extra
Sun vizors	Two flexible vizors, universally pivoted

Instruments: Speedometer with decimal trip distance recorder, clock, fuel contents gauge, coolant thermometer.

Warning lights: Dynamo charge, oil pressure, headlamp main beam, direction indicators.

Locks:		
With ignition key		Ignition/starter switch
With other key		Either front door, luggage locker
Glove lockers		None
Map pockets		None
Parcel shelves:		Shallow map shelf below left side of facia, deep parcel well behind rear seat.
Ashtrays		One on facia panel, two on rear doors
Cigar lighters		One on facia panel
Interior lights:		One above windscreen, with courtesy switches on front doors.
Interior heaters:		Fresh air type with screen de-misters and two-speed fan.
Car radio		Optional extra
Extras available		Safety harness, also any proprietary accessories
Upholstery material		P.V.C. plastic-coated fabric
Floor covering		Rubber mats
Exterior colours standardized		4 (all with grey top)

Alternative body styles: None (2-door saloon with similar power unit made in Sweden but not marketed in Britain).

Maintenance

Sump: 4¾ pints plus 1½ pints in filter, S.A.E. 20 (below freezing, S.A.E. 10; above 90° F., S.A.E. 30).

Gearbox	1¾ pints, S.A.E. 80 gear oil	
Rear axle	2¼ pints S.A.E. 80 hypoid gear oil	
Steering gear lubricant	S.A.E. 80 hypoid gear oil	

Cooling system capacity (including heater): 15 pints (3 drain taps)

Chassis lubrication: By grease gun every 3,000 miles to 8 points.

Ignition timing: 4° b.t.d.c. static (or set by stroboscope to 21° b.t.d.c. at 1,500 r.p.m., with vacuum advance disconnected).

Contact-breaker gap 0.016-0.020 in.

Sparking plug types (for hard driving): AC 43 com, Auto-lite AH4, Bosch W 225 T3, or Champion J-6.

Sparking plug gap 0.028-0.032 in.

Valve timing: Inlet opens 32° before t.d.c. and closes 72° after b.d.c.; exhaust opens 70° before b.d.c. and closes 34° after t.d.c.

Tappet clearances (hot):

Inlet and exhaust		0.020 in.
Front wheel toe-in		0 to 0.157 in. unladen
Camber angle		0 to ½° positive, unladen
Castor angle		0 to 1° positive, unladen
Steering swivel pin inclination		8°

Tyre pressures:

Front		20-23 lb.
Rear		24-27 lb. according to load
Brake fluid		Heavy duty
Battery		6 volt, 85 amp.hr.

ROAD TEST VOLVO 122-S

A newer and more attractive product of superb Swedish engineering

THE VOLVO 122-S, called the Amazon in Europe, has been built and sold in Sweden for over two years, but it is new to the U.S. market.

And a refreshing new car it is, too: pleasant looking, easy (and fun) to drive, economical and durable in the extreme. It is also refreshing to find a company that actually does something to make its product safe for the occupants, and does it without making asinine statements that the public won't buy safety.

The safety-conscious Swedes have inaugurated many features we've long advocated for cars and would like to see incorporated, in some form, in all passenger vehicles. Here we have the padded instrument panel, a dished steering wheel that is attached to a column built to collapse under pressure, and a plastic package shelf on the passenger's side which folds under impact. The sun visors are of thick foam rubber construction, and seat belts (diagonal straps that extend from the floor in the center across the passenger to the door post) are available on order.

You can, of course, put safety belts in any car, and many vehicles have padded instrument panels to protect the passenger, but most manufacturers then install a row of projecting knobs, handles or switches below the padding which nullifies its effectiveness. The only deterrents to the effectiveness of the Volvo's padded panel were the two radio control knobs.

The newest import is a handsome car in a reserved way, with no evident ostentation or gaudiness. Its excellent over-all proportions carry it well past the average medium-sized sedan in appearance. From some angles, notably the side and rear, it resembles the smaller Simca Aronde (nothing wrong with that). It is quite easily lost in the traffic shuffle, not being an outstanding example of something new in the styling department. This, of course, will keep the car from getting old as quickly as some other contemporary vehicles, so it can mean money for its owner at trade-in time.

A close examination of the car, along with many miles behind the wheel, brought favorable comments from every tester and rider. Design, construction and general quality are obviously excellent, and there is a pervasive feeling of durability. Yet there is no indication of the luxury that we expected of a 4-cyl sedan in this price bracket.

With 3 less horsepower, the same gear ratios, and over-all size and frontal area similar to those of the PV-544, the performance would be expected to be about the same on both models, and so it is. The 122-S weighs

Under the hood, a fine 4-cyl, ohv engine and a fine heater, so necessary for Sweden's cold climate.

COLOR PHOTO BY RAY HALIN

PHOTOS BY POOLE

18

The pleasing proportions are evident in this side view of the 122-S.

65 lb more, and acceleration is reduced proportionately. It is our feeling that the last Volvo we tested (the PV-444, October 1958) was in slightly better tune. If so, the difference in performance was a little more than would be indicated by the two cars' specifications:

	PV-444	122-S
Weight	2160 lb	2225 lb
BHP	85 @ 5500	88 @ 5500
Torque	90 @ 3500	90 @ 3500
Gear ratio (over all)	4.55:1	4.55:1
0–30 mph	4.2 sec	4.7 sec
0–60 mph	13.0 sec	16.2 sec
Top speed	93.5 mph	91.9 mph

Road & Track has not tested a 159 PV-544 because of its similarity to the 444, but the above figures can be assumed to be accurate for this year's model.

The 1600-cc, 4-cyl engine is extremely flexible and, as we've said before, one of the most free-revving rocker-arm engines we've seen. It also seemed to run smoother than other Volvo engines.

Even low gear is synchronized in the 4-speed transmission. Still controlled by a long floor-mounted lever, as in more familiar Volvo models, it operated magnificently every time. There is still too much gap between the ratios of 2nd and 3rd, but the bright side of this is that the car starts easily in 2nd gear. Those who shift more traditionally will find that the shift lever has so strong a spring that it wants to go directly from low gear to 4th instead of 2nd. Yet the engine is so amenable that it pulls

calmly (albeit with little power) from about 15 mph on up.

Both the sturdily vinyl-covered seats and the suspension of the 122-S appear at first to be a little too stiff for comfort. Longer excursions in the car emphasize the wisdom of fairly firm seat cushions (ultra-soft seats are fine on a sofa, but not in a car) but also point up the need for a more bucket-like treatment of the seats themselves; the backs are fine. The suspension grows to feel a little softer. Bad bumps, however, catch it napping.

Foot room for both front- and rear-seat passengers is good, but the knee room of rear-seat passengers is somewhat limited when the front seats are in a position from the center of travel on back. The front-seat adjustment handle itself is the biggest single annoyance in the entire car. It seems mechanically sound, but it's the poorest we've seen for safety. We strongly advise purchasers of the car to exercise extreme caution when adjusting the seat (especially rearward), lest their cut or pinched fingers require medical aid. The rear door window handles are difficult to use when the front seats are back.

The emergency brake is handily located on the driver's left and a clever protective ring prevents its accidental release. Several times during the test we were moved to wonder why they didn't mount it on the driver's right, between the front seats.

Visibility is excellent, due to the well placed window areas (not overly large) and the high seating position. After several years in cars with low builds, the seating

A pleasant but unexciting interior is functional.

The neat luggage compartment is easily accessible . . .

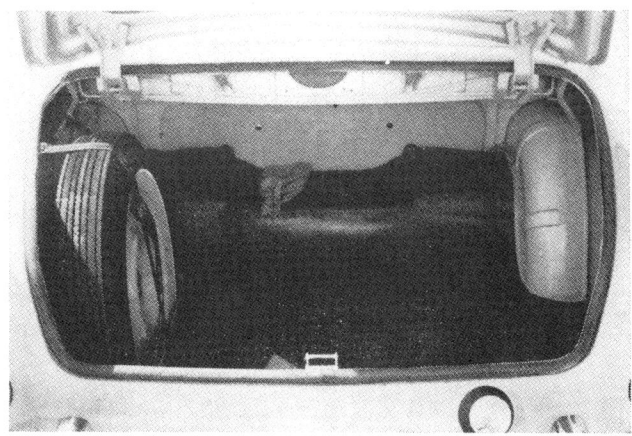

Double grille treatment is different and very well done.

position of a Volvo reminds us of that in a pickup truck. The upright seating, more practical than it is stylish, helps to get comfortable room for 4 passengers in the 102-in. wheelbase. The attractive appearance of the car is even more impressive in view of the excellent head room.

The unit body/frame construction, in addition to being a safer and stronger design than the separate frame and body, offers a very solid-feeling, rattle-free ride. Flinging the car around mountain curves, bumping along over secondary roads, crawling through city traffic, and cruising at 65–70 mph on the highway are all done with the greatest of ease. Corners can be taken with gusto, though with considerable body roll and squealing of tires (added pressure is advisable for faster driving). The ZF steering is precise and gives a good feel of the road to the driver. It seems slower than its 3.2 turns lock to lock.

Like the PV-444 and 544, the 122-S has coil springs and telescopic dampers all around with forged wishbones in front, coupled with a large-diameter anti-roll bar, and trailing arms at the rear with lateral axle location by Panhard rod.

The brakes are British Girling drum type, with 2 leading shoes in front. No brake fade was noticed during our tests, but considerable odor was evident after hard usage.

It has been claimed by many observers that the 122-S body is the same as that of the 1953 Willys sedan. Volvo supposedly purchased the Willys dies at the termination of production of the Willys car, but a comparison of the two makes reveals little or no resemblance. ◙

. . . Under the large, heavily ornamented trunk lid.

ROAD & TRACK ROAD TEST 215

VOLVO 122-S

SPECIFICATIONS

List price	$2895
Curb weight	2390
Test weight	2715
distribution, %	52.3/47.7
Dimensions, length	173
width	63.5
height	59.2
Wheelbase	102.4
Tread, f and r	51.7
Tire size	5.90-15
Brake lining area	165
Steering, turns	3.2
turning circle, ft	33
Engine type	4 cyl, ohv
Bore & stroke	3.125 x 3.15
Displacement, cu. in.	96.6
cc	1586
Compression ratio	8.20
Bhp @ rpm	85 @ 5500
equivalent mph	92.0
Torque, lb-ft	87 @ 3500
equivalent mph	58.5

GEAR RATIOS

O/d (n.a.), over all		
4th (1.00)		4.55
3rd (1.31)		5.97
2nd (2.18)		9.93
1st (3.45)		15.7

CALCULATED DATA

Lb/hp (test wt)	32.1
Cu ft/ton mile	72.6
Mph/1000 rpm (4th)	16.7
Engine revs/mile	3590
Piston travel, ft/mile	1885
Rpm @ 2500 ft/min	4760
equivalent mph	79.6
R&T wear index	67.6

PERFORMANCE

Top speed (4th), mph	91.9
best timed run	92.9
3rd (6450)	82
2nd (6500)	50
1st (6500)	31

FUEL CONSUMPTION

Normal range, mpg	24/27

ACCELERATION

0-30 mph, sec	4.7
0-40 mph	7.4
0-50 mph	11.4
0-60 mph	16.2
0-70 mph	22.8
0-80 mph	32.5
0-90 mph	
0-100 mph	
Standing ¼ mile	20
speed at end, mph	66

TAPLEY DATA

4th, lb/ton @ mph	175 @ 55
3rd	230 @ 50
2nd	360 @ 37
1st	450 @ 22
Total drag at 60 mph, lb	114

SPEEDOMETER ERROR

30 mph	actual 31.5
40 mph	41.0
50 mph	50.5
60 mph	58.9
70 mph	67.2
80 mph	77.6
90 mph	
100 mph	

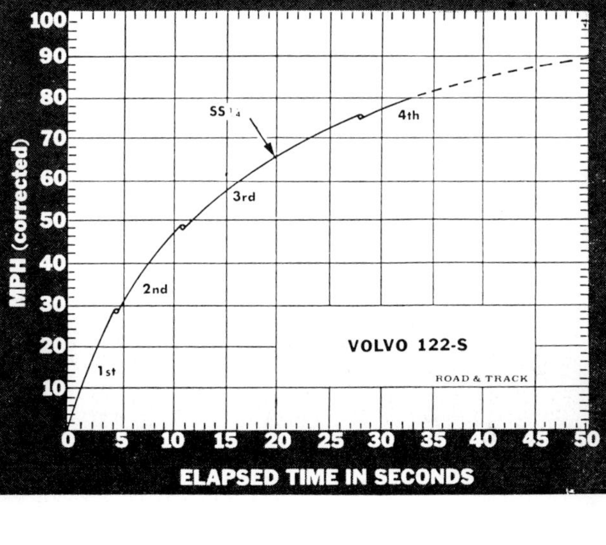

[Acceleration graph: MPH (corrected) vs ELAPSED TIME IN SECONDS, showing 1st, 2nd, 3rd, 4th gear curve with SS ¼ marked. VOLVO 122-S, ROAD & TRACK]

ROAD TEST

The Volvo 122S

by Roy Salvadori

DURING the past two years or so the Volvo has come to be recognised in this country for the very good car that it is. The model which preceded the 122S was the reputation-maker, but its uninteresting appearance prevented it from catching the public's enthusiasm. With the introduction of the Amazon, which was quickly re-christened the 122S, the inherent good qualities of the car were dressed up in a much more acceptable body shell.

By most standards the 122S is a very handsome car. It is also extremely practical. For a 1½-litre car it might seem on first impression to be over-bodied: however, the weight has been kept down so well and the very ordinary looking engine puts out such a rugged 85 b.h.p. there is no question of this. The kerb weight of the car is only 21 cwt. and yet there is ample room for four adults inside, with good leg and head room, and a very capacious luggage boot. The quality of the finish both outside and in is very high indeed. Although not using the traditional luxury appointments of the British car, everything about the Volvo has been well thought out and well executed. On the exterior the paintwork is good and nearly all the bright parts are stainless steel, as befits a car coming from the home of stainless steel manufacture.

Mechanically there seems very little of the unusual about the Volvo. The 1·58-litre engine is an overhead valve four-cylinder unit—which was originally a boat engine—with the high power output of 85 at 5,500 r.p.m. on an 8·2 to 1 compression ratio. Somewhat surprisingly twin S.U. carburettors are employed on cars for Britain, and these probably contribute to quite good fuel consumption. Transmission is through a four-speed gearbox with synchromesh on all gears, and it is operated by a floor-mounted gear lever. Nor is there anything unusual about the suspension

The simple elegance of the 122S is uncluttered by embellishment

which is by coil springs and wishbones at the front, and at the rear by coil springs and locating arms with a conventional hypoid axle. Construction is of the integral type and appears to be exceptionally stiff and rattle-free.

Detail-work

So much for the details of the car. They give no indication of what a very pleasant and purposeful vehicle this is, and I think the main reason for this is that the engineering has been meticulously done and the overall resultant design is extremely well balanced. I would not say that the car is outstanding in any single respect, but due to a remarkable combination of virtues it emerges as a really fine car. I have already touched on the spaciousness of the interior, and when you sit in the driving seat you are immediately impressed by the good visibility and the handiness of all the controls. The speedometer is of the moving band type, and the rest of the instruments are grouped below it in an elongated panel in front of the driver, well-shielded to protect the windscreen from reflections. The long-stalked gear lever comes readily to hand, and the column-mounted indicator switch is also used for flashing the headlamps during day or night driving. Pedals are well placed and there is a foot dip switch. Seats are of the individual type in the front, but not quite as curved as they might be. Although this allows three people to be carried in the front in an emergency it does not give such good location for the driver and passenger under fast cornering.

On the road

As soon as you move off from rest the liveliness of the engine becomes apparent. By taking the revs well up 30 m.p.h. can be achieved in bottom gear although a more normal speed would be between 20 and 25 m.p.h. The gear change is precise and fast despite the rather long travel necessitated by the elongated shank of the gear lever. Accelerating through the gears 70 m.p.h. is reached very quickly indeed for a saloon car with this modest engine capacity. Thereafter, as might be expected, the acceleration is much slower, but a maximum speed of approaching 90 m.p.h. is available given the right road conditions. The steering is delightfully light and precise, and the suspension a good combination of softness for bumpy roads and roll resistance for fast cornering. The large drum brakes I found quite adequate for anything that they were called upon to do, although I must reiterate my preference for discs. In these days of congested roads, if there is to be any possibility of driving quickly with absolute safety there must be no danger of brakes fading after repeated applications. In my experience the disc is the only method that gives this guarantee. In the wet the car is pleasant to handle and has good qualities of adhesion, while, as might be expected with the healthy power and torque available, it is quite easy to slide the back if the throttle pedal is used too vigorously.

Among a number of small features which I appreciate on the Volvo are the remarkably powerful heater, the radiator blind, the provision for safety belts, the very sensible padding on the facia (this is sufficiently strong to provide some real protection in the event of a hard impact), the laminated glass in the windscreen, and the very good sound deadening. On the test car the worst fault was a distinct reluctance to start first thing in the morning. Con-

The simple yet sturdy construction and layout of front (above) and rear (below) suspension systems

Good rearward visibility and commodious luggage accommodation are shown in this view

sistent experiment failed to reveal a method whereby the car could be started short of giving the engine a very long run over on the starter.

Carburettor roar was very noticeable under hard acceleration, but it became unobtrusive when cruising on a light throttle opening. Although, as I mentioned above, the brakes were good there was rather a lot of play in the pedal. I have found this to be fault on all Volvos—not just the test car—and it appears that nothing has been done to improve it.

As I mentioned before the predominant impression left by the car is undoubtedly one of perfect balance: balance not only in the sense of good design but also in respect of general handling. Lively performance combined with good roadholding, spaciousness and well-thought-out detail work must give the car great appeal for the enthusiastic family motorist. Unfortunately, of course, the incidence of duty and purchase tax make the car more expensive than it would otherwise be in this country. Even so, the person looking for those qualities which I have defined will recognise in the Volvo a very worthwhile investment.

The Volvo engine, while no glamour unit, now boasts a polished rocker cover. Twin S.U.s as fitted to the British models can be seen

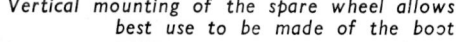

Vertical mounting of the spare wheel allows best use to be made of the boot

SPECIFICATION

ENGINE:

Four cylinders; bore 79·4 mm. (3·12 in.), stroke 80 mm. (3·15 in.). Cubic capacity, 1583 c.c. Overhead valves operated by pushrods. Compression ratio, 8·2 to 1. Maximum b.h.p. (gross), 85 at 5,500 r.p.m. Maximum torque, 87 lb. ft. at 3,500 r.p.m. Fuel system: twin S.U. carburettors, AC mechanical fuel pump. Capacities: fuel tank, 10 gallons; cooling system, 13 pints; oil sump, 6½ pints. Electrical system: 6-volt, 85 amp/hr. battery.

TRANSMISSION:

Single dry plate clutch. Four-speed gearbox with synchromesh on all ratios. Overall ratios: 1st, 4·6; 2nd, 6·0; 3rd, 9·9; top, 15·7 to 1. Centrally-mounted gear lever. Hypoid bevel rear axle.

CHASSIS:

Suspension: front, independent by coil springs, transverse wishbones and anti-roll bar; rear, live axle with coil springs and radius arms. Brakes: Wagner hydraulic; total lining area, 153·7 sq. in. Steering: cam and roller; two-spoke steering wheel; turning circle, 32 ft. 6 in. Tyres: 5·90 × 15, whitewall, tubeless.

DIMENSIONS:

			ft.	in.
Wheelbase	...	...	8	6½
Track: front	...	...	4	3½
rear	...	...	4	4
Overall length	...	...	14	5½
Height	...	...	4	11
Width	...	...	5	4
Ground clearance	...	...		7¼
Weight	...	...	21 cwt.	

PERFORMANCE:

Acceleration through gears:

m.p.h.	sec.
0—30	5·6
0—40	7·6
0—50	12·8
0—60	15·2
0—70	22·8

Speed in gears:

1st	30 m.p.h.
2nd	45 m.p.h.
3rd	70 m.p.h.

The  in detail

1. Rear air cleaner

2. Front air cleaner

3. Float chamber (front carburetter)

4. Fuel line

5. Cylinder head

6. Cylinder block

7. Ignition setting mark (T.D.C.)

8. Dynamo

9. Engine mounting

10. Oil pressure relief valve

11. Oil filter

12. Type and number designation plate (early production)

13. Starter motor

14. Solenoid

15. Flywheel housing

16. Exhaust manifold

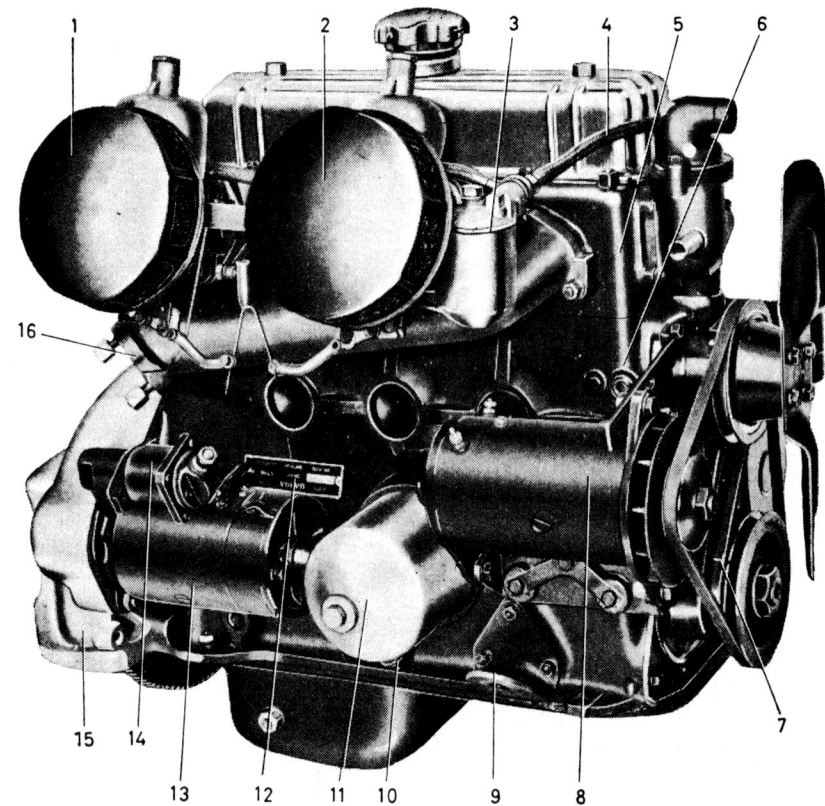

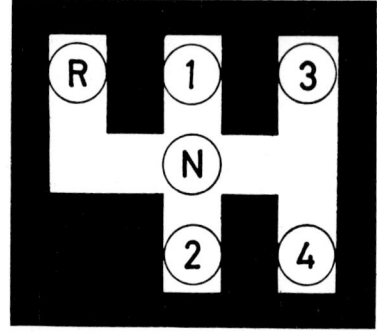

Gear positions.

All four forward ratios

are equipped with

synchromesh.

The independent front suspension system incorporates coil springs, wishbones, anti-roll torsion bar and double-acting hydraulic dampers.

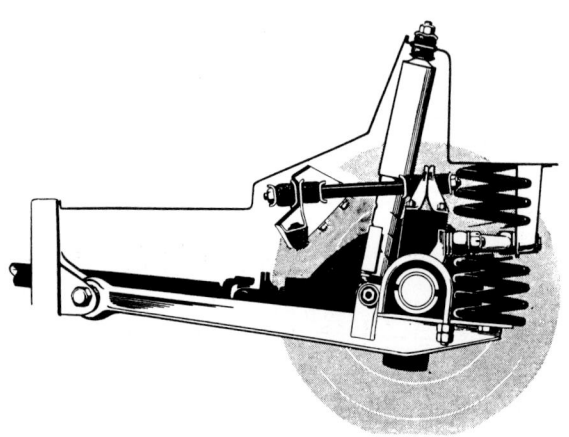

The rigid rear axle is positively located by torque arms and Panhard rod; coil springs and double-acting hydraulic dampers are employed.

The Volvo 122S saloon

IF it is true, as someone has said, that motor-cars are divisible into three classes — cars for men, cars for gentlemen and cars for gents — we must, without hesitation, place the Volvo 122S (the former "Amazon" by which the car was known has been dropped) in the first category. Here, indeed, is a man's car. The solidity of its construction, the deep-chested roar of its motor under acceleration, the positive way in which it handles and in which all its controls respond, the effortlessness of its manner of high-speed cruising, the excellence of the illumination provided by its headlights — even minor items such as the "feel" of the precision-turned pieces of steel which make up the brake- and gear-levers — all these aspects of the Volvo proclaim it to be of that particular breed of car which will delight any man who derives pleasure from sound mechanical workmanship and design.

Not that the Volvo's layout incorporates any radical departures from what has come to be regarded as conventional in contemporary motor-engineering practice. The four-cylinder, o.h.v. motor (admittedly equipped with twin carburetters) is water-cooled and mounted in front; the compression ratio of 8.2 to 1 at which this motor operates is not high by to-day's standards; rear suspension is not independent and, in appearance, the car is not noticeably different from the mass of 4/5-seater saloons common on South African roads. The interior layout reveals nothing really out-of-the-ordinary

beyond full instrumentation which lacks only a revolution counter. Items such as ash-trays, courtesy lights, neatly-upholstered seats, grab handles are featured in many other cars of lesser price and in themselves do not distinguish the Volvo in any way.

But exceptional the car certainly is. As our performance figures show, the 122S has powers of acceleration and speed far in advance of what is regarded as usual. The obvious care and thought which have gone into the design of the car and its assembly indicate, also, that its term of useful life will also exceed what might be expected. The Volvo's road behaviour is excellent — more that of a sports-car than a medium-sized family saloon — and we can only assume that the manufacturers have set about improving what has come to be regarded as normal practice to produce as good a car as possible in its price range.

That this is a quality machine is immediately evident. Before the driver is a full array of instruments, lacking, as we have said, only a rev. counter. Windscreen wipers are two-speed and park themselves automatically. Lighting of the instruments is rheostat-controlled and although there is no "cubby-hole" in the accepted sense, there is, opposite the front-seat passenger, a wide and deep parcel shelf under the dash-board illuminated by a light separate from the instrument lights.

Differing from the general run-of-the-mill cars (and

(continued overleaf)

Restrained, smooth lines are evident in this three-quarter front view of the Volvo 122S. The entire under-carriage is rubberized.

proclaiming its Swedish origin where every car must be fitted with a heater) the Volvo is standardly-equipped with a heater, defroster and fresh air system, the controls for which being grouped in the centre of the instrument panel. This system is fitted with a two-speed fan to enable the interior temperature to be adjusted to a nicety in cold weather.

This installation, we must point out, was of no real help in really hot weather. Fresh air, it is true, is ducted to the car's interior through an intake in front of the windscreen — an intake, that is, away from exhaust fumes of cars ahead — but the air itself was not cooled in any way.

The seats, both fore and aft, were firm and comfortable. At first, it was thought that they were too firm but in the course of a longish run into the country on a particularly hot day, no discomfort was experienced by driver or passengers. A parcel shelf in the rear compartment is located behind and below the top of the seat; thus, items stowed there were not in the way of the driver's vision through the rear-view mirror and were not thrown forward under emergency braking. (At speed, the car's rear-view mirror tended to vibrate somewhat, giving the driver an "out-of-focus" impression of what was going on behind the car.).

The boot, in which the spare wheel is stowed vertically, is large without being cavernous. It has enough room for the reasonable travelling needs of a family of four or five. At night, incidentally, when the boot lid is lifted, the interior is lit by the lights which illuminate the rear registration plate. A complete tool kit and a jack are supplied as normal equipment.

A sign of the Volvo's sporting qualities is evident from the twin carburetters which are seen when the bonnet is raised. Another hint of the car's cold-country origins is provided by the radiator blind which is operated by a chain running over the top of the motor from the instrument panel.

The Volvo's sports-car qualities are noticed as soon as the motor is started. (This was accomplished, by the way, without the use of the choke on three succcessive

The engine compartment of the 122S: behind the motor and next to the battery is the heater-defroster-fresh air unit. The chain to operate the radiator blind extends across the motor and above the twin SU carburetters.

Below the dash-board and in front of the floor-mounted gear lever is the heater-defroster-fresh air unit. All foot controls are pendant, the gear-lever being on the right of the driver's seat. All 122S models sold in South Africa are fitted with right-hand drive.

days although it did take a few minutes running before the engine was warmed up.) The gears — all four are synchromesh — engage instantly and positively. The engine, quite quiet at tick-over speeds or when idling through traffic, suddenly develops a pronounced power roar when the accelerator is depressed and hurls the car forward in a manner eminently satisfying to an enthusiast.

This power roar is only audible when the motor is under load. Once the car reaches a point where pressure is removed from the accelerator — be that point 35 m.p.h. or 85 m.p.h. — the roar disappears and although the motor is not quiet, it is certainly not noisy. With quarter-lights open, there is some wind roar, but this is never excessive and at no time did passengers have to shout to make themselves heard.

At high speed, the 122S is vibrationless, apart, that is, from the already-mentioned rear-view mirror. The car holds the road exceptionally well, cornering under power being accompanied by very little tyre howl. As we have said, the car's controls respond positively. Gear-changing, to the enthusiast, is a delight although, in the car under test which had less than 2,000 miles on the clock, pressure had to be exerted to overcome some stiffness. The trip distance recorder calibrated in tenths, clicked over from one digit to the next without moving across slowly as do most instruments of this sort. Without going into unnecessary details, let it suffice merely to say that everything about the Volvo behaves as it should do, the speedometer being exceptionally accurate.

In one aspect only can we fault the car's behaviour on the road. During a high-speed run on, we hasten to point out, an extremely gusty day, the Volvo's stability was slightly impaired by cross-winds. This was particularly noticeable by rear-seat passengers when pressure was removed from the accelerator pedal. The rear of the car, under these conditions, tended to sway slightly and necessitated correcting action being taken by the driver. Under extremely heavy acceleration from standstill on rough roads, the rear axle tended to tramp about. We do not regard this is a real fault as no careful owner would go to such extremes of trying to get away in a hurry as we did.

The test car was fitted with a non-standard all-wave Elra radio equipped, in turn, with two loud-speakers. One was mounted under the dash-board and the other behind the rear seat — the boot conveniently acting as a sort of "sounding-box". This item, with its aerial which extends and disappears as the radio is switched on or off costs an extra £47.15.0 fitted, plus £18.18.0 for the automatic aerial.

As regards fuel consumption: the local distributors have informed us that the 122S returns a figure of 32 m.p.g. at 60 m.p.h. After our series of acceleration tests, we achieved a figure of 24.5 m.p.g. over 113 miles. Fuel consumption for a further mileage of 193 miles was 29.2 m.p.g. Much of this latter mileage was covered at speeds in the 80's — in fact, over a 30-mile stretch of national road we overtook every car ahead of us excepting one large American saloon — and we have no doubt that a steady 60 m.p.h. would result in 32 m.p.g.

We were fortunate — as far as this road test was concerned — in that during our temporary ownership of the 122S we were able to drive for some miles over exceptionally dusty roads and also through a heavy rainstorm. Sealing of the car against dust was found to be excellent; apart from a very slight layer which formed on the padded instrument panel, dust was absent. In the rain, we found that the wipers cleared the windscreen effectively and quickly; not a drop of moisture entered the car at all.

The ride over macadam was, as we have said, vibrationless. Over rougher terrain, the car showed up well — sudden bumps and rises on an undulating gravel road being effectively damped. Rear-seat passengers were not thrown up and down too much when these rises were taken at speeds higher than normal, leg-room at the rear being adequate for six-footers.

Much attention has been paid to the safety of the Volvo's occupants. The car, itself of heavier-than-normal steel gauge construction, is fitted with a laminated windshield which will not shatter or become opaque if hit by stones thrown up by other cars; sun visors are padded and safety belts are an optional accessory. With a total lining area of 165 sq. ins., the brakes

Vertical installation of the spare wheel can be seen in this photograph of the boot; the rear of the radio loudspeaker unit can be seen in the centre below the lid.

proved to be fade-free and brought the car to an even halt at all times.

Because the name Volvo is comparatively new to South African motorists, we include in this road test some brief notes about the company and its products. The first Volvo car was produced in 1927. This was an open four-seater with a top speed of 37 m.p.h. During the first year of operations, the company turned out 297 cars. In 1958, total production was more than 70,000, the grand total of all cars produced being, at the end of that year, more than 450,000. Rear axles for the cars are made by Bofors, bodies by Svenska Stalpressnings, engines by the Volvo Pentaverken subsidiary and gearboxes by its Kopingverken.

Non-Swedish items in the car tested included instruments by Vdo, electrics by Bosch, brakes by Girling and carburetters by SU.

Costing nearly £1,200 at the S.A. coast, the purchase of a 122S will involve for most people a heavy capital outlay. But this, in our view, is worth while in view of the car's very evident reliability and its consequent relatively small depreciation costs. The Volvo enhances in every way the enviable reputation which other Swedish products have built up in all parts of the world and will satisfy the sporting instincts of the enthusiast as well as the transportation needs of his family. ●

SPECIFICATION AND PERFORMANCE

BRIEF SPECIFICATION

Make ... ... VOLVO
Model ... ... 122S saloon
Style of Engine In-line, 4-cylinder, o.h.v., water-cooled, front-mounted.
Bore ... 3·125 ins. (79·37 mm.)
Stroke ... 3·15 ins. (80 mm.)
Cubic Capacity 97 cu. ins. (1,580 c.c.).
Maximum Horse-Power 85 b.h.p. at 5,500 r.p.m. (Compression ratio 8·2 : 1).
Brakes Self-centering, hydraulic, 10 in. drums, total lining area, 165 sq. ins.
Front Suspension Independent with coil springs and control arms and stabilizer bar.
Rear Suspension Coil springs, rear axle carried on two rubber-mounted support arms, two torque arms and a track bar.
Transmission System S.d.p. clutch with four manually-controlled forward gears, all synchromesh.
Gear Ratios (overall) 1st 15·7 to 1
2nd 9·94 to 1
3rd 5·97 to 1
Top 4·56 to 1
Rev. 15·84 to 1
Final Drive Ratio ... 4·56 to 1
Steering Cam and roller, 3¼ turns, lock-to-lock.
Turning Circle ... ... 32 ft.
Overall Length ... 14 ft. 7·2 ins.
Overall Width ... 5 ft. 3·8 ins.
Overall Height ... 4 ft. 11·5 ins.
Ground Clearance (with 4 passengers) ... ... 6·7 ins.
Price ... £1,197 (at the S.A. Coast)
Weight ... ... 2,400 lbs.
Annual Licence £8
Basic Insurance Premium (in Cape Town) £26

PERFORMANCE

Acceleration 0-30 m.p.h. 4·9 secs.

0-40 m.p.h. 7·9 secs.
0-50 m.p.h. 11·8 secs.
0-60 m.p.h. 17·0 secs.
0-70 m.p.h. 21·7 secs.
0-80 m.p.h. 32·5 secs.
In top gear from a steady 20 m.p.h. to 40 m.p.h. ... ... 10·0 secs.
In top gear from a steady 30 m.p.h. to 50 m.p.h. ... ... 10·3 secs.
In top gear from a steady 40 m.p.h. to 60 m.p.h. ... ... 10·4 secs.
In top gear from a steady 50 m.p.h. to 70 m.p.h. ... ... 11·9 secs.
In top gear from a steady 60 m.p.h. to 80 m.p.h. ... ... 18·8 secs.
Maximum Speed ... 91 m.p.h.
Reasonable Maximum Speed in
1st gear ... ... 25 m.p.h.
2nd gear ... ... 45 m.p.h.
3rd gear ... ... 73 m.p.h.
Fuel Consumption ... 29·2 m.p.g.
Test Conditions Sea level, warm, extremely gusty, dry road. 90-octane fuel.

by John Christy

I F YOU ARE AMONG THOSE who believe the Volvo 122-S is nothing more nor less than a Swedish Willys, forget it. The relationship of the Aero sedan to Volvo's latest confection is, as the fiction publishers say, purely coincidental. After getting the word straight from Volvo that the four-door is not now and never has been thumped by a Willys die, we — suspicious souls that we are — made a direct comparison between the two cars, parking one next to the other. The Willys is wider by inches, higher, shorter in the belt line and deeper in section from window sill to rocker panel. Fender and deck lines are entirely different. The only resemblance, and admittedly a striking one, is in the roof line and the rear window cut-out. All of which leads us to believe that if a fib has been told and Willys dies were indeed used, the men of AB Volvo are the finest disguise artists extant.

The lack of resemblance between America's earliest modern compact and the Gothenburg offering doesn't end with the sheet metal, either. The 122-S is, for a 96-incher, a going, handling fool. The late Mike Hawthorne put it very succinctly when he stated that the 122-S gets underway "with

Twin SU side-draft carburetors help put 85-hp muscle in the mild appearing 96.5-cubic-inch engine of Volvo's newest creation.

VOLVO
122-S ROAD TEST

Instrumentation shows fine quality but lacks in quantity. Gauges are used for fuel and water temp, warning light for oil pressure.

The Swedes have a word for it—"SPORTS SEDAN," which means sedan-smooth and sports car-stable

the vigor of a man who discovers that the log he's sitting on has two bloodshot eyes and enormous teeth."

This performance comes from a mild appearing but exceedingly muscular four-cylinder engine that is the Volvo firm's main prime mover in various guises. Originally a 1500cc or 91-cubic-inch unit, this has been bored out in the last three years to give a displacement of 1582cc or 96.5 cubic inches. As used in Volvo cars this engine develops an honest 85 bhp at 5500 rpm and 87 lbs. ft. of torque at 3500. Power falls off very slowly after peak power has been reached, particularly in the special equipment versions, and even showroom stock models can be twisted to 6500 rpm in third gear. Outwardly the engine is similar to the last of the pre-BMC MG engines, the XPEG, though a little bulkier. In many respects it is similar in interior design as well except

that the intake ports are individual, not siamesed as in the late lamented British engine, and in dimension it is bigger in bore and shorter of stroke.

Volvo, in the guise of the PV-444 and PV-544 two-door sedans, have been noted for sporting characteristics and have cleaned house in virtually every sedan race in this country and abroad, to say nothing of a disconcerting ability to hold their own against more than a few outright sports cars. One might be justified in supposing that the 122-S might be a somewhat calmer cousin; with a goer like the PV-544 in the stable, Volvo might consider producing a calmer version for the more faint of heart. Not so — the 122-S is every bit the sizzler its homely cousin is. It is every bit the handler the two-door is, too, but in a somewhat different way.

The difference is rather hard to pinpoint in a hard and fast manner but one experience was enough to prove the point. Our first introduction to the 122-S was in the summer of 1959, the car being one of the first of the breed on these shores. In company with Volvo dealer and team driver Art Riley we set off to take a look at New England. With us we took Art's two-time Little LeMans-winning PV-444. The 122-S was new and dead stock while Art's machine was equipped with the optional power kit. In a fast day's run over a twisting Berkshire road (which will remain nameless for obvious reasons), we discovered several interesting facts.

VOLVO

Outstanding among these was that under certain circumstances the 122-S could outcorner the 444 without even working hard at it. Fast downhill bends, for instance, were duck soup for the four-door, as were medium-fast level bends. Esses that had the taller 444 whip-sawing were taken almost level with the 122-S. Uphill turns and tight corners remained the exclusive property of the 444 although not without some intimidation from the 122-S. The major difference between the two cars seemed to be one of weight transfer, the outboard front wheel of the 444 taking a tremendous part of the load on fast turns. While this weight shift occurred in the 122-S it didn't do so to the same shocking degree as in the 444. Weight shift or no, the 444 was quicker by seconds around the Lime Rock Park course. Art got the 444 around the normal (as opposed to the Little LeMans) course at a steady 1:17 to 1:18 while the best the 122-S could do the circuit in was 1:21 and up. Undoubtedly part of this was due to the preparation the two-door had received and part definitely to Art's notoriously heavy right foot and skill in the car; it remains to be seen what the 122-S could do given the same preparation and the same driver. As it is, anything in the low 1:20's at Lime Rock is good scooting for a sedan.

It takes a somewhat practiced eye and a long look underneath the car to discover why it handles as well as it does. At first glance there is nothing underneath to give a clue to the roadability — a fairly ordinary ladder-type frame of heavy-appearing channel is supported by four coil springs. At the front there is the expected independent suspension with unequal length A-arms and a torsion anti-roll bar; at the rear there is a perfectly ordinary rigid rear axle housing. This last however is located very firmly both longitudinally and laterally by a pair of very large trailing arms, a pair of torque rods, and by a hefty Panhard rod that effectively prevents any side-to-side motion. Most of the roll stiffness is in the front while the rear is permitted considerable wheel-travel. The result is not only a smooth ride but a complete reluctance to pick up a rear wheel on corners although the stability of the car is such that this nuisance can be provoked if one wants to try hard enough. There is a payoff in another direction, too, in that the car is one of those vehicles that can be barreled quite rapidly and in complete comfort over gravel or dirt roads with a feeling of great stability and without the feeling that the car might fall apart from the beating.

What else does the 122-S have to offer for the extra $400 price tag? Plenty. In common with the 544 it has the all-synchro four-speed box that is a pleasure to operate although a bit widely split between second and third. It features a

Raised trunk lid reveals compact but adequate space for most traveling needs. Spare tire is mounted upright and to the extreme left side for maximum room. The trunk is tightly sealed against moisture.

Outward appearance of 122-S engine is similar to the last of the pre-BMC MG powerplants. Although appearing to be a calmer version of Volvo's two-dooor PV-544, the 122-S is every bit the sizzler its cousin is. This four-cylinder ohv was originally a 1500cc engine, but has been increased to 1582cc.

low-gear ratio that seems to be more designed for climbing walls and uprooting stumps than anything else. Coupled to this is the same 1580cc engine that puts the sizzle into the two-door. While some won't appreciate the engine's slightly loping idle, the minor idling-speed roughness indicates cam timing designed more for performance than for coin-balancing smoothness at low speeds. So quick-starting is the engine that, if one wants to impress passengers or play at LeMans starts, the car can be gotten under way merely by twisting the key and hitting the throttle with the transmission previously placed in low gear and *without using the clutch*. It's not recommended but it can be done. Response is instantaneous and as the engine winds up there is an eminently satisfying strong snore from the carburetors, indicating huge amounts of air being gulped in by the deep-breathing four-barrel.

Getting under way is a cinch. The clutch takes hold with a strong but not sudden bite and shifts are made quickly and smoothly through a set of synchronizers in the gearbox that none but the most callously brutal hand can beat, either up or down. Once under way it would be wise for the man who wants to keep his license unsullied to pay attention to the speedometer from time to time. This red ribbon has a tendency to march unobtrusively up to quite illegal speeds if an unconsciously heavy foot is used. Cruising speed is in the 65-to-75 range and so slight is the wind noise — or all other noises for that matter, except for the carburetor snore — that you're there before you realize it.

As a car goes so should it stop. The Volvo stops as well as it goes. Halt after halt at .75 G can be made with little or no pedal loss or apparent fade. The brakes are self-centering duo-servo units with 10-inch drums and 239 square inches of swept area. They're smooth, require only moderate pedal pressure even on a wet day and have no tendency to grab either together or alternately.

Volvo advertising refers to this car and the 544 as "sports sedans" and the description is an accurate one. Steering is by a cam-and-roller unit with a ratio of 15 to 1, giving a lock-to-lock figure of 3⅓ turns, sporting enough in any car. There's just enough return and just enough road feel through

it to give the driver an accurate idea of the status of the front wheels. Overall handling is likewise "sporty" but in no way stiff. The unique Volvo suspension system damps out any hopping or skipping tendencies on turns even over the most abominable washboard roads. The overall effect is that one feels one could drive the car all day, and fast, without emerging fatigued at the end of the run, a feeling borne out by experience.

If it weren't for this smooth, secure ride, one could get jarred in the 122-S. Despite the overall appearance of luxury, the seats in this car are *firm* unto the point of spartanism. One of the few complaints we can register is this hard seating, for with it goes a lack of lateral support that under certain circumstances can lead to acute discomfort, causing one to have to hang on to the wheel for the support that should have been given by the seat. Room that would be welcome with a more form-fitting seat becomes just something to slop around in when the car is pushed in a hurry over a twisting road.

One other complaint needs to be registered regarding instruments. This is not so much concerning quality as quantity — there just are not enough. Aside from the speedometer, an extremely accurate instrument by the way, there are only a fuel gauge and a water temperature gauge, neither of which is marked in numbers. The rest of the indicators are warning lights for oil pressure, generator and high beam. While gauges for current output and oil pressure may not be vital they are handy to have in a car such as this.

One nice thing about the trip mileage indicator, though, is that one needn't twiddle for eons with a little knob to reset it — a simple pull ratchets the meter back to zero in one fast motion. Like z-z-zip. Other nice things: All controls are adequately marked and conveniently placed and, best of all, decently lit so there's no fumbling in the dark for a seldom-used gadget. Other complaint: There is no lockable glove box or stowage compartment, a simple package shelf being placed under the dash on the passenger's side which means one keeps cameras and other valuables in the trunk if parking lots are to be used. Personal property insurance might also be a good idea.

COMFORTABLY DESIGNED INTERIOR PLUS UNIQUE SUSPENSION SYSTEM OFFERS A FIRM BUT UNTIRING RIDE TO DRIVER AND PASSENGERS.

It might be expected that a car from Scandinavia would be fairly weather-tight and well heated. The Volvo is almost hermetically sealed — to the point, in fact, where it takes a strong push to shut the door if everything is closed tightly, unless the ventilator is open. The heater is controlled by three toggles and a switch on the dash, and when turned on full blast it can literally fry the polish on one's shoes. The toggles control, from left to right, the amount of outside air admitted, the proportions of warm air to floor and defroster, and finally the degree of heat from nil to broiling.

VOLVO

The Swedes are a practical race that obviously has no use for extraneous tomfoolery like vast shrouds of plastic or metal around vital parts. Everything mechanical is eminently get-at-able from the rear end up through the engine compartment which is as uncluttered as that of any modern vehicle. In fact so accessible is everything that the afore-mentioned Art Riley was able on a rainy afternoon to rebuild a seized engine blown during a race at Lime Rock. It was his only transportation and had been driven there. Between two o'clock in the afternoon and dark, Riley had the engine field stripped, the offending piston removed and the engine back together. Try that one on your merry Oldsmobile.

The final question is whether, in the light of the presence of the American compacts which approximate the size of the Volvo, the car is worth the $2600 price tag. The answer, at least to this reviewer, is a definite yes. The price of the car includes every normal accessory including a radio — in other words, you pay your money and you walk out the door own-ing a complete car. While the factory or p.o.e. prices on other compacts and imports may be lower, actual purchase price of, say, a Corvair, Falcon, Valiant, A-90, Fiat 2100 or similar car with the same equipment will be virtually the same if not more. All things considered, the Volvo 122 S is an awful lot of car for the money. •

Volvo emblem cleverly disguises car's trunk lock and latch. Lid pops open simply by raising lower portion of the emblem.

Hand brake is located on left side of front seat, in typical European fashion. Note ample legroom and recessed steering.

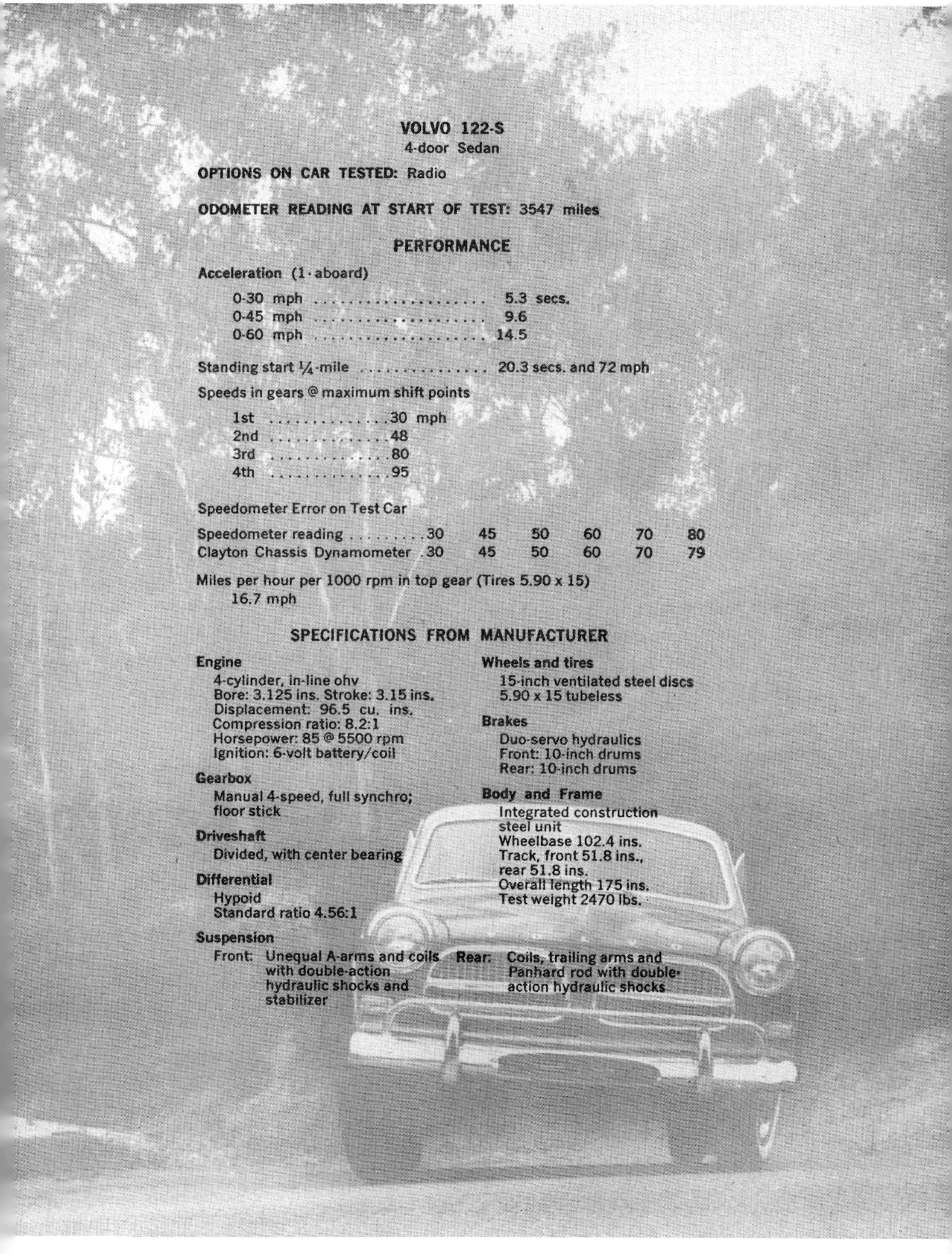

VOLVO 122-S
4-door Sedan

OPTIONS ON CAR TESTED: Radio

ODOMETER READING AT START OF TEST: 3547 miles

PERFORMANCE

Acceleration (1·aboard)

0-30 mph	5.3 secs.
0-45 mph	9.6
0-60 mph	14.5

Standing start ¼-mile .... 20.3 secs. and 72 mph

Speeds in gears @ maximum shift points

1st	30 mph
2nd	48
3rd	80
4th	95

Speedometer Error on Test Car

Speedometer reading	30	45	50	60	70	80
Clayton Chassis Dynamometer	30	45	50	60	70	79

Miles per hour per 1000 rpm in top gear (Tires 5.90 x 15)
    16.7 mph

SPECIFICATIONS FROM MANUFACTURER

Engine
4-cylinder, in-line ohv
Bore: 3.125 ins. Stroke: 3.15 ins.
Displacement: 96.5 cu. ins.
Compression ratio: 8.2:1
Horsepower: 85 @ 5500 rpm
Ignition: 6-volt battery/coil

Gearbox
Manual 4-speed, full synchro;
floor stick

Driveshaft
Divided, with center bearing

Differential
Hypoid
Standard ratio 4.56:1

Suspension
Front: Unequal A-arms and coils
with double-action
hydraulic shocks and
stabilizer

Wheels and tires
15-inch ventilated steel discs
5.90 x 15 tubeless

Brakes
Duo-servo hydraulics
Front: 10-inch drums
Rear: 10-inch drums

Body and Frame
Integrated construction
steel unit
Wheelbase 102.4 ins.
Track, front 51.8 ins.,
rear 51.8 ins.
Overall length 175 ins.
Test weight 2470 lbs.

Rear: Coils, trailing arms and
Panhard rod with double-
action hydraulic shocks

VOLVO

VOLVO 122-S

SPECIFICATIONS

List price	$2495
Price, as tested	2495
Curb weight, lb	2400
Test weight	2750
distribution, %	52/48
Tire size	5.90-15
Tire capacity, lb	3420
Brake lining area	165
Engine type	4 cyl, ohv
Bore & stroke	3.125 x 3.15
Displacement, cc	1586
cu in	96.6
Compression ratio	8.2
Bhp @ rpm	85 @ 5500
equivalent mph	92.6
Torque, lb-ft	87 @ 3500
equivalent mph	58.4

GEAR RATIOS

4th (1.00), overall	4.55
3rd (1.36)	6.18
2nd (1.99)	9.05
1st (3.13)	14.2

DIMENSIONS

Wheelbase, in	102.4
Tread, f and r	51.7/51.7
Over-all length, in	175
width	63.3
height	59.2
equivalent vol, cu ft	382
Frontal area, sq ft	20.9
Ground clearance, in	7.7
Steering ratio, o/a	n.a.
turns, lock to lock	3.3
turning circle, ft	32
Hip room, front	.53
Hip room, rear	.52
Pedal to seat back	.41
Floor to ground	13
Luggage vol, cu ft	n.a.

PERFORMANCE

Top speed (4th), mph	90
best timed run	87.4
3rd (6500)	79.8
2nd (6500)	54.5
1st (6500)	34.8

FUEL CONSUMPTION

Normal range, mpg	24/27

ACCELERATION

0-30 mph, sec	4.8
0-40	7.9
0-50	12.0
0-60	16.6
0-70	23.0
0-80	33.1
0-100	
Standing ¼ mile	20.4
speed at end	66.3

PULLING POWER

4th, lb/ton @ mph	160 @ 55
3rd	220 @ 48
2nd	340 @ 36
Total drag at 60 mph, lb	124

SPEEDOMETER ERROR

30 mph, actual	30.9
60 mph	60.7
90 mph	

CALCULATED DATA

Lb/hp (test wt)	32.3
Cu ft/ton mile	73.2
Mph/1000 rpm	16.7
Engine revs/mile	3600
Piston travel, ft/mile	1890
Car Life wear index	68.0

ACCELERATION & COASTING

(Graph: MPH vs ELAPSED TIME IN SECONDS, showing SS¼, 4th, 3rd, 2nd, 1st curves)

BUICK SPECIAL/185 BHP

SPECIFICATIONS

List price	$2529
Price, as tested	2941
Curb weight, lb	2560
Test weight	2880
distribution, %	55.7/44.3
Tire size	6.50-13
Tire capacity, lb	3980
Brake lining area	130
Engine type	V-8, ohv
Bore & stroke	3.50 x 2.80
Displacement, cc	3524
cu in	215
Compression ratio	10.25
Bhp @ rpm	185 @ 4800
equivalent mph	100.8
Torque, lb-ft	230 @ 2800
equivalent mph	58.8

GEAR RATIOS

4th ( ), overall	n.a.
2nd (1.00)	3.36
1st (1.58)	5.30
1st (1.58 x 2.50)	13.27

DIMENSIONS

Wheelbase, in	112.1
Tread, f and r	56.0/56.0
Over-all length, in	188.4
width	71.3
height	52.8
equivalent vol, cu ft	410
Frontal area, sq ft	20.9
Ground clearance, in	4.9
Steering ratio, o/a	20.8
turns, lock to lock	4.2
turning circle, ft	38.1
Hip room, front	53.0
Hip room, rear	53.0
Pedal to seat back	39.2
Floor to ground	10.3
Luggage vol, cu ft	28.0

PERFORMANCE

Top speed (2nd), mph	110
best timed run	
2nd ( )	
1st (5200)	69
1st (5250)	28

FUEL CONSUMPTION

Normal range, mpg	17/21

ACCELERATION

0-30 mph, sec	3.7
0-40	5.4
0-50	7.5
0-60	10.0
0-70	14.1
0-80	20.0
0-100	
Standing ¼ mile	17.4
speed at end	76.0

PULLING POWER

4th lb/ton @ mph	@
2nd	340 @ 42
1st	off scale
Total drag at 60 mph, lb	144

SPEEDOMETER ERROR

30 mph, actual	25.6
60 mph	51.7
90 mph	79.3

CALCULATED DATA

Lb/hp (test wt)	15.6
Cu ft/ton mile	123.5
Mph/1000 rpm	21.0
Engine revs/mile	2860
Piston travel, ft/mile	1335
Car Life wear index	38.2

ACCELERATION & COASTING

(Graph: MPH vs ELAPSED TIME IN SECONDS, showing SS¼, 2nd, 1st curves)

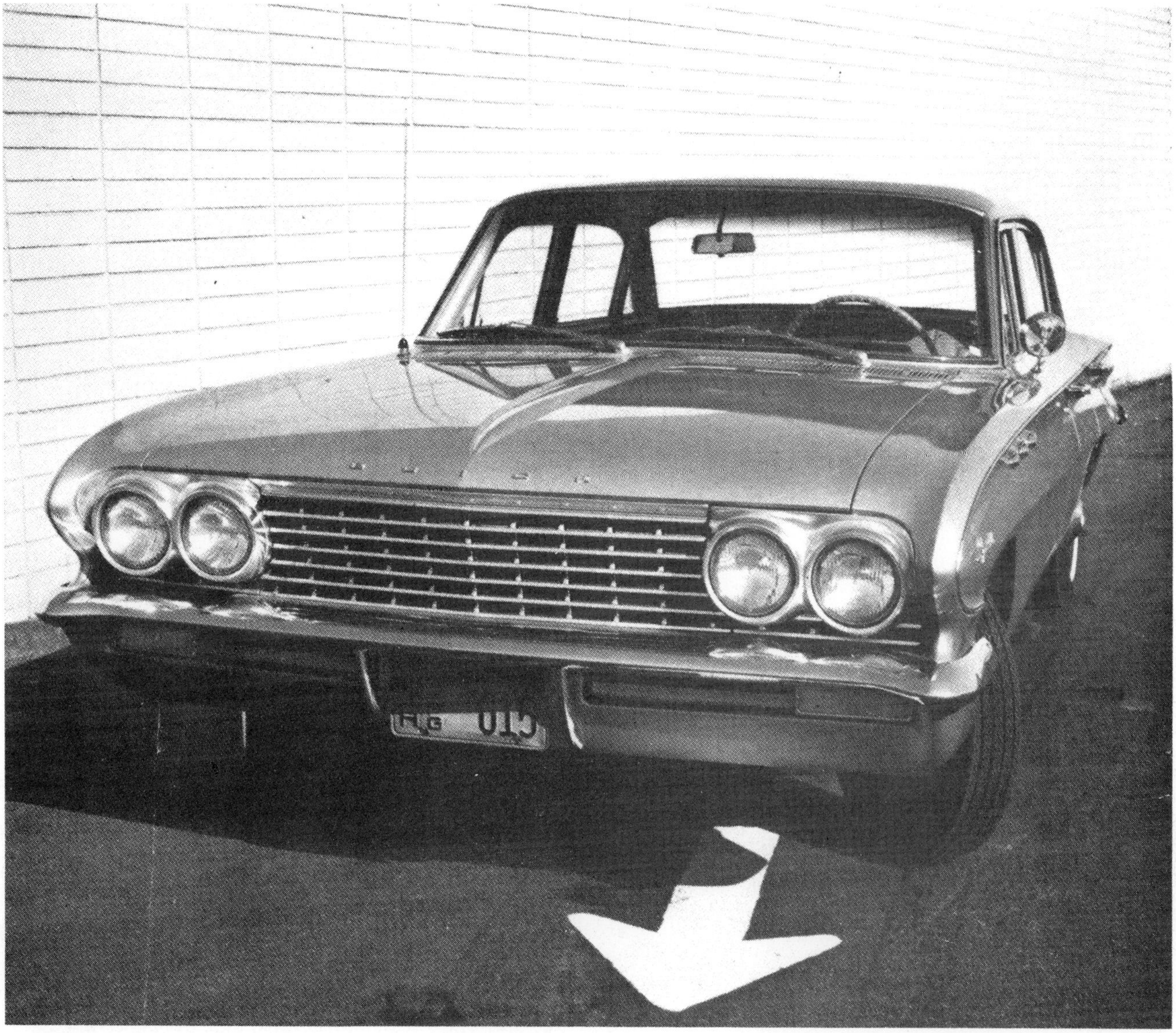

CAR LIFE ROAD TEST

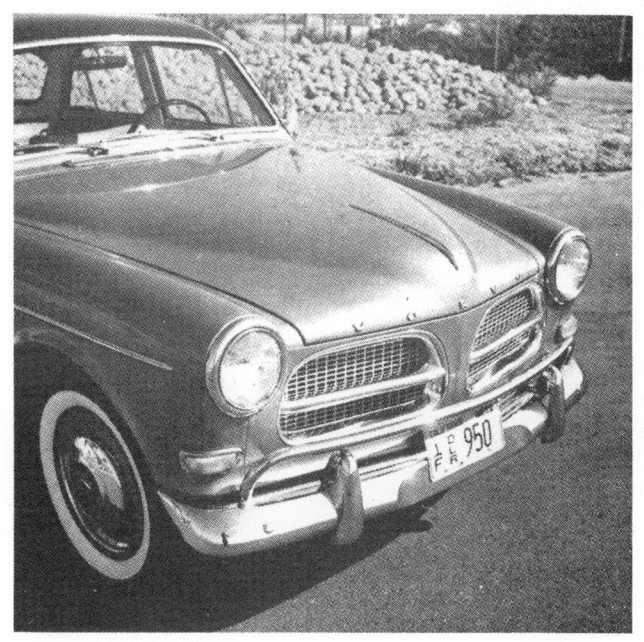

BUICK SPECIAL versus VOLVO 122-S

Variety is the spice of the current buyer's market in both imported and domestic cars; witness the differences between the Buick Special and the Swedish Volvo 122-S.

Both sell for about the same price, as nearly as we can find out. (See this month's "Outlook" for more on prices.) One thing we're sure of, and that is the $2495 the Volvo costs (FOB Port of Entry). The Buick figure supplied by a dealer was $2529, certainly close enough to the Volvo figure to invite comparison by anyone seriously considering purchase of a "smaller than standard" (we're tired of saying "compact") sedan.

Actually, the fact that the buyer *has* such a choice is indicative of the car market we have today. American manufacturers are bringing out new models so fast the dealers can't possibly have a full stock of every available combination. Simultaneously with this profusion of latest offerings, the European manufacturers are bidding ever higher for the Yankee Dollar, either with drastic price cuts (Fiat, Alfa Romeo) or with new models of their own, such as this Volvo.

Intended to sell alongside of, not instead of, the more familiar Volvo PV-544 (you know, the one that looks like the '48 Ford), the 122-S is a Swede of another complexion. While the PV-544 is long on go (for its engine size) and short on room, the 122-S is less lively than roomy, with a much more livable passenger compartment—large enough so that we dare compare this "biggest" Volvo with the "smallest" Buick.

Starting from the back and working toward the front, we find the first surprise in comparing these two cars is with regard to useful, usable luggage capacity. Although

Volvo; a little higher, narrower . . .

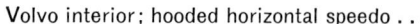

Volvo interior; hooded horizontal speedo . . .

at first glance the Buick Special appears to have a far larger trunk than does the Volvo 122-S, it only works out that way in practice if you're hauling golf balls instead of suitcases. The Buick trunk is too low to stand a standard-sized case upright, so the first restriction appears immediately. In attempting to use the full capability of the compartment with suitcases horizontal, the irregularity of the Buick trunk floor makes itself known, and a further reduction in room is felt. When all is said and done, the apparently smaller compartment in the 122-S will actually hold more luggage.

Around front it's another story, this time with the nod going to Buick all the way. We're referring to the engine size. The Buick engine looks bigger, and is—more than twice as big to be more specific. While the Volvo gets by with not quite 97 cubic inches of piston displacement, the Buick gets its urge from 215. The Volvo divides its inches among four cylinders, with the four's inherent roughness, while the Buick has eight cylinders arranged in the now-traditional American pattern ohv Vee. Those of you who favor economy over performance may take issue with us for inferring that Buick's bigger engine makes it a more desirable car than the Volvo—but look at it another way: with more than twice the engine size, the Buick uses but one-third more fuel. In exchange for this increase in consumption, the Buick out-performs the Volvo by a considerable margin, as the accompanying data panels show. Roughly, it takes the Buick about two-thirds as long to accelerate from a standstill to any given speed it can attain as it does for the Volvo to accelerate through the same range.

So you have a choice: 0 to 60 in 16.6 with 27 miles per gallon (Volvo), or 0 to 60 in 10 flat with 21 miles per gallon (Buick). However, with the pace of U.S. traffic being what it is (fast) and the cost of U.S. gas being what *it* is (low) you would be foolish, we think, to trade that much performance for that small an increase in mileage.

For instance: if you drive 10,000 miles per year using 30¢-per-gallon gas, the Volvo's gas bill will be $111.06,

Buick; lower, wider, sleeker, too.

while for an equal distance in the Buick, you will spend $142.86. This results in a difference of $31.80, which is less than some of us have paid to get de luxe trim on a new car—an item far less functional than the dramatic margin of performance that exists between the Buick and the Volvo.

It has been our contention for many years that the greatest thing the makers of foreign cars could do to make their products more appealing to the American buyer would be to increase the displacement of their engines for better performance, even if accompanied by a proportionate reduction in operating economy. Since the U.S. market is life itself for many of these firms, and special models just for U.S. consumption are not unheard of, the reason this obvious move has yet to be made is a mystery to us.

The real cost of operating a car in the U.S. is not gasoline, but time—mechanic's time, that is. A car's freedom from automotive hypochondria is far more significant in determining annual cost of operation than is its fuel consumption. On this level, the Volvo may score over the Buick if its owner happens to be the do-it-yourself type. The simple, four-cylinder engine in the 122-S is the kind a mechanically inclined owner can work on himself—if he has the time and inclination to do so. Whether this simplicity will keep the Volvo out of the shop more or less than the Buick remains to be seen, as neither car has had time to prove itself.

Cost of ownership, as contrasted to cost of operation, is determined largely by annual depreciation. Although this varies from area to area, the introduction of the American compact cars has forced the resale of imports down in most sections of the country. In view of this, it's difficult to predict which would score better on resale value.

So we have covered trunk space, performance, cost of operation, and finally cost of ownership on these two cars. Thus far it looks as though the Buick's main down-check is in the practicality (or lack of it) in its trunk. But

perhaps you don't select the car you buy for any of the aforementioned reasons. As a matter of fact, none of the automotive experts we know do so. They, like most of us, buy a car on the basis of what it will do and whether they happen to like it or not. Cost enters the picture only by determining whether or not they get the car of their choice, or instead, must lower their sights and settle for something a little less expensive.

In the area of car characteristics that, while not intangible, are immeasurable insofar as they influence the potential purchaser, the Volvo starts rating higher than in the more practical comparisons. For instance, consider finish of the cars: the Buick once again amazed us with the apparent ease with which GM turns out mass produced cars with excellent matching of panels, trim strips, etc., and with a minimum of paint runs, file marks, body lead cracking out, and that sort of thing.

The Volvo was more of the same —in fact, its quality was, if anything, even finer, although when the relative quantity of cars produced by both factories is considered it's no wonder. Just this slight edge in solidness, the firm feel of the doors, the lush-leather quality of the leatherette seats, the faultless paint job, may sway the buyer from the Buick to the Volvo. The Volvo has that indefinable aura of custom craftsmanship which is one of the primary causes of foreign car sales getting as good a toe hold in this country as they had up until the time of the compact's introduction.

Buick interior; hooded horizontal speedo (!).

Buick engine compartment; compact can be a synonym for crowded.

As far as interior room is concerned, it's obvious that the Buick has more of it—but to the man who uses only a fraction of the room his car has to offer, what does it matter if he has 30% more room than he actually needs, or just 10%?

One of the intrinsic things about foreign cars is their ability to handle. In fact, it has gotten to the point where otherwise knowledgeable Americans assume that just because a car is foreign-made, it must handle better than the domestic products—all of them. This is not necessarily so, and in the case of the Special-Volvo comparison under way here it is not applicable at all.

Said one test driver, returning from a trip of several hundred miles involving both cars, "In the Buick, I feel that if I should enter a turn too fast to make it around, I would leave the road and overturn. In the Volvo, under identical circumstances, I think I would overturn and leave the road." In other words, the pronounced understeer inherent in all American cars ("pushing" the front end in corners) has its counterpart in the Volvo's top-heaviness. Neither situation is ideal. The ideal would be a car that understeered ("pushed" the front end) up to a point where it was about to leave the driver's intended path (the road) at which time it would switch over to what is known as final oversteer; that is, the rear end would break loose and come around the outside of the curve, forcing the driver to reverse the wheel to keep the turn radius from being decreased.

The writer has personally experienced this type of car, and found that it saves one from running out of road if the speed into the turn isn't unreasonable but still rather fast; then a spin (still on the road) results, with no harm done except to the ego. High speed runs disclosed that as far up as it went, the Volvo's handling characteristics re-

Buick styling; the teardrop bowed to the razor's edge.

Volvo styling; beltlines aren't so high in Scandinavia.

Volvo engine compartment; ". . . and everything in its place."

mained the same. All was well except that it felt a little "tippy"—but then so does the Super Chief on the stretch of tracks near Tucumcari, New Mexico, and we've never heard of one rolling off the rails yet.

On the other hand, the Buick's good handling went by the board at an indicated 100 mph (about 85, actual time). From there on up, it felt as if directional stability was something it just wasn't interested in, and it would just as soon go right as left if you let it have its head. In a car this fast, it's criminal to drive it wide open; but we feel it's equally culpable to sell a car that has an engine that's faster than its chassis, if you follow our meaning.

Part of this sad state of affairs is the Special's ridiculously slow steering (4⅕ turns) which effectively damps out road feel at high speeds. However, too-soft (for high-speed travel cross country) shock absorbers are equally responsible. Modifying the steering is beyond the scope of the average owner, but having a better set of shocks installed is not, and we heartily recommend this.

Finally, the topic of gear shifting must be explored, as these cars exemplify two entirely different approaches to the question. The Volvo is for the man who wants to shift for himself, and have a ratio for every occasion, so they give him four forward.

The Buick, on the other hand, has the best darn two-speed plus torque converter type box we've yet encount-

Buick trunk; less for bags than for golf balls.

Volvo trunk; less cubic inches, more capacity.

ered—and this was a real surprise to us. We have always looked down our nose (technically speaking) at this particular type.

Both boxes are well suited to the torque curves of their respective engines, with the Buick getting the nod on convenience, of course. We honestly feel that more buyers, pondering whether to go for an import or a domestic "small car," will choose on the basis of the availability (or lack) of an automatic transmission than for any other reason. If we're right, the imports that are big enough to have any one of the American compacts for competition are in real trouble. They haven't had automatics before, in many cases because their engines don't have a sufficient margin of power to overcome the automatic's power loss.

It's a pity, we think—there's nothing better for the consumer than competition between manufacturers after his dollars, and one of the nicest things about the Volvo, for instance, is that we think it makes an excellent example for our domestic manufacturers to strive to better. ∎

There's no greater width difference in back than in front yet vertical taillights and horizontal chrome succeed in clouding the issue completely.

ENTER THE VOLVO

Like the SAAB we tested last month, the Volvo 122S hails from Sweden. Its comfort and performance are outstanding, reports Bryan Hanrahan

modern **MOTOR** ROAD TEST

SWEDEN has several rather curious features . . .

The liquor laws make Victoria's six o'clock closing seem a miracle of broadmindedness and sophistication. Don't know if there's any connection, but Swedes commit suicide at a faster rate than any other nation in the world.

And — just to point-up the validity of their outlook — the crazy blighters drive on the left-hand side of the road in cars that have LEFT-HAND DRIVE.

Presumably this is to make absolutely sure that the front passenger cops it in a prang instead of the driver. Parking must be a bit easier, too.

Any other good reason for this state of affairs I fail to see.

It seems that the double-left system started before the war, when there was only a small local car industry. The Swedes bought left-hand-drive American cars but stuck to driving on the left.

Somehow everyone got used to it all. After the war, with an expanding local industry that could have put the steering wheel either side, they insisted:
● The steering MUST be on the left.
● The left-hand side of the road is the PROPER side on which to drive a car with a left-hand drive.

This was decided in a referendum. And so matters stand today.

The wonder of it all is that the cars the Swedes build are not crazy, too. Their Volvo 122S saloon is one of the most outstanding bits of machinery I've yet laid hands on.

ENGINE, of modest 1580c.c. capacity and completely conventional—as is the rest of the car—puts out a solid 85 horses with help of twin SU carburettors and 8.2 to 1 compression. Genuine maximum is 92 m.p.h.

This doesn't mean it's perfect — but it IS to my tastes.

Body shape has a slightly old-Simca look. Long and narrow, but with good vision (which wasn't the old Simca's strongest point).

Decor is restrained, with little chrome. The two-hole radiator grille is reminiscent of Fiat's 1100TV Spyders of a couple of years ago.

Inside there's room for only four really — but four in luxury. Three adults will fit into the rear seat, but it's a bit of a squeeze.

Generously Equipped

The test car had form-fitting twin bucket seats in front, adjustable to any rake, height or reach for four-footer or six-footer. These go with the floor gearshift; if the shift's on the column, you get a bench seat up front.

The back seat is a nicely contoured bench, but restricted by wheel-arch encroachment. A folding central armrest is provided here as standard.

All seats are covered in luvverly leather.

The heater is designed to cope with blizzardly Scandinavian conditions. Watch out if you use the two-speed fan — you could get first-degree burns!

The electric screen-washers are almost instantaneous in action — rare for this breed.

Two-speed, self-parking electric wipers clean every vital bit of the flattish screen. There's a cigarette lighter (or should I call it a cigar lighter, since the car is expected to sell here at around £1700, tax-paid?).

All four doors have pulls and armrests. The courtesy light works off all doors.

Luvverly carpet on the floor.

Full seat harness — not just lap straps — is fitted to each front seat. Each harness snaps neatly away on a button on the doorpost; it is equally easy to snap into place on a hefty lug on the floor, between the seats. Both lap and shoulder straps join at the snap.

It's the neatest, safest and least involved belt set-up I've seen. If more were designed like this, it might do a lot to help belts become more popular.

I found you could tailor the driving position to suit yourself. Just set the seat cushion and squabs where you want them, and the wheel falls nicely to hand. The feet connect with pendant brake and clutch pedals that aren't set too high off the deck.

The well-padded dash holds the instruments in front of the driver; there's a speedo with both total mileage and trip recorders, and fuel and water-temperature gauges.

Minor controls are below: the cig-

BOOT is roomy, with every bit of space usable—and the side-mounted spare can be extracted without shifting luggage. Well dustproofed, too.

DOORS open wide, seating is luxurious. You get twin front seats with floor shift, a bench with column lever.

arette lighter, choke knob, light switch and panel rheostat on the left of the column; key-start ignition lock and heater-fan switch on the right.

The other heater controls are central. There's an ashtray to the left of them, and a radio in front of the passenger; if you don't want to buy a radio, the aperture for it will serve as a tiny glovebox. A small parcels shelf is also provided under the left corner of the dash.

Bonnet release catch is below the column.

Boot has 20 cu. ft. of usable space, with the spare tucked away at one side in an upright position. Tools are essentials only.

The petrol filler is wide and means business—it will take the full blast from any pump.

Small Engine, Big Results

The Volvo's mechanics are the story of the conventional doing the highly unconventional. The almost-square 1580c.c. four-cylinder engine is mounted with acres of room to work around it. Output is 85 b.h.p. at 5500 r.p.m. on a compression ratio of 8.2 to 1 And does that donk hop into it!"

But its characteristics won't suit everyone—the man who doesn't really care for driving will probably reckon he has to work too much.

Maximum urge (87lb./ft.) doesn't come in till 3500 r.p.m. You have to really rev this car; and when you do, the results are more than gratifying — they are sensational.

Zero to 50 m.p.h. in just over 10 seconds, to 60 in 16.5, and to 70 in 24.2; a true top speed of 92.1 m.p.h. —over 30 in first, 54 in second and 72 in third!

But if you don't pile on those revs she'll just crawl around like a two-stroke lawnmower with an impediment in the carburettor.

As you'll judge by the figures, the gear ratios are beautifully chosen. All forward cogs are synchronised and controlled by a long but highly positive floor gear-change lever.

This refuses to work at all unless the clutch pedal is floored. When you get used to it, you are really operating precise mechanism.

Our test car, by the way, was the only Volvo in Australia at the time of writing. It had been in four trials and had led a hard life as a hack for over 11,000 miles. Ray Fleming, service manager for the importers in Victoria (Regent Motors, of Melbourne), swears that no oil has been added between oil-changes.

I believe him. The car went perfectly, however hard it was thrashed on test—and it's the sort of machinery that just asks to be thrashed.

Maximum revs do not fuss the engine — nor is it noisy. The two big inclined SU carburettors hardly gurgle through a combined silencer-aircleaner.

Only time the engine does anything it shouldn't is when you switch off. The 8.2 compression — in this particular engine design, at any rate —is a bit high for our fuel and sometimes you get a bit of running-on.

But I found that by using another essential bit of equipment on the car — a radiator blind — and keeping the temperature steady at 170deg. F. the trouble was minimised.

The thermostat has a very wide range by our standards, to cope with Scandinavian cold. At low speeds she tends to run hot; at high speeds, cold.

A sort of lavatory chain below the dash works the blind smoothly. But don't expect to be able to adjust it at over 70 m.p.h. — slipstream pressure is too great.

By using the blind to keep temperature steady, I managed to get within coo-ee of the factory's claimed fuel-consumption figure of 23 m.p.g.

The drive is taken from the engine by a hydraulically operated single dry-plate clutch with 52.7 sq.

Continued on page 89

Left: Keith and Lorna Gamble have already achieved success in trials with the Volvo.

Below: Twin air intakes give the front of the Volvo a distinctive, purposeful air.

Right: For those who can afford it the beautiful P1800 coupe will be available.

Pat Hayes drives the latest new car for the Australian market —

SWEDEN'S

VOLVO 122 S

"Big in area but thinly populated, Sweden has some fast highways, a lot of very rough dirt roads and a climate which can include extremes of cold, so cars that spend most of their lives out of doors must be well built: elsewhere in the world also, it is thought, there are many places where a car needs to be tougher than are many 1961 designs intended primarily for smooth going."
— Joseph Lowry — "The Motor".

WHEN GLOWING REPORTS about a Swedish "Peoples' car", the Volvo, started to filter through from overseas. I began to hope that here might be a car which would suit the Australian enthusiast as much as his European brothers.

Only "might" because, like everybody else, we've been caught before. Only too often in the past have English and American motoring journalists raised our hopes with reports of cars with dream-like cornering and performance figures which seemed almost outside the realms of imagination. What's more, these cars lived up to their overseas reputations in these fields when they came to Australia.

But because of the conditions under which overseas journalists test cars, there were many things they did not tell us. This was not done through any malicious intent or because the makers treated them to a fine champagne and chicken lunch during the test, but because most of their readers are not interested in such features.

And so it happens that when the cars eventually came to Australia we found a few flies in what was said to be the perfect ointment.

On dirt roads dust poured in to leave the occupants choking and gasping for air and their clothes looking as though they's been worn by trans-Nullabor trekkers.

Ground clearance, which on paper looked to be enough for a run through an uncleared paddock, became useless when the springs bottomed crossing a spoon drain and an overhanging tail section caused the exhaust to be ripped off. Dust and jolts from Australia's potholes (among the best in the world) rendered window handles and door locks inoperable and dust, pouring in through air filters which would only just keep out rabbits, gave owners one of the fastest "while you wait" rebores they had ever experienced.

As a final blow, the heat caused beautiful veneer cappings on the fascia panel to lift, plastic fittings to crack and break and fibreglass, finely finished when new, to craze and roughen.

Add to this a few alloy metal parts which became fatigued and cracked after a few thousand miles and your mouth watering car became, categorically, a heap.

Despite this knowledge, hope springs eternal and I was anxious to get my hands on a Volvo just in case it did turn out to be at least a little better than it's European cousins. Reading Joseph Lowrey's description of Sweden (from an article in which he describes the Volvo factory) made me keener still.

The car which Regent Motors, Melbourne, provided turned out to be a revelation. Not only did it live up to what overseas journalists said about it but it did not fall down in the departments about which they never say anything.

Before we get too lyrical in our praise it might be best to point out here that there is nothing fantastic about this car. It is a car produced for the ordinary motorist who wants a sporting flavour in his everyday motoring and who expects a car to last. Any other manufacturer could do the same thing if he wanted to. The Volvo will NOT do 100 m.p.h., it will NOT corner like a Ferrari (or a Maserati or an E type Jaguar nor will it out-accelerate them) and it will NOT give you £3,000 worth of motoring for its middle class price tag of around £1,700.

What will it do then ?

All that a reasonably sane driver would expect it to and possibly a little more. So let's get down to business.

The Volvo looks like the majority of continental cars: Styling is not futuristic, or even modern and is reminiscent of Borgward and Simca with, if you look hard enough, a little dash of Mercedes Benz. Like these three it has distinctive but conservative lines designed for simple beauty and efficient manufacture. In keeping with these edicts chrome plating is sparse but what there is looks solid enough to last. Something, of course, which all chrome plating should do, but today many examples show rust spots after only one night in the open.

The four doors open wide and the lack of a drastic wrap-round windscreen allows easy entry to the front seats. The doors close with a soul-satisfying "clunk" that speaks for solid construction and old-style craftsmanship. Inside, the finish is a mixture of business and luxury. Seats are fine-grained leathercloth, tall, well-upholstered buckets in front and a bench with rounded corners in the rear, which, with a central armrest, converts to what amounts to another two buckets.

Trim is metal where most manufacturers use plastic and while the plastic manufacturers claim their products are sufficiently durable there is no more re-assuring sight than solid metal fascia panels and window surrounds. The gear lever is a solid, chromed stick which goes direct to the selectors. It looks solid enough to coerce the most elusive cogs to mesh.

The seating position is excellent. The driver sits high enough to get a commanding view over the bonnet and with the steering wheel held in classical "straight arm" style the pedals and gear lever are perfectly positioned. One fault could be the fact that pedals are slightly offset and it is possible for the foot to slip off the pedals at the most inopportune moments. This fault would probably disappear when one became better acquainted with the car.

Instruments are grouped in a binnacle directly in front of the driver. A strip type speedometer extends the full width of the binnacle and under it are four windows housing a temperature gauge, a trip meter graduated down to tenths of a mile, the usual mileage indicator and a fuel gauge. Generator charge, oil pressure, high beam (headlight) and direction indicators signify their operation or non-operation by means of the usual warning lights.

A stalk on the side of the steering column operates the traffic indicators and will also flash the headlamps in continental fashion if pulled toward the driver.

Standard equipment includes two-speed windscreen wipers, a cigarette lighter, two ashtrays for rear seat passengers, a radiator blind, a heater demister unit and safety belts for front seat passengers.

Of these the radiator blind and safety belts are probably two items which most motorists meet only rarely. The blind is controlled by chain under the fascia panel. Pulled out to its fullest extent the chain raises a blind in front of the radiator. The consequent lack of cold air through the cooling system enables the car to warm up to its proper operating temperature within minutes of starting on cold mornings. Our test was carried out during one of Melbourne's well-known frosts and while other cars were still spluttering when we joined them on one of the main highways we were already operating at full efficiency.

The safety belts consist of the normal lap strap coupled with a diagonal strap running from the door pillars across the body to a lug between the front seats. The number of times that passengers forgot they were wearing the harness and attempted to leave the car without dis-engaging the simple locking mechanism was ample proof that the harness did not restrict comfort or normal movement while travelling.

Apart from the safety harnesses, Volvo have paid greater

Interior is well designed for comfort and safety. Safety belts are standard.

Engine accessibility is good. Equipment at top right is the efficient heater unit.

attention to safety than most car manufacturers. There are no projecting metal pieces anywhere inside the car and any part against which one would be thrown in an accident (fascia panel, sun visors and the front section of the front window ledges) has been provided with thick padding. For the same reason, roof pillars are thicker than those on the majority of modern cars and should prevent the roof from caving in if a driver is unlucky enough to roll the car.

The heater demister is one of the most efficient we've ever experienced. It can be regulated to a nicety through a thermostat unit and with the two-speed fan at full boost and the temperature control opened to its fullest extent the heater provides more heat than would be required in a blizzard.

An overhead valve, 1580 c.c. four cylinder engine of almost "square" dimensions powers the Volvo. In the 122S this engine has been fitted with two horizontal type SU carburettors and developed to provide 85 brake horsepower at 5,500 r.p.m.

The four-speed all-synchromesh gearbox is fast and efficient. At first, when one tries to disregard the strong spring loading between second and third gear it seems slow and awkward. Then it is found that by using this loading the lever can be swapped from second to third gear in one swift, sweeping movement. It is virtually impossible to beat the synchromesh.

For a 2,405 lb. car powered by a 1600 c.c. engine the Volvo has quite a respectable road performance. Maximum speeds in each gear are: 35 m.p.h. (1st), 55 m.p.h. (2nd), 80 m.p.h. (3rd) and 94 m.p.h. (4th). The standing quarter mile comes up in just over 20 seconds and 60 m.p.h. in just under 17 seconds.

The ride is firm but not annoyingly so and the steering, in true continental style, is light and beautifully precise.

Driven hard into corners the Volvo tends to understeer at first but when approaching the limit this changes slowly to oversteer causing the tail to drift out slowly if full power is not maintained. Using these handling characteristics it is possible to take varying lines through corners without much fear of getting into serious trouble. The front suspension is independent with coil springs and wishbones and at the rear the rigid axle is sprung by coils with massive radius arms and an anti-roll bar to hold it in position.

Although trials driver, Keith Gamble, had competed in several gruelling events with the car loaned to us, it showed no signs of the wear and tear it must have experienced. Road noise was fairly high but there were no body rattles or slackness in any of the controls — a good recommendation if one

has any inkling of the type of terrain covered by most trial routes.

Fuel consumption on our test was only about 20 m.p.g. but this is accounted for by the fact that the car was under full acceleration for most of the time and hardly ever in top gear. In normal driving conditions the car should return a more moderate 27 m.p.g. or better.

Incidentally, readers who have never heard of the Volvo before will be interested to know that last year the company produced as many cars as Mercedes Benz as well as many commercial vehicles. The company is part of a large, well-organised group so spares should not cause any difficulty.

Volvo tell us they built this car to be "a speedy, safe and durable means of transportation but, at the same time, extremely smart and eminently representative" — in our opinion they have succeeded.

Victorian Distributors: Regent Motors.
N.S.W. Distributors: Antill Ranger.

VOLVO SPECIFICATIONS
Model: 122S
Engine: 4 Cylinders in line, o.h.v.
  Bore and Stroke: 79.37 x 80 m.m.
  Capacity: 1580 c.c.
  Compression ratio: 8.2 to one.
  Output: 85 b.h.p. at 5500 r.p.m. (S.A.E.)
  Torque: 86.8 lb. ft. at 3500 r.p.m.
  Carburettor: Twin horizontal SU.
Transmission:
  Gearbox: Four speed with synchro on all.
Suspension:
  Front: Wishbones and coil springs.
  Rear: Rigid axle with coil springs, radius and anti-roll bar.
Brakes: Hydraulic drum.
Steering: Cam and roller, 3½ turns lock to lock.
Tyres: 5.90 x 15.
Chassis: Integral.
Bodywork: Saloon, four seats, four doors.
Dimensions:
  Wheelbase: 102.4 ins.
  Track 51.77 ins. front and rear.
  Length: 175 ins. Width: 64 ins.
  Hight: 59.75 ins.
  Ground Clearance (four occupants): 7.3 ins.
  Dry Weight: 2,400 lbs. Turning Circle: 32.5 ft.

VOLVO B-18 *P-1800 Punch for the Whole Volvo Line*

THE CAR

Volvo AB is quietly dropping the 1.78-litre P-1800 sports engine into its Amazon and PV 544 series. Along with this change comes a grab-bag full of other improvements which on the Amazon include disc brakes in front, a 12-volt electrical system, modified exhaust layout, a more robust front axle member and new bushings for the lower control arm shaft.

The Amazon (now known, at least to Volvo, as the P 120 E) uses much the same mechanical components and running gear as its humbler PV 544 brother, under a roomier and more stylish four-door exterior.

TECHNICAL

Volvo's new engine nomenclature deserves a brief explanation. On all Amazons and PV 544's equipped with the gutsier four-cylinder B-18 engine there is a small badge attached to the grille. This engine, officially numbered B-18 D by the factory, differs slightly

from the unit used in the P-1800 and known in that form as the B-18 B. It's milder by about 10 horsepower due in part to white metal instead of lead bronze main bearings ,slightly lower compression ratio, and the absence of the P-1800's high-lift cam. Two SU carburetors are supplied, however, and despite its detuning the B-18 D is still a beefy, thoroughly capable powerplant.

The rest of the Amazon is traditional Volvo, conventional in design and rugged in operation. Transmission is a four-speed, all-synchromesh unit with a robust floor shift. Suspension is by coil springs at all four corners.

Our test car was a P 120 E Amazon equipped with the B-18 D engine and accompanying disc brakes.

STYLING

The Amazon is pure Nordic in appearance, sober and simple. Volvo's aim with its deluxe model was

apparently to achieve a feeling of dignity and restraint in keeping with the car's polished character. For our taste the car might have been a little less restrained; it's infinitely livelier than it looks.

Heavy gauge metal, flush-fitting panels and thorough assembly were evident. This is true of all Volvos, but extra care seems to have been taken with the Amazon. Paint finish is excellent.

INTERIOR

Seating room in the Amazon doesn't appear to be drastically better than the PV 544 despite this car's two extra doors and larger body. In fact those narrow rear doors offer very skimpy legroom, and the seat is rounded off severely at the corners to help passengers squeeze in and out. Seats themselves are definitely on the firm side, moreso in the separate front two than in the bench-type rear seat, divided by a folding armrest. Doors are unobstrusively well trimmed and carry ashtrays in the rear. Grey rubber matting is used on the floor front and back.

A solid-feeling shift lever pokes up from the high transmission tunnel and is placed just right for brisk shifts in any direction. The handbrake, mounted on the floor to the left of the driver's seat, got in our way more than once while entering and leaving the car and seems a curious placement.

A 120-mph speedometer dominates the instrument console, running beneath a lightly padded cowl and carrying fuel and temperature gauges and lights for oil pressure and amps. There is also a trip mileage recorder apart from the odometer. Black plastic knobs control choke, wipers, lights, cigar lighter and the heater fan. The main heater control, illuminated at night, sits in the centre of the instrument panel, and to its right beneath the panel is a small padded tray in lieu of a locking glove compartment. A handy item on this as on all Volvos is a radiator blind pull-chain for cold starting.

Due to the high seating position and adjustability of front seats both for rake and fore-and-aft position, driving the Amazon shoud be pleasant for almost anyone.

DRIVING

The key-started B-18 D engine snaps into action

rapidly, and though a tap of the throttle indicates plenty of zip there is almost no noise or vibration at idling speeds. Gear ratios in the Amazon have apparently been chosen with a consistent power curve in mind rather than low-gear acceleration alone. Yet one needn't excuse this car's takeoff by any means; 0-30 mph was achieved consistently in very little more than four seconds, by which time the exhaust had gained a delightfully husky sound. First gear can be held right up to 25 mph and second is good until 50 without the feeling that an explosion is imminent. This is, as we reported in our P-1800 test, an almost incredibly strong engine which somehow avoids the penalty of feeling rough at any revs. For maximum performance it would help greatly to have a tachometer.

The Amazon rides firmly, without waver and with the feel of a much heavier car at cruising speeds. Steering felt slightly heavy in tight spots but is neither sluggish nor overly sensitive in normal driving. Cruising at 70 mph, there is no indication of engine strain. In truth it would be all too easy to slip up to flagrantly illegal speeds before the car gave any indication of being extended.

Fast cornering is a cinch once you lose the fear of that tire squeal and learn to roll with the car. There seems to be a lot of bulk to throw around, yet the Amazon will do precisely what you want it to in the handling department if throttle and wheel are used effectively to control its tendency to understeer. Brakes on our test car needed a strong tromping to halt things in a hurry, but medium pressure serves adequately in all but panic situations.

ECONOMY

Average consumption (premium fuel) in combined city and open-road use worked out to 25 mpg. A good five mpg more is quite possible at steady speeds.

STORAGE SPACE

A well-finished trunk, with the spare tire mounted upright to the left and rubber carpeting on the floor, should suffice for the family style use for which the Amazon is intended.

HEATER AND VENTILATION

The Amazon's heating system works up warm air and delivers it where it's wanted in a hurry. It's one of the best such rigs on any imported car and should serve admirably in the worst Canadian winter weather.

LAST WORD

The Amazon was always a solid if not spectacular buy; with the additional kick of its new B-18 D engine it's very nearly spectacular too. Some people might not like its typically uncompromising Volvo ride, and despite its four doors this is not the roomiest import. But it's certainly among the best-handling and most responsive sedans available today.

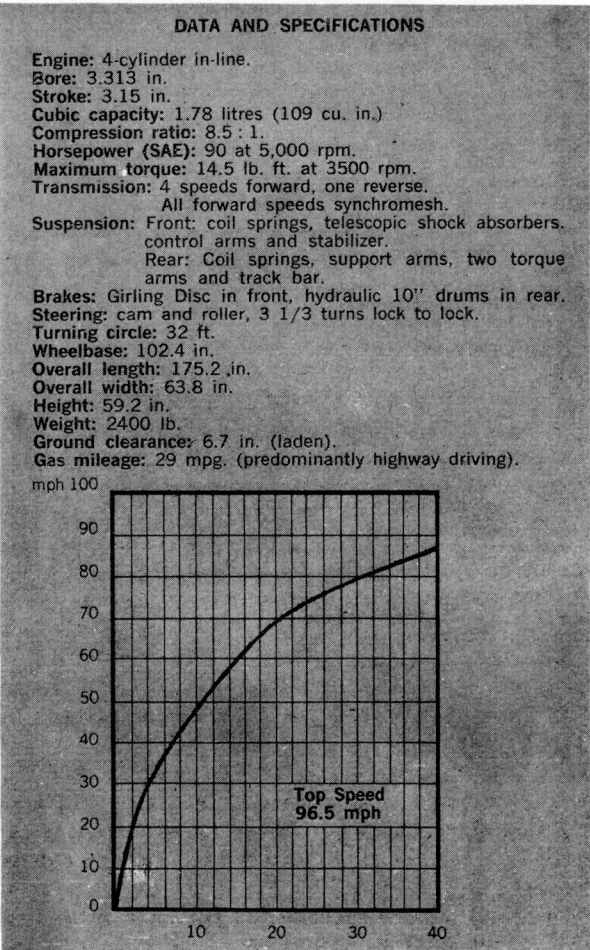

DATA AND SPECIFICATIONS

Engine: 4-cylinder in-line.
Bore: 3.313 in.
Stroke: 3.15 in.
Cubic capacity: 1.78 litres (109 cu. in.)
Compression ratio: 8.5 : 1.
Horsepower (SAE): 90 at 5,000 rpm.
Maximum torque: 14.5 lb. ft. at 3500 rpm.
Transmission: 4 speeds forward, one reverse. All forward speeds synchromesh.
Suspension: Front: coil springs, telescopic shock absorbers. control arms and stabilizer. Rear: Coil springs, support arms, two torque arms and track bar.
Brakes: Girling Disc in front, hydraulic 10'' drums in rear.
Steering: cam and roller, 3 1/3 turns lock to lock.
Turning circle: 32 ft.
Wheelbase: 102.4 in.
Overall length: 175.2 in.
Overall width: 63.8 in.
Height: 59.2 in.
Weight: 2400 lb.
Ground clearance: 6.7 in. (laden).
Gas mileage: 29 mpg. (predominantly highway driving).

Top Speed 96.5 mph

VOLVO 122-S

Quality is its most important asset

OF ALL THE CARS that are brought to our shores, one of the best-suited to American needs and driving habits is the Volvo 122-S. Not because the Volvo is a copy of any American design, but simply because it is sufficiently roomy and has the performance required to cope with our brisk traffic conditions. Moreover, it has the sort of high-speed cruising potential that is absolutely essential for touring on this immense continent of ours.

Of course, there are reasons; all cars reflect the driving conditions that prevail in the countries where they are built, and the Volvo is an excellent mirror for Sweden. Sweden still has a lot of the "wide open spaces" that we fondly imagine to be our exclusive property, and Swedes customarily take extended trips by automobile. At a pretty fierce clip, too, as anyone who has driven there will attest. In fact, they expect much the same sort of service from their cars as we want from ours, and the Volvo is built with that in mind.

In exterior appearance, the 122-S is rather "Ameri-can" in flavor. It would probably blend right into the general rush and crush of our traffic, were it not for the fact that it lacks the staggering visual impact of some of our stylists' gaudier creations. And, too, in a crowd of the new American cars the Volvo would look a trifle tall. But, though the height of the 122-S may not be modish, it does lend certain advantages to the car. Getting in and out when parked by a high curb can mean a heroic struggle when a contemporary "Detroit" product is involved. But, due to the generous vertical dimensions of the Volvo, one can pop in or out very handily.

Another plus for the car's height is the commanding view it gives in traffic. Naturally, it really isn't all that much higher than any other car, but the extra bit does give an edge in threading through traffic. As a final note on the Volvo's height we would like to say that it is somewhat more apparent than real, the illusion being created by the depth of the unit-structure body. There are no frame members underneath to occupy space, and the car's underpan is also the flooring, which makes the interior seem very deep and the seat very high indeed.

While the styling of the Volvo may have been open to some criticism, the standard of workmanship—in either over-all or detail finish—was not. Our test car was a light pearlescent-grey with a black roof and this gave an over-all effect of subdued elegance. This impression was carried over into the car's interior, where a grey leather-grained plastic covered the most comfortable small-car seats that we have ever seen. The only sour note was provided by the floor mats, also of a grey plastic but very definitely lacking the luxury look of the seats.

The instrument and control layout was very good, with a couple of exceptions. The driving controls, the ones that make the car go, stop and turn, were well arranged and performed their functions precisely and without undue reaching on the part of the driver. And lesser controls, such as the choke, light switch and the like, were also neatly and effectively placed. The instruments, however, although grouped into a neat cluster and very easy to see, gave complete information regarding only the temperature and fuel. Oil pressure and generator charging are covered, *very* loosely, by warning lights. The speedometer was the most uncertain of the lot; it has a horizontal red band to indicate speed and, as this band is slashed off at an angle, one has a fairly broad range of speed, from the point to the heel of the slash, from which to choose a road speed. We used the point, and discovered that it gave very close to the true speed, being just a few tenths slow.

Despite the minor aggravations produced by the instruments (which are much the same in 9 out of 10 cars being sold today) we liked the 122-S. The back seat is, as is the case in all medium to small cars, a bit cramped for 3 persons, but with the armrest pulled down in the middle there is almost armchair comfort for two. The front seats are just about perfect; they are shaped as though the designers had human beings in mind and are adjustable for rake, and for leg room.

Our test car was a bit too new for our liking, being a trifle stiff but, even so, was a real delight. The steering was particularly good; the wheel is placed where it can be cranked around very smartly and the steering gear (cam and roller) gave remarkable precision. Our joy

over the excellence of the steering was, at first, restrained somewhat by the top-heavy feeling of the machine and by the self-steering effects of the rear-axle layout. However, after a time we stopped struggling against these characteristics and learned to use them. Once past this point we had no further difficulties and, in fact, found that the car could be driven very forcefully without the results becoming untidy.

A substantial part of the Volvo's sporting side is supplied by the new 4-speed transmission, which is also used in the new P-1800 sports car, as described in our technical analysis of that car (March 1961). The ratios are much improved over the earlier 4-speed unit—being near perfect—and the synchromesh is absolutely unbeatable. Synchromesh is supplied on all gears and its action is strong enough to allow the driver considerable clumsiness before it makes any audible protest. One thing is certain: what the driver commands, the transmission will have to do, for the shift lever is both long and strong enough to overcome any resistance on the part of the transmission.

Visibility from the driver's seat was good, better than from the older PV-444, but not as good as it might be if it were not for the presence of the unusually thick

window posts. In this respect, however, we can't complain too much; while the thick posts *do* obstruct one's vision slightly, the effect is not so severe as if the top were flattened down level with the hood—which is precisely what may happen when one of those other slender-post lovelies inverts itself. Therefore, any feelings of claustrophobia that we may have had were well mixed with a comfortable sense of being surrounded by protecting structure. This impression of solid strength was materially heightened by the fact that the Volvo never showed the slightest tendency to rattle—even after being subjected to some fast running over a really atrocious road.

Unfortunately, though the body was very solid and little road noise was to be heard, engine noise was very much in evidence. The 122-S is powered with the "sports" version of the PV-series engine and has a pair of SU carburetors (a single Solex is used on the standard model). These are capped with a pair of "gravel-strainer" air cleaners and, while they will prevent large rocks and small animals from being aspirated, they don't hold back the carburetor noise a bit. Therefore, the engine noise varies with the throttle setting, and if the car is being driven very hard, the sound level is entirely too high for comfort. This is an unhappy characteristic and we found it most objectionable. We have been assured that an

effective intake-air silencing system is in the works and will be along shortly.

These criticisms notwithstanding, we were very favorably impressed with the Volvo 122-S. It is a solidly constructed machine, a spirited performer, and handles well. In the area of pure practicality it offers very comfortable seating for four and has a large usable, cube-shaped trunk. Its brakes are good—though not outstanding—and its turning circle allows a U-turn in any but the most narrow streets. Under the hood there is no confused mass of plumbing—everything can be reached and routine service should be easy to perform. And one touch that we liked very much was the radiator-blind; this can be pulled up (by means of a small chain) from inside the car and it gives an extremely rapid warm-up on cold mornings. This last point will be important in far-north areas where the car's occupants will want the heater to start doing its job as soon as possible—and the Volvo's heater/ventilator system will deliver a scorching blast if necessary.

Even though the 122-S has some very tough competitors for the American compact market, Volvo's prospects are quite good. Any shortcomings the 122-S may have are equaled by flaws in its competition, and the Volvo does offer a sporting flavor—perhaps the best quality of all.

DIMENSIONS

Wheelbase, in.........102.4
Tread, f and r....51.7/51.7
Over-all length, in.....175
    width............63.5
    height...........59.2
    equivalent vol, cu ft....382
Frontal area, sq ft....20.9
Ground clearance, in....7.7
Steering ratio, o/a......n.a.
    turns, lock to lock......3.3
    turning circle, ft......32
Hip room, front.........53
Hip room, rear.........52
Pedal to seat back....41
Floor to ground...........13

CALCULATED DATA

Lb/hp (test wt)........32.3
Cu ft/ton mile........73.2
Mph/1000 rpm (4th)....16.7
Engine revs/mile........3600
Piston travel, ft/mile....1890
Rpm @ 2500 ft/min....4760
    equivalent mph.......79.4
R&T wear index........68.0

SPECIFICATIONS

List price............$2495
Curb weight, lb.........2400
Test weight...........2750
    distribution, %.....52/48
Tire size.........5.90–15
Brake swept area........166
Engine type.....4 cyl, ohv
Bore & stroke....3.125 x 3.15
Displacement, cc........1586
    cu in.............96.6
Compression ratio.......8.2
Bhp @ rpm......85 @ 5500
    equivalent mph.....92.6
Torque, lb-ft......87 @ 3500
    equivalent mph.......58.4

GEAR RATIOS

4th (1.00)..............4.55
3rd (1.36)..............6.18
2nd (1.99)............9.05
1st (3.13)............14.2

SPEEDOMETER ERROR

30 mph...........actual, 30.9
60 mph..................60.7

PERFORMANCE

Top speed (4th), mph......90
    best timed run.......87.4
3rd (6500)...........79.8
2nd (6500)...........54.5
1st (6500)...........34.8

FUEL CONSUMPTION

Normal range, mpg....24/27

ACCELERATION

0-30 mph, sec..........4.8
0-40....................7.9
0-50...................12.0
0-60...................16.6
0-70...................23.0
0-80...................33.1
0-100...................
Standing ¼ mile.......20.4
    speed at end.........66.3

TAPLEY DATA

4th, lb/ton @ mph..160 @ 55
3rd...............220 @ 48
2nd...............340 @ 36
Total drag at 60 mph, lb...124

ENGINE SPEED IN GEARS

4th
3rd
2nd
1st

2000 3000 4000 5000
ENGINE SPEED IN RPM

MPH

ACCELERATION & COASTING

90
80
70
60
50
40
30
20
10

SS¼
4th
3rd
2nd
1st

5 10 15 20 25 30 35 40 45
ELAPSED TIME IN SECONDS

ROAD TEST

VOLVO
122-S

A Unique Import That Offers Distinctive Competition To The Domestic Compacts

ALTHOUGH IT WAS INTRODUCED to U.S car buyers only a little more than five years ago as a make of imported cars, Volvo is now in the front rank both in reputation and in popularity. There are many reasons for this, among them high quality and high performance at a reasonable price, but one factor that is relatively new is the increasing variety in the Volvo lines of cars.

The bread-and-butter cars of the Volvo line are, of course, the various versions of the 544 series — which have been previously reported on in recent issues of MOTOR TREND. There are two additional important vehicles bearing the Volvo name which should not be overlooked. One, obviously, is the new P-1800 sports car (see September MOTOR TREND), and the other is the impressive and more luxurious sedan, designated as the 122-S.

A road test of the Volvo 122-S over an extensive distance of more than 2,000 miles was just conducted by the MOTOR TREND staff. Many of its characteristics are similar to those of the PV-544 series, and to a limited extent even the P-1800 sports car. But the 122-S is a distinctive car in its own right.

Much of the Volvo reputation has been built on the high performance of its cars in comparison with competing makes in the same class. The 122-S continues this tradition, although its general performance ability is not as startling as the Volvo PV-544 in the 85-hp version.

The 122-S has substantially the same engine, drive line and running gear as the companion 544's. The engine is the four-cylinder, running 8.2 compression and rated at 85 hp. There are dual SU carburetors feeding the 97 cubic inches, and power is transmitted through an all-synchromesh four-speed gearbox to the rear axle, which has a final ratio of 4.56.

The accelerating time from 0 to 60 mph for the 122-S is about 16½ seconds on an average. This makes it somewhat slower than the PV-544, which hits 60 mph in approximately 13½ seconds, but it is faster than the PV-544 Special Deluxe (the lowest-priced Volvo with same engine but lower compression and single carburetor to yield 60 hp), which posts a time of just over 18 seconds.

The slower acceleration of the 122-S is, of course, directly the result of the heavier body, nearly 300 pounds more than the hot PV-544. It is a coincidence, but nonetheless interesting, that the 122-S also costs about $300 more than the PV-544, or almost exactly $1 per pound for the luxury and better finish.

The Volvo engine, used in all Volvos with the exception of the P-1800, is really a small economy engine, but it is so

The Volvo engine is a rare type, in that it combines true gasoline economy with excellent performance. Although it uses dual SU carburetors which add to accelerating ability, in average driving it will yield 21 to 25 mpg.

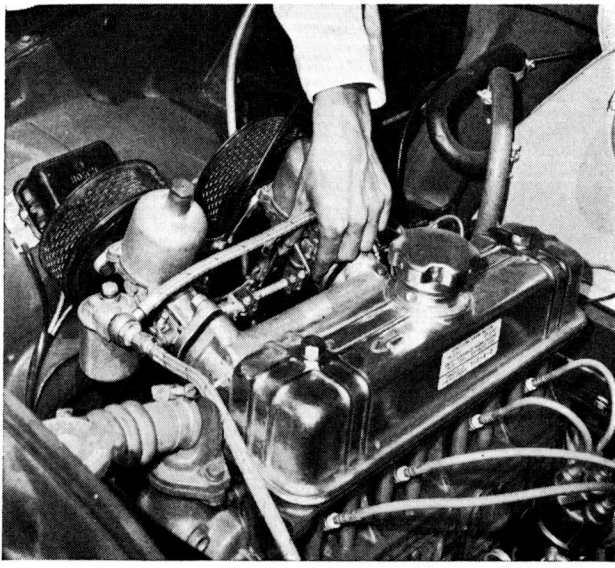

The cockpit of the 122-S has major controls, including the steering wheel and floor gear lever, in excellent positions. Major flaw of the instruments is a ribbon-type speedometer, which is harder to read than a dial.

ruggedly designed and constructed that it is capable of functioning under the greater strain of performance as well. As might be expected from a small-displacement, high-revving hot engine, the gears must be given close attention if maximum performance is to be realized. Yet the four-speed transmission, which feels rough at idle, is extremely quiet and works beautifully. It undoubtedly adds to the enjoyment of many Volvo drivers.

The 122-S fuel economy is very good. The average gas mileage rarely falls below 20 even under great stress, and sometimes climbs as high as 27 or 28. The average range determined by this road test was 21 to 25 mpg. The 85-hp PV-544 will average 22 to 27 mpg, while the 60-hp PV-544 has a range of 23 to 28 mpg. An interesting bit of evidence on the penalty of higher performance and increased weight.

While the quick and precise steering and the very short turning radius of the 122-S make it a car of good maneuverability, its general handling is not up to the very high standards of the PV-544. The reason for this is the heavier car body and the softer ride offered in the attempt to provide luxury. These characteristics result in a little more lean when cornering, combined with a slight tendency to oversteering.

The 122-S in action on all types of roads is notable for the absence of rattles and vibrations and a very low level of wind noise. Such quietness, however, makes the engine roar all the more noticeable, so the car has the traditional Volvo bark in its exhaust and considerable intake roar. Some of this engine noise can be muffled, if so desired, by the installation of an optional air cleaner silencer unit which recently has become available.

It has been pointed out repeatedly in the past two years that the 122-S, originally called the Amazon, has a striking resemblance to the last Willys sedan manufactured in the United States. A similar resemblance is often noted between the PV-544 and the fastback Fords of the mid-1940's. No official explanation has ever been offered for this by Volvo, but according to informed sources in car design circles, the Volvos were not copied from American styles, nor were obsolete dies involved. Instead, it was a matter of certain stylists themselves who conformed to particular theories of design.

Whatever the origin of the body lines, the Volvos generally are well laid out and the 122-S adheres to this policy. There's no need to stoop or crouch when entering or leaving, since the car height is adequate and the doors open wide and have ample clearance over curb lines. The individual semi-bucket front seats are chair high and fully adjustable, even to the rake of the seat back. The rear seat accommodates two riders, with more than enough room for comfort, plus the added luxury of a center armrest which can fold out of the way to make room for a third passenger. All four window cranks are easily reached from the driver's seat.

Volvo long preceded Detroit in the installation of anchor points for seat belts, introducing them more than five years ago as standard equipment in all cars. Now all Volvos sold in Europe have the belts as a standard item, although the cars imported into the U.S. have only the attachment points. The optional belts include a combination chest and lap type, which is much more effective in case of impact.

The Volvo 122-S occupies a unique position among the imported cars. It is almost without an exact counterpart in that class, and it is best compared with the domestic compacts. It matches Detroit's smaller cars very well in price, performance and durability and exceeds nearly all in gasoline mileage, quality and in luxury for a moderate price. The 122-S appeared on the U.S. scene at the very same time the new compacts were first revealed. Chances are it'll go on unchanged a lot longer than most of them. /MT

MOTOR TREND TEST DATA

TEST CAR:	Volvo 122-S
BODY TYPE:	Four-door sedan
BASE PRICE:	$2495
ENGINE TYPE:	Ohv Four
DISPLACEMENT:	97 cubic inches
COMPRESSION RATIO:	8.2-to-1
CARBURETION:	Twin SU
HORSEPOWER:	85 @ 5500 rpm
TRANSMISSION:	Four-speed, all synchro manual
REAR AXLE RATIO:	4.56
GAS MILEAGE:	21 to 25 miles per gallon
ACCELERATION:	0-30 mph in 4.9 seconds, 0-45 mph in 8.8 seconds and 0-60 mph in 16.5 sec.
SPEEDOMETER ERROR:	Indicated 30, 45 and 60 mph are actual 30, 45 and 59 mph, respectively
ODOMETER ERROR:	Indicated 100 miles is actual 98 miles
WEIGHT-POWER RATIO:	28.2 lbs. per horsepower
HORSEPOWER PER CUBIC INCH:	.88

VOLVO spells speed and comfort and the new B.18 shows it better than ever. Recently Volvo have been dropping a 1.78-litre engine into their 122S. This engine is, with some very slight modifications, the engine fitted to the P.1800, and as well as this Volvo have included a whole load of other ameliorations in their new car.

Perhaps the most important of these improvements are disc brakes on the front. But the 12-volt electrical system now used was also long overdue. And the stronger front axle member and new bushings for the lower control arm shaft are items adding greatly to the robust qualities of this Swedish car.

Ken Rudd of Worthing decided that it was worth while making modifications to the already improved car, and the one I tried had a hot engine and modified suspension.

Due to adverse weather conditions I was fortunate enough to have the car for twice the normal period of test, and when the time came to return it was rather reluctant to let it go. For of all the medium saloons I have tried this was by far the best. One of the most pleasant things about a Volvo is that it is so unostentatious; its looks are sober yet purposeful and the finish is of high standard.

The 1.78 engine is a very flexible unit in standard form, and coupled to a Ruddspeed cylinder head with a 9:1 compression ratio it was even smoother and just as flexible. Oversize valves giving considerably more valve area to the inlets are fitted, and the ports hand finished for better gas flow. The capacity of each combustion chamber is balanced, this adding greatly to the flexibility. A modified exhaust system was fitted which made the car sound a good deal healthier, the exhaust note sounding rather like that of an o.h.c. engine.

The suspension had been altered fairly radically and included exchange springs. which lowered the car a trifle, and Koni adjustable shock absorbers. These alterations, coupled with Pirelli Cintura tyres. were the only mods carried out and the transformation was surprising, to say the least.

As for performance, the acceleration times of 0-30 m.p.h., 3.7 secs.; 0-50 m.p.h., 8.3 secs.; and 0-60 m.p.h., 12.1 secs., speak for themselves. The Konis and stiffer suspension permit little or no tramp when making fast getaways, a vast improvement over a standard Volvo. These acceleration times were recorded in mediocre conditions, and backed up the driver's earlier impressions of the potential of the car. A top speed of over the ton (102.5 m.p.h.) was also very surprising, but as this speed could be attained in most conditions, a higher speed was obviously possible with wind and gradient in the car's favour. Whilst on the performance figures, it is worth mentioning that the speedometer was approximately 6 per cent. fast all through the range.

The cruising speed of the car was up in the nineties, as the overdrive which was fitted was fairly high geared. The speeds through gears for optimum figures proved to be: first, 30 m.p.h.; second, 50 m.p.h.; third, 70 m.p.h.; fourth, 90 m.p.h., the increased speeds in the gears afforded by the greater rev. range being offset by the lower geared axle ratio. The low rear axle ratio was one of the features which made the acceleration times so good.

The performance was matched by the excellent brakes. Not before time, Volvo have decided to fit discs on the front as standard equipment, and these make the excellent performance of the car very usable. For the brakes never gave me a moment's doubt as to their intentions. which were always to pull up the car powerfully and progressively, with never a thought of fade.

Matching the performance was the enthusiastic roadholding, which again inspired confidence. The rear end, with Konis, was very predictable in its movements, and the change over from understeer to oversteering in extreme conditions was always unhurried. Ice and snow were encountered during the test. and here, of course, the Volvo was in its element. The balance of the whole car made icy roads easy meat, and judging by the antics of the cars which I passed the Volvo was in a class of its own. It was also a great help to have proper heating and demisting, as both hands could be kept on the wheel, and were not employed feverishly trying to stop the screen from misting up.

The car had a slight tendency to understeer, but the power available made this more of an advantage than the contrary. The steering might be described as heavy by some people, but it was certainly well up to its job and made it difficult to overcorrect rear-end break away. The Pirelli tyres added greatly to the adhesion, and during the test were run at 30 and 32 p.s.i., several pounds higher than recommended pressures. These pressures were maintained even during the icy conditions without adverse effects on handling.

The safety harness fitted as standard equipment was the best I have ever tried, as it was so simple to put on and adjust. The trouble with many harnesses is that they are so uncomfortable people don't bother to wear them.

The instrumentation is excellent, all controls coming easily to hand, the speedometer is the strip type and the gauges easily read. The car is a full five-seater, which can easily take six, and the luggage space in the boot is enormous.

Economy was another good feature, the test car returning some 28 miles per gallon under average conditions. Driven very hard indeed the petrol consumption would rise to 25 m.p.g., but on long journeys a figure nearer 30 m.p.g. might be obtained.

It is difficult to explain why the Volvo was always used in preference to other faster cars at my disposal. Perhaps because journeys were made in identical times with far less effort and considerable saving in bad language. All passengers, whether enthusiasts or elderly people, had good words for the comfort, and the latter were not aware of the speed they were travelling—a great advantage!

Sober—
Yet Purposeful

PATRICK McNALLY

Tests the New

Volvo B.18 with

Ken Rudd

Modifications

PERFORMANCE DATA

Acceleration Times: 0-30 m.p.h., 3.7 secs.; 0-50 m.p.h., 8.3 secs.; 0-60 m.p.h., 12.1 secs.

Standing $\frac{1}{4}$ mile: 18.6 secs.

Maximum Speed: 102.5 m.p.h.

Fuel Consumption: 28 m.p.g.

SPECIFICATION

VOLVO 122S B.18 SALOON—5 DPO

Ruddspeed cylinder head. Compression ratio 9:1. £29 10s.

Stage I exhaust system. £12 10s.

Ruddspeed Suspension. £10.

Koni Shock Absorbers. £18 10s. (When fitted on new car £14 10s.)

Pirelli Cintura tyres on exchange on new car. £27 10s.

ROAD TEST/8-62 VOLVO 122S

An already desirable Swedish beauty is now even more so.

VOLVO'S FOUR DOOR SEDAN, the 122, known in Europe but not here as the "Amazon," has been around for several years with the hopped-up 1600 cc engine, but due to small distribution, this saucy package has not been as familiar a sight as its kissin' cousin, the PV-544. For 1962, however, the quota has been upgraded and so has the car. The highly stressed 1600 engine has given way to the larger (1800 cc) B-18 series engine, similar to that in the P-1800 Grand Touring car, but with a milder cam timing and 10 fewer horses under the hood. Power alone is not the only improvement but detail changes have been made all down the line to make an already desirable package even more so.

We picked our test unit up from Volvo Western, the factory-branch importer/distributor for the western half of the U.S. and launched into a discussion of availability of these hard-to-come-by cars. It's been a year since the distribution was taken over by the factory and six months since those in charge here were able to report that they had dealership and organization well enough set up to handle a flow of vehicles approaching the percentage that should sell in this country. At present, there's still a long waiting list for all Volvos, but the situation is improving on a day-to-day basis.

The 122S, selling for just under $2700 in Los Angeles, is a deluxe package with four doors, full interior appoint-

ments, and a pleasant, modern body. It holds its passengers about a foot higher from the ground than our late-model, domestic products do, but this becomes obvious only from the interior. The construction technique, gauge of metal, and quality of workmanship are that of the PV-544 on which the firm built its reputation. Extra monies spent on finish materials and sound-deadening make it one of the most solid-feeling units we've ever sat in — not completely quiet, but most certainly solid. It is a four-passenger sedan, with two buckets up front and a comfortable bench-type seat in the rear. Five adults can be squeezed in, but it's just that — a squeeze. A folding armrest is provided between the two rear passengers and, while footroom is a bit sparse, the rear section is more than adequate for long trips. The bucket seats in front have easy-to-operate levers for three-position adjustment of the backrest and the fore-&-aft adjustment is good. A nice-sized wheel and hung footpedals put the driver in a good, relaxed position for maximum control. Instrumentation won't win any prizes — at least not from us — as it includes one of those damnable ribbon-type speedometers with the mitered end. When you're stretching speed limits it requires a constant, close watch. Other gauges are adequate and legible. The dash knobs are well designed and we especially liked the bayonet-handle for the choke and its location, to the left of the wheel.

Equally likeable was the lever-type emergency brake, located on the floor and to the left of the driver's seat, easily reached yet not interfering with entry.

The weather-proofing controls — wipers, heater, and defroster are first-rate, as would be expected from a car made in Scandinavian climes. The usual window-shade radiator shroud, operated by a chain under the dash, allows the driver to select his engine temperature for both efficient operation and maximum heater output. Venting is on the weak side; the vent windows in the front doors are small and emit a wind-whistle at high speeds even when closed. Opening the door windows will turn the whistle to a roar with accompanying air-blast. The single under-dash vent is operated with the heater controls and is inadequate to provide cool air by its lonesome in warm weather.

In terms of ride, the 122S is excellent, with a firmness than lets you feel the road but not jolt your insides on rough bumps. We took a 1000-mile journey over all types of roads without any discomfort in this department. Handling is likewise excellent, with some reservations — these are confined to the rear section of the car. Well-driven, the sedan can negotiate twisty, mountain roads at sports-car velocity, but it requires careful control. First, there is a tendency for the body to shift some on the rear suspension. Second, if choppy bumps are encountered in a hard corner there is an immediate loss of power to the road as the inside rear wheel bounces high off the road. Amazingly, this isn't accompanied by the expected loss-of-adhesion in the rear, but the ability to get the power to the road is as much a handling factor

Rugged, roadable, and reliable, the Volvo 122S is a neat and likeable compact sedan with enough performance to be spirited and enough deluxe appointments to be comfortable in a large variety of applications. Shown here being "wrung out" during our 200-mile test, it emerged unscathed and delivered mpg's consistently in the high twenties. Cruising speeds up to 90 mph were easily attained and gradeability under fully-laden conditions was good. There were only minor flaws.

PHOTOS: RANDY HOLT

VOLVO 122S

as anything else. On the open highway stability is good except when gusty crosswinds are encountered. The body float in the rear is again felt with a slight-re-aiming and subsequent correction required. Though these corrections were a constant and conscious effort, they didn't reach the annoying stage but did slightly detract from our otherwise fine impression of the car. Handling characteristics, though dependent on a number of variables, are near neutral with some understeer in tight turns and oversteer on the fast ones.

Performance garnered from the 1.8-liter engine puts the sedan very definitely in the "peppy" category, despite a curb weight of 2400 pounds. Fully laden, we negotiated healthy grades at 75 mph and extended cruising above 90 mph was both possible and practical — though not exactly legal. What was really impressive for this kind of speed was the gas mileage garnered: 22 mpg! In our normal commuting we averaged 27 mpg and in steady cruising at 65-70 mph netted 25 mpg. The 122S is a real miser in this department.

Behind the rugged engine is a floor-shift, four-speed, full-synchromesh transmission. It's coupled to the former by a smooth, hydraulic-actuated clutch of more-than-adequate strength. The shift handle comes a long way out of the floor, making it easy to reach from the high seat. Reverse is a bit of a chore to engage and both First and Second are occasionally elusive, but we'd far rather have this than a shift on the wheel — the only other solution. From a ratio standpoint the top three gears are well-chosen for the type of car, just a shade below the close-ratio category. First cog, however, feels a trifle on the high side and acceleration from a standing start is slow because of this. Mid-range acceleration is excellent, with good power for passing up to 65 mph.

The brakes, a disc-front/drum-rear combination, similar to that on the P-1800, really anchor down the car at any speed and combine with the inherent stability to accomplish it without any fuss. Power-assist is an available option but pedal pressure is moderate, so we don't see the need for it. The braking is good enough to allow the sedan to go about a third farther into a corner approach than would normally be expected and, over extended applications, we couldn't incur any fade.

Noise level, much of it contributed by outside air as we mentioned earlier, was above what we consider the normal for a conventional sedan. The remaining portion was almost entirely a result of harmonic amplification within the unit-constructed body. At speeds above 50 mph the engine, transmission and differential make their presence known to a noticeable degree. At speeds above 30 mph a partial turn of the wheel will transmit a growl into the front suspension that apparently is caused by the wavy edge of the tire tread. It doesn't seem that the components themselves are noisy, just that the body amplifies what little noise they do produce. We mention this in detail because some owners are more critical about the noise level than others. We didn't find it really objectionable, but definitely noticeable.

An outstanding feature of the car, as typical of Volvo products, is the quality of materials used — steel, paint, and fabric. Ditto for the caliber of workmanship expended. In summing up our impression of the car, these factors rated high. The 122S is not exactly an economy sedan, nor is its price category, but it *is* economical to operate and maintain. We see no reason for it not to be as rugged and reliable as the PV-544, with more luxury and *much* better styling. It's a pleasant and flexible machine to operate and its many attributes make it well worth the price.
— *Jerry Titus*

VEHICLE	Volvo	MODEL	122S
PRICE (as tested) $2695.00 (POE, L.A.)		OPTIONS	None

ENGINE:

Type	4 cyl., 4-cycle, water-cooled, 5-main crankshaft (B18B)
Head	Cast iron, 4-port, removable
Valves	OHV, pushrod-rocker actuated
Max. bhp	100 @ 5500 rpm
Max. Torque	110 lbs. ft. @ 3800 rpm
Bore	3.313 in., 84 mm.
Stroke	3.15 in., 80 mm.
Displacement	109 cu. in., 1780 cc.
Compression Ratio	9.5 to 1
Induction System	2 Sidedraft SU carburetors
Exhaust System	Cast manifold into single pipe
Electrical System	12 Volt

CLUTCH:	Single disc, dry	DIFFERENTIAL:	Love, hypoid
Diameter:	8.5 in.	Ratio:	4.1 to 1
Actuation:	hydraulic	Drive Axles (type):	enclosed, semi-floating
TRANSMISSION:	4-speed, full-synchromesh	STEERING:	Cam & roller
Ratios: 1st	3.13 to 1	Turns Lock to Lock:	3¼
2nd	1.99 to 1	Turn Circle:	32 ft.
3rd	1.36 to 1	BRAKES:	disc front, drum rear
4th	1.0 to 1	Disc diameter	10.87 in.
		Drum diameter	9 in.
		Swept Area	317 sq. in.

CHASSIS:

Frame and Body:	Integral, welded steel, subframe front end
Front Suspension:	I.F.S., coil springs, tube shocks, stabilizer bar
Rear Suspension:	Live, coilsprings, tube shocks, trailing stabilizer arms, Panhard rod
Tire Size and Type:	5.90 x 15 Goodyear tubeless

WEIGHTS AND MEASURES:

Wheelbase	102.5 in.	Ground Clearance	6.5 in.
Front Track	51.7 in.	Curb Weight	2405 lbs.
Rear Track	51.7 in.	Test Weight	2642 lbs.
Overal Height	59.25 in.	Crankcase	5 qts.
Overal Width	63.75 in.	Cooling System	n.a. qts.
Overall Length	175 in.	Gas Tank	12 gals.

PERFORMANCE:

0-30	4.0 sec.	0-70	19.2 sec.
0-40	7.2 sec.	0-80	26.6 sec.
0-50	10.1 sec.	0-90	36.0 sec.
0-60	14.5 sec.	0-100	— sec.

Standing ¼ mile .... 20.6 sec. @ 73 mph
Top Speed (av. two-way run) .... 101 mph

Speed Error	30	40	50	60	70	80	90
Actual	30	40	50	59	69	79	88

Fuel Consumption Test: .... 22 mpg   RPM Red-line .... 6000 rpm
Average: .... 27 mp   Speed Ranges in gears:
Recommended Shift Points: 1st .... 0 to 28 mph

Max. 1st	30 mph	2nd	10 to 48 mph
Max. 2nd	50 mph	3rd	18 to 68 mph
Max. 3rd	70 mph	4th	30 to top mph

Brake Test: .... 70 Average % G, over 10 stops
No fade encountered.

REFERENCE FACTORS:

BHP per Cubic Inch	0.91
Lbs. per bhp	24
Piston Speed @ Peak rpm	2886 ft./min.
Sq. In. Swept Brake Area per Lb.	0.123

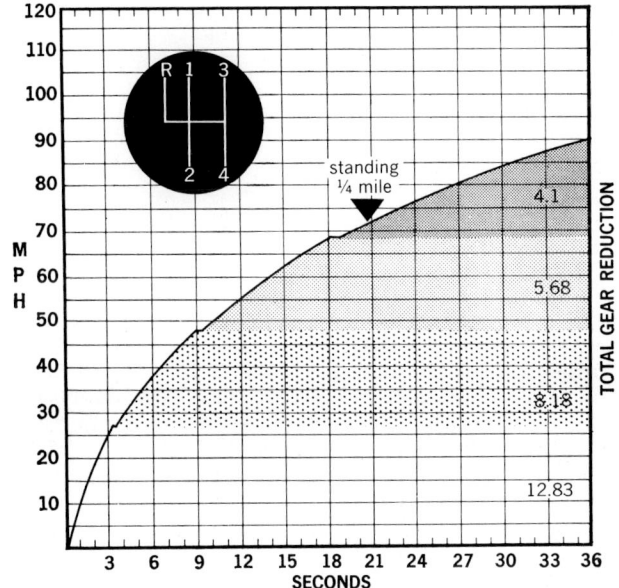

Road Test: Volvo 122S

This new model is improved in both comfort and performance

Since we last tested the Volvo 122S (August, 1960, page 54), this four-door sedan has been given a more spacious interior, a more powerful engine, a gearbox with more evenly spaced ratios, and disc brakes on the front wheels, along with some detail modifications. The 1962 model is called the B-18 series, and takes its designation from the 1.8-liter engine derived from the P-1800 sports car. This engine, in the sports and two less highly tuned versions, now becomes Volvo's only passenger-car unit, the last of the big side-valve sixes having been made some time in 1959 and the B-16 unit being wholly superseded. In developing the new engine, the Volvo engineers were aiming at increased low-range torque, greater rigidity and reduced vibration, improved silence and better fuel economy. This has been achieved by giving the engine five main bearings and completely revised porting, giving separate intake ports to each cylinder, while SU H6 carburetors replace H4 units. The valves are larger and the cam lift lower than in the B-16, and while maximum power is up only five bhp, maximum torque has been raised from 87 to 105 lb./ft. This engine is installed in the familiar four-door body-cum-chassis structure originally known as the Amazon, now with a few exterior trim changes in order to distinguish the B-18 series. Single-tone color schemes have been adopted exclusively and really suit the car much better than some of the two-tone combinations previously used. The extra space inside has been obtained by lowering the rear seat, retreating and curving the seat back, and giving a different profile to the front-seat backrests. Two-door bodies are already being made for delivery in the home market, but plans announced in the summer of 1960 for

a station wagon using the same basic body shell have not yet materialized. While Reutter seats remain optional in the 122S, the standard front seats now have three positions of backrest tilt, and long fore-and-aft travel. They are also dished in the seat and shaped in the back to give lateral support. This is much appreciated, as body roll has not been eliminated. The suspension, in fact, remains unaltered, and a heavier anti-roll bar would be an improvement. The spring rates are more than an acceptable compromise, giving a certain firmness when needed and providing a soft ride with plenty of wheel travel on rough surfaces. The steering is very light and accurate in normal driving, but in spite of the large-diameter steering wheel, considerable muscular effort is called for when parking. The controls are sensibly placed and easy to reach, handbrake on the left and gear lever on the right.

With the new engine goes a larger and stronger clutch, which has been successfully tested at 9,000 rpm. The gearbox has the same ratios as the P-1800, with really excellent synchromesh on all ratios, including first. The gear lever is rather long, but the movements are so precise that its length is hardly noticed. The rear-axle ratio has been changed from 4.56 to 4.10, raising top speed somewhat. Volvo has sensibly adopted disc brakes on the front wheels. The Girling brake units are the same as those used on the P-1800. It is interesting to note that power assistance has not been found necessary, and it was almost surprising that in normal use the brake pedal pressures were so light that it was hard to tell whether servo brakes were fitted or not. The elimination of a booster certainly contributes to keeping the total car price down. The rear brakes are nine-inch drums with one leading and one trailing shoe. Both brake and clutch pedals are of the pendant type and operate hydraulic cylinders. Even the accelerator is pendant, in lieu of the previous organ-type pedal used on this model, facilitating heel-and-toe operation. The legroom on the left of the clutch is very generous and the clutch foot can rest easily and be used for bracing the driver whenever extra support is needed.

The instrumentation is poor for a car of this class, with non-precision gauges and warning lights instead of proper dials. The Volvo P-1800 has large-diameter circular instruments, and it is not unimaginable that even the fastest Swedish-made sedan should in time be made available with a set of instruments of similar layout to replace the current thermometer-strip speedometer and the other gauges. There are door pockets for maps and other travel documents but no glove box. There is a shelf, however, under the dashboard, where cigarettes, sunglasses, etc., may be stored, but which is too low for comfort and cannot be reached when the safety harness is in use. The space normally occupied by the glove box has been reserved for the radio, while a large ashtray is mounted in the middle of the dashboard. The sun visors are padded for safety as well as the top of the dashboard, which extends backwards into both front door panels.

The safety harness is of a type designed by Volvo, with a belt across the waist as well as the single shoulder strap, both fitting on the same lock which engages on a clip on the central floor tunnel. When not in use, they are clipped on a rubber button on the door pillars. These belts are really comfortable, so that if you would not wear them for reasons of safety, you would put them on because they help keep you in your seat during hard cornering yet give your arms complete freedom of movement. This was much appreciated during the test, most of which took place on snow- and ice-covered roads. The controllability of the car on slippery surfaces is commendable, for it can be corrected into the intended course from unusually advanced broadslides. The front wheels retain a good directional grip under most road conditions although, with standard tires, rear-wheel traction seemed only about average. The value of Scandinavian winters was evident not only in its road-holding characteristics, but also in the heater. Both heater and defroster are highly efficient and warm-up was especially rapid. There was no defroster for the rear window, but the car's interior is so tight that after initial warmup, there is no trouble with mist on any window.

Over-all performance of the car belies not only its unpretentious appearance, but also the engine size. It will pull away from 20 mph in fourth gear, but also is capable of exceeding 50 mph in second. Naturally, there is a substantial difference between what the car will do with only two people in it, and what it will do fully laden. But thanks to the new-found lower-end torque, this difference is no longer so accentuated as in the previous model (with the B-16 engine). What is still more impressive is that the improved performance is realized with a degree of silence which approaches the standards of engines twice the size, although only owners of the B-16-B-engined Volvos would call it complete. Maintenance is just as simple as for the 1961 model, with oil change and lubrication at 3,000 miles and only eight grease fittings. In its native Sweden, Volvo (counting all passenger-car models) outsells even Volkswagen, although the price difference is substantial. About 50% of the production is exported to more than 50 different countries throughout the world. The 122S model sells at $2,595 POE New York.

VOLVO 122S

Price as tested: $2,595 POE New York

Importer:
Volvo Distributing, Inc.
452 Hudson Terrace
Englewood Cliffs, New Jersey

ENGINE:

Displacement	108.6 cu in, 1,780 cc
Dimensions	4-cyl, 3.31-in bore, 3.15-in stroke
Valve gear:	Pushrod-operated overhead valves (vertical, in-line).
Compression ratio	8.5 to one
Power (SAE)	90 bhp @ 5,500 rpm
Torque	105 lb-ft @ 3,500 rpm
Usable range of engine speeds	1,250-6,000 rpm
Corrected piston speed @ 5,500 rpm	2,680 fpm
Fuel recommended	Premium
Mileage	22-32 mpg
Range on 12-gallon tank	265-385 miles

CHASSIS:

Wheelbase	102.5 in
Tread	F 51½ in, R 51½ in
Length	177 in
Ground clearance	7 in
Suspension:	F, Ind., wishbones and coil springs, anti-roll bar R, Live axle; radius arms and torque rods, coil springs, Panhard rod.
Steering	Cam and roller
Turns, lock to lock	3¼
Turning circle diameter between curbs	32 ft
Tire and rim size	5.90 x 15, 15 x 4J
Pressures recommended	F 24, R 26
Brakes; type, swept area:	Girling 10⅞-inch discs front, 9-inch drums rear, 350 sq in
Curb weight (full tank)	2,390 lbs
Percentage on the driving wheels	48

DRIVE TRAIN:

Gear	Synchro?	Ratio	Step	Overall	Mph per 1000 rpm
Rev	No	3.25	—	13.15	-5.7
1st	Yes	3.13	37%	12.80	5.9
2nd	Yes	1.99	46%	8.16	9.6
3rd	Yes	1.36	36%	5.58	13.6
4th	Yes	1.00	—	4.1	18.4

Final drive ratio: 4.1.

ACCELERATION:

Zero to	Seconds
30 mph	3.8
40 mph	6.5
50 mph	9.7
60 mph	14.6
70 mph	19.9
80 mph	28.1
Standing ¼-mile	19.9

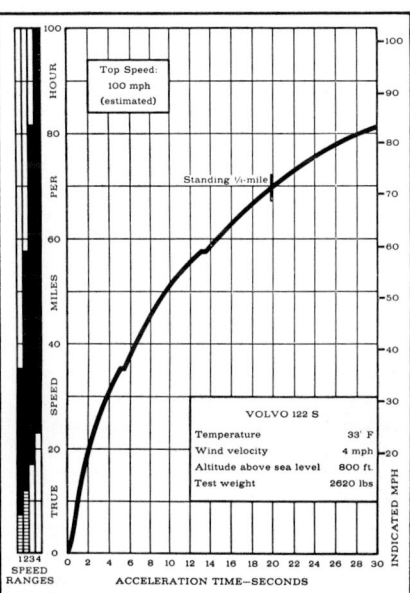

Top Speed: 100 mph (estimated)

Standing ¼-mile

VOLVO 122 S

Temperature	33° F
Wind velocity	4 mph
Altitude above sea level	800 ft
Test weight	2620 lbs

ACCELERATION TIME—SECONDS

ROAD TEST
VOLVO 122-S

SCALE: 10" DIVISIONS

DIMENSIONS

Wheelbase, in	102.4
Tread, f and r	51.7
Over-all length, in	175
width	63.5
height	59.2
equivalent vol, cu ft	382
Frontal area, sq ft	20.9
Ground clearance, in	7.7
Steering ratio, o/a	n.a.
turns, lock to lock	3.3
turning circle, ft	32
Hip room, front	2 x 19
Hip room, rear	52
Pedal to seat back, max	44
Floor to ground	13

CALCULATED DATA

Lb/hp (test wt)	30.8
Cu ft/ton mile	74.2
Mph/1000 rpm (4th)	18.4
Engine revs/mile	3260
Piston travel, ft/mile	1715
Rpm @ 2500 ft/min	4760
equivalent mph	87.5
R&T wear index	55.9

SPECIFICATIONS

List price	$2695
Curb weight, lb	2410
Test weight	2765
distribution, %	52/48
Tire size	5.90-15
Brake swept area	339
Engine type	4 cyl, ohv
Bore & stroke	3.31 x 3.15
Displacement, cc	1780
cu in	108.5
Compression ratio	8.50
Bhp @ rpm	90 @ 5000
equivalent mph	92.0
Torque, lb-ft	105 @ 4000
equivalent mph	73.6

GEAR RATIOS

4th (1.00)	4.10
3rd (1.36)	5.57
2nd (1.99)	8.16
1st (3.13)	12.8

SPEEDOMETER ERROR

30 mph	actual, 29.4
60 mph	59.4

PERFORMANCE

Top speed (4th), mph	93
best timed run	93.5
3rd (5400)	73
2nd (5400)	50
1st (5450)	32

FUEL CONSUMPTION

Normal range, mpg	21/26

ACCELERATION

0-30 mph, sec	3.8
0-40	6.3
0-50	9.5
0-60	14.5
0-70	20.2
0-80	28.3
0-100	
Standing ¼ mile	19.5
speed at end	69

TAPLEY DATA

4th, lb/ton @ mph	205 @ 48
3rd	290 @ 42
2nd	425 @ 37
Total drag at 60 mph, lb	118

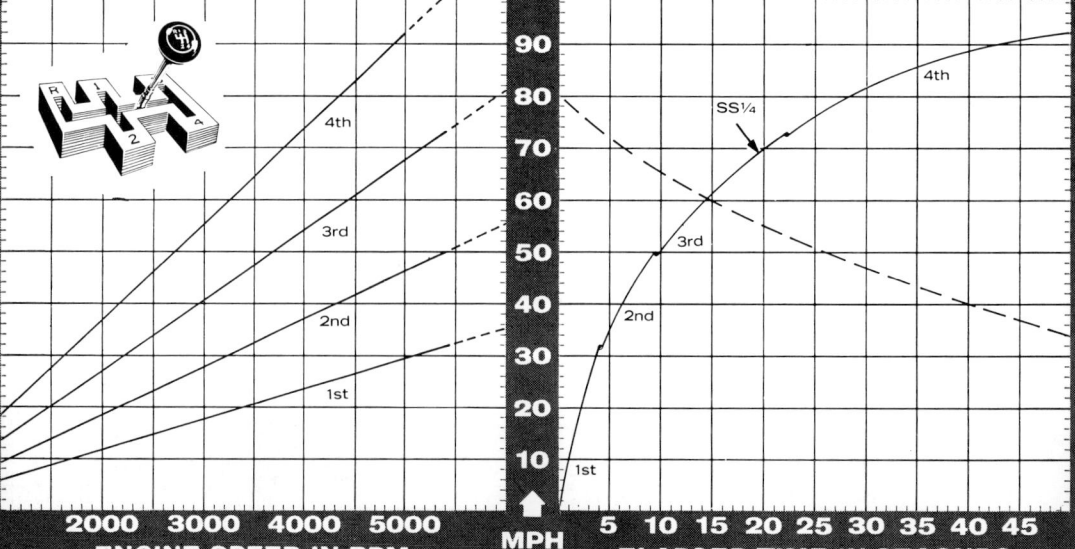

ENGINE SPEED IN GEARS

ACCELERATION & COASTING

ENGINE SPEED IN RPM

MPH

ELAPSED TIME IN SECONDS

VOLVO 122-S B-18

It may look the same on the outside, but a new engine with 5-main bearings and disc brakes on the front wheels are now standard

VOLVO THREE YEARS AGO, when we made our initial acquaintance with the Volvo 122-S 4-door sedan, it impressed us very favorably. Here was a roomy, comfortable, 1.6-liter automobile that was solidly built. And, more important (at least to us), it had very lively performance and a genuine top speed of 92 mph. In fact, that earlier sedan would out-perform most of the new crop of American compacts that were introduced in the fall of 1959.

But the 1959 Volvo had a few minor faults. Specifically,

VOLVO 122-S B-18

the smallish engine had to be thrashed a bit (through the gears) to get full performance from its very-high-output, sports-car-like engine. Though the engine proved that it could really take it, the net result was a certain amount of "rowing" with the gear lever and, at full throttle, a considerable volume of engine noise.

Now, for 1962, all Volvo models including the PV-544 2-door sedan, have a new and larger engine known as the B-18 unit. This powerplant is simply the new 5-main-bearing job originally announced for the P-1800 sports coupe, but de-tuned just a few degrees; down from 100 bhp to 90. With 1780 cc (up 12%), the significant factor is much more torque; now 105 lb-ft—an increase of 20.7%.

This higher torque means much greater flexibility in high gear at low speeds even though the axle ratio has been lowered by 10%. The sum of these changes means less gear-shifting, even better performance and much quieter high speed cruising. Hill climbing ability also improves—our Tapley meter readings show a clear gain of 17% in high gear ability, which simply means that steeper hills can be climbed without shifting down into 3rd gear.

Translating all of these changes into acceleration, nearly a full second has been trimmed from the standing-start ¼ mile time and the car pulls much more strongly in the 40-70 mph speed range usually maintained in touring. An especially attractive feature was the ease with which the car would attain 75 mph in 3rd gear. Needless to say, this makes 3rd a very brisk passing gear: only slightly more than 10 sec are required for the new 122-S to accelerate from 50 to 70 mph. Such performance can certainly be useful in tight situations, even though the B-18 engine still develops a power roar as the engine revs go up. The noise isn't nearly as noticeable as with the old engine, but for those who might object to it, Volvo has an accessory air-intake silencer system, part number 279891.

The 4-speed transmission retains the synchromesh on 1st gear and the ratios are very well chosen. The gears are quiet and, once you've mastered the longish wobble-stick control, it is very easy to select the proper gear for any situation.

The increased performance of the B-18-engined version

of the 122-S has been matched by an improvement in the car's braking power. Dunlop disc brakes, like those on the P-1800, are now used on the front wheels of the sedan. Unlike the sports car, however, the 122-S has no power booster—and apparently none is needed. The pedal pressure required is a trifle higher than before but not enough to be bothersome, and the car will pull down from its top speed in a way that is equaled by few sedans available at any price.

Vehicle code regulations in Sweden are stiffer in some respects than those in the U.S. All cars operating in Sweden, for example, are required to have mud flaps behind each wheel to prevent mud and gravel from being thrown behind the car. The safety conscious Swedes have inaugurated many features we've long advocated for cars and would like to see incorporated, in some form, in all passenger vehicles. The Volvo has a padded instrument panel that is attached to a column built to collapse under pressure, and a plastic package shelf on the passenger's side which folds under impact. The sun visors are of thick foam rubber construction, and safety belts (diagonal straps that extend from the floor in the center across the passenger to the door post) are available on order. The attachment points are installed in every car.

As for the niceties, the early 122-S sedans had somewhat skimpy individual seats in front, but plenty of head and leg room to accommodate the large American physique. Now these 4-door sedans all come through with beautifully upholstered and improved semi-bucket type seats in front and the interiors are definitely more luxurious in appearance as well as in fact. There is room for 5 adults, with a bit of a squeeze for 3 in the rear, but this is a comfortable 4-seater for all practical purposes. In addition to the fine seating, the driver will find that his personal comfort has been considered in the positioning of the controls. The steering wheel is exactly where it should be, there is sufficient room around the pedals to eliminate foot-fumbling, and all knobs and switches are clearly marked and within easy reach. Driving the car is a real pleasure; it is, of course, not up to sports car standards, but relative to other sedans it must be rated as very good.

Seldom does our entire staff reach unanimity of opinion, but in the case of the 122-S it happened. Everyone agreed that this new Volvo really was "superb Swedish engineering."

Volvo 122 1,780 c.c.

IN reporting on our industry's overseas rivals one is faced with two problems that do not exist so far as British cars are concerned. The first is to overcome—without appearing to do so—an inherent prejudice against foreign cars that is possessed by many British motorists. The second is to ensure that, in doing so, one does not appear guilty of having an in-built bias in the opposite direction—an extremely difficult compromise to strike if the "foreigner" in question is a particularly good one. It is perhaps best first to outline the salient points, before going into details of the road test findings, and locate the Volvo 122 in the general motoring picture, so that readers can assess the car themselves, comparing its specification and price with those of known makes and models.

The car falls into the 1,600 to 2,000 c.c. category for competition purposes, with its engine capacity of 1,780 c.c. This larger engine was introduced last October as an alternative to the 1,582 c.c. unit of the 122S model, the re-engined version then becoming the 122S B18. This has recently been redesignated the 122. With a compression ratio of 8·5 to 1, and output of 90 b.h.p. at 5,000 r.p.m. it is, in fact, a somewhat detuned version of the 100 b.h.p. unit used in the P1800 G.T. coupé. The new engine gives the car a mean maximum speed of 94·7 m.p.h., and a best in one direction of 99·5 m.p.h., a fair turn of speed for a 4-5-seater family saloon, with no pretensions to being a sporting car.

Girling disc brakes are fitted to the front wheels of the 122, and other improvements include a changeover from 6- to 12-volt electrical system, a larger clutch to deal with the extra power, a new headlamp flasher, right-hand drive for the British market, and an optional Laycock-de Normanville overdrive, operating only in top gear.

In general conception, there is nothing out of the ordinary about the Volvo's unit construction, coil-spring and wishbone independent front suspension, coil-spring and live axle at the rear, and its straightforward pushrod overhead valve four-cylinder engine. In its basic form, however, the car incorporates a great many features which one has grown to look for as optional extras on cheaper cars and as standard equipment only on the considerably more costly.

As might be expected of a firm which pioneered the standardization of such equipment, seat belts are provided, of a particularly convenient design embodying the lap and diagonal strap layout. Cigar lighter, heaters, headlamp flasher, two-speed (and particularly good) windscreen wipers, screen washer, map-reading lamp, radiator blind, rheostat control to the panel lights, and front seats with three-position adjustment to the backrests' rake (as well as the normal fore-and-aft movement of the seats as a whole), all these come with the car. The sole items listed as optional extras are the Laycock-de Normanville overdrive, the radio and a brake servo.

The basic price ex-works in its native Sweden is £855, which, with the addition of import duty and British purchase tax, increases to £1,293 10s 3d (£1,376 0s 3d with overdrive) in this country.

First impression of the car—and an impression that grows as one gets better acquainted with it—is that everything about it is extremely serviceable, strong, and well finished. There is no polished timber, no leather, and there are no pile carpets, but there is an excellent moulded-rubber floor covering which fits perfectly, is almost indestructible, and is very easy to clean; the facia is topped with a matt-black non-reflecting plastics material, the instruments being grouped directly ahead of the driver beneath a hood. The driving position is good, the wheel being more nearly vertical

PRICES		
Basic (with 4-door saloon body)		£940
Purchase Tax		£353 10s 3d
Total (in G.B.)		**£1,293 10s 3d**
Extras	Basic	Inc. Tax
Laycock-de Normanville overdrive ...	£60	£82 10s 0d

Manufacturer : A.B. Volvo, Gothenburg, Sweden
Concessionaires : Volvo Concessionaires Ltd., 28 Albemarle Street, London, W.1

Test Conditions
Weather .... Dry and sunny, 5-10 m.p.h. wind
Temperature .... 32-36 deg. F. (0-2 deg. C.)
Barometer 29·85in. Hg.
Dry concrete and tarmac surfaces.

Weight
Weight (with oil, water, and half-full fuel tank)
21·4cwt (2,394lb—1,085·6kg).
Front-rear distribution, per cent: F, 54; R, 46.
Laden as tested ...... 24·4cwt (2,730lb—1,238·6kg.)

Turning Circles
Between kerbs .... L, 34ft 5in.; R, 34ft 0in.
Between walls ...... L, 36ft 4in.; R, 35ft 11in.
Turns of steering wheel from lock to lock...... 3·5

Performance Data
Overdrive top gear m.p.h. per 1,000 r.p.m. ... 21
Top gear m.p.h. per 1,000 r.p.m................. 16·2
Engine revs. at mean max. speed............... 4,520
Mean piston speed at max. power 2,760ft/min
B.h.p. per ton laden .... 65·7

FUEL AND OIL CONSUMPTION

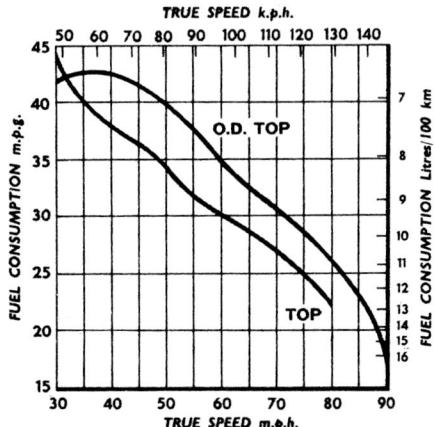

FUEL .... Premium grade
(97 octane RM)
Test Distance .... 1,342 miles
Overall consumption .... 25·3 m.p.g.
(11·16 lit/100 km.)
Normal Range .... 24 to 32 m.p.g.
(12·8-8·83 lit/100 km.)
OIL: SAE 30 ...... Consumption, 10,000 m.p.g.

HILL CLIMBING AT STEADY SPEEDS

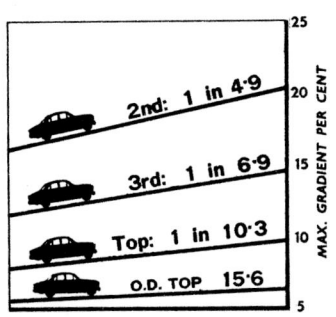

GEAR PULL	O.D. Top	Top	3rd	2nd
(lb per ton)	146	215	320	453
Speed range (m.p.h.)	36—40	35—40	32—37	27—33

MAXIMUM SPEEDS AND ACCELERATION (mean) TIMES

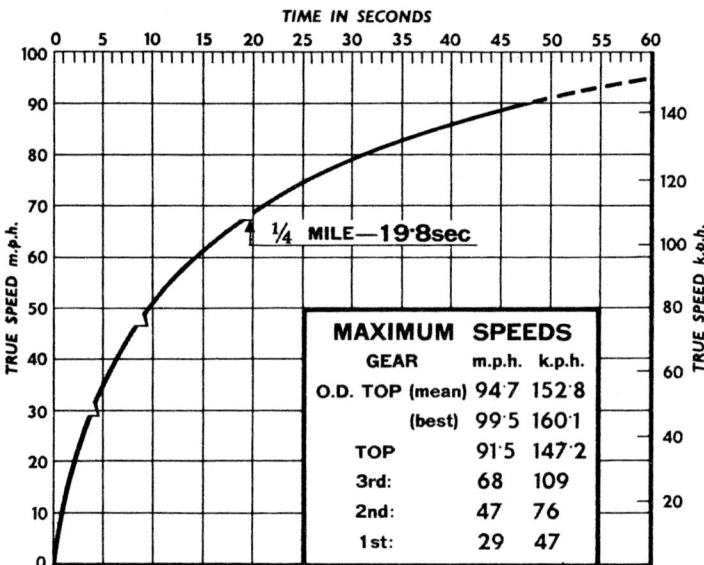

¼ MILE—19·8sec

MAXIMUM SPEEDS

GEAR	m.p.h.	k.p.h.
O.D. TOP (mean)	94·7	152·8
(best)	99·5	160·1
TOP	91·5	147·2
3rd:	68	109
2nd:	47	76
1st:	29	47

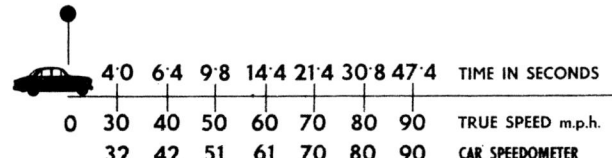

	4·0	6·4	9·8	14·4	21·4	30·8	47·4	TIME IN SECONDS
0	30	40	50	60	70	80	90	TRUE SPEED m.p.h.
	32	42	51	61	70	80	90	CAR SPEEDOMETER

Speed range and time in seconds

m.p.h.	O.D. Top	Top	Third	Second	First
10—30	—	—	7·3	4·4	—
20—40	13·5	8·8	6·5	4·6	—
30—50	14·1	8·9	6·4	—	—
40—60	14·5	9·4	8·0	—	—
50—70	17·8	11·4	—	—	—
60—80	—	14·9	—	—	—
70—90	—	26·0	—	—	—

BRAKES
(from 30 m.p.h. in neutral)

Pedal load	Retardation	Equiv. distance
25lb	0·14g	216ft
50lb	0·33g	90ft
75lb	0·50g	62ft
100lb	0·74g	40ft
150lb	0·83g	36ft
Handbrake	0·37g	80ft

CLUTCH   Pedal load and travel—50lb and 4·75in.

Passengers are particularly well cared for in regard to comfort, leg-room and protection in the event of a sudden stop. Floor coverings are of strong, moulded rubber; front seats are separately adjustable; seat belts are standard; instruments and controls are all within comfortable reach of the driver

than usual; and the range of seat adjustment is enough to give the ideal semi-straight-armed "stance" to drivers of widely differing shapes and sizes.

Throughout the interior is evidence that the manufacturers have had passenger safety very much in mind. All along the scuttle edge, for example, there is a very firmly padded roll which seems strong enough to keep the occupants off the underlying metal even in cases of severe impact. Over the front passenger's feet is a useful parcels shelf which, too, has a padded edge, and is made of a firm but collapsible material. The sun vizors are soft and pliable, and the steering wheel, with its single, broad, horizontal spoke, is designed to present the largest possible flat area to the driver's chest on impact. Finally, the car clearly has been designed from the start with seat belts in mind; there is not a switch or control normally used in driving that cannot be reached easily when one is held in the driving seat.

Door handles and window winders are well placed, high up on the doors; and the swinging ventilator panels in the front doors—"Point of entry No. 1" in the car thief's *vade mecum*—have locking safety catches. There are perhaps too many warning lights of one sort or another disposed along the instrument panel—ignition, main beam, traffic indicators, oil pressure and overdrive. It is odd that the overdrive light should be red, a colour which is universally accepted as a signal that something is wrong. Until one remembers that the Volvo is by no means a sporting car, one finds oneself feeling that the horizontal, moving-column speedometer is strangely out of character. This instrument proved to be 100 per cent accurate at speeds from 70 to 100 m.p.h.

Quick to Warm Up

Throughout the test period the engine started immediately, usually without the use of the choke, though once or twice after a night in the below-freezing open it required momentary use of this control. It warms up quickly, helped if necessary by the radiator blind, and develops its full power very soon. It is one of those rare cars in which one feels at home immediately, driving it well from the start, without having to go through the usual period of acclimatization.

The engine is smooth and reasonably silent at all normal speeds and degrees of throttle opening, though it becomes somewhat more obtrusive when the throttle is opened fully or at high speeds. Presumably this is due to the lack of silencing by the small pancake-type air filters on the two

S.U. carburettors, compared with the more usual, large cleaner-silencer found on family saloons. The car responds immediately to the throttle, and will pull away perfectly happily from a standstill in second gear; first gear, in fact, is superfluous save in very hilly countries, or where a great deal of dead-slow traffic crawling is required, for it is very low.

Apart from this, the four ratios are well spaced, giving maximum speeds of 29, 47, 68, 91·5 m.p.h. There is synchromesh for all four gears, and even during the particularly harsh conditions of the test the synchromesh was unbeatable. The substantial central gear lever is spring loaded into the right-hand plane of the "gate," so that provided only fore-and-aft pressure is used it will select only top or third gears. The gear change is excellent, being clean and precise, the slight extra pressure required to move the lever over to the second-first plane being negligible; yet it is sufficient to prevent one selecting first gear when trying for third, a possibility with synchromesh on first.

Stronger Clutch

The clutch pressure is light, and with the increase in plate diameter to 8·5in. introduced with the larger engine, it will stand full throttle gear changes without superfluous slipping; it has no trouble whatever in moving the car away on a 1-in-3 test gradient without any undue revving of the engine. There were no vibration periods in the engine, though there was a light rumble, or tremor, from the transmission at all speeds.

Without the optional servo assistance, the brake pedal load was somewhat on the heavy side, particularly for a woman driver, and it would improve driving ease if this extra were fitted. Apart from this, the brakes provided plenty of "feel" through the pedal, showed no signs of fade within the relatively demanding conditions of use during the test period, and gave powerful straight-line stopping power that was fully up to the car's performance. However, the rear wheels would lock when the brakes were applied really hard at low speeds, pedal pressures of over 100lb being required to produce this. The rear wheels would bounce or patter, and the impression that the rear damping may have been at fault is confirmed by the car's performance over a single artificial hump, the rear wheels bucking, and being thrown clear of the road.

The handbrake lever is very well placed, horizontally to the right of the driving seat. A sensible guard to the release button prevents it being touched accidentally into the "off"

The rear window's shape is repeated in that of the driving mirror, so that the rearward view is first class. Mud flaps protect following traffic in wet weather. A reversing lamp is not included as standard in the specification

position; and the handbrake held the car on a 1-in-3 test gradient, facing either uphill or down. It is also powerful enough to pull the car up reasonably quickly should the main braking system fail.

Suspension is fairly firm at low speeds and over the smaller bumps; yet at high speeds and over long-frequency irregularities the ride is unexpectedly soft and very comfortable. There is, however, rather excessive road noise on certain types of surface. At 25 m.p.h. over a washboard surface the ride was rough, as in any car, but at 45 m.p.h. it levelled off to the extent that the passenger was able comfortably to write perfectly legible notes. This was also the case over the extremely rough *pavé*, on which the suspension behaved better than most. Driven as a family saloon, the Volvo provides extremely effortless motoring; and even driven as a sports car it requires very little more concentration.

Light, even at manoeuvring speeds in heavy traffic, the steering is very positive and precise and transmits almost no road shocks, though one can feel through the steering wheel exactly what the car is doing, and when the tyres are, nearing the limit of adhesion. There is little roll on corners, and to all intents and purposes the steering is

Underbonnet accessibility is excellent, as this photograph shows. Attention has been paid to the finish throughout the engine compartment, the engine being painted scarlet, with aluminium valve cover

neutral, with perhaps a shade of understeer that is not normally apparent. Curiously, the Firestone tyres fitted as standard do not give the grip in wet weather that one would expect after sampling the performance in the dry; a driver who is proposing to treat the car as more than a workaday family saloon might do well to take advice on other tread patterns.

Though a child can be carried in relative comfort in the middle of the rear seat, or a fifth passenger on short journeys, the bodywork is essentially for four, those on the rear seat having ample leg room by virtue of the big recesses beneath the separate front seats. The upholstery is deep and comfortable, both at the back and front, though sometimes one gets the impression on cornering that the backrests of the front seats are scarcely secure enough. Closer inspection reveals that the framework is perfectly rigid, but that the backrests' side-pads flex sideways and thus one loses support in cornering, unless using the seat-belts. There is no doubt, however, about the comfort and quality of the upholstery; the squab shape is excellent, and after a long run occupants have no signs of stiffness.

Fuel Economy

Despite a performance that is nothing short of amazing—completion of the standing start quarter-mile in 19·8sec, and reaching 30 m.p.h. from rest in 4·0sec and 60 m.p.h. in 14·4sec—the fuel consumption is very reasonable indeed. On the M1 Motorway, at cruising speeds between 75 and 85 m.p.h., the car gave a figure of 26 m.p.g., and at a fast-touring average over give-and-take roads this figure increased to almost 30 m.p.g. With the 10-gallon tank capacity, a range of some 250 miles between refills could be expected. For the 1,500-mile road test period the overall consumption worked out at 25·3 m.p.g.—much better than many a smaller car with a lower performance can achieve.

The Laycock overdrive on the car tested was well adjusted, and engaged remarkably smoothly whether under power or not. It is a true overdrive, in the sense that it is a fuel- and engine-saving gear for cruising, and not merely a higher top gear. As a result, the car's performance in this ratio is influenced very much by winds and gradients. In still air, for example, it will reach 94·7 m.p.h. in overdrive top, and 91·5 in normal top. Against a slight headwind, however, the time taken from 70 to 90 m.p.h. in overdrive was 97sec—compared with 27·3sec in the opposite direction; a while later, when the wind force had increased, the car would not even reach 90 m.p.h. in overdrive top, whereas it would do so in normal top.

Those who preserve their copies of *Autocar* Road Tests will no doubt be as surprised as were the road test staff to find that the 1,583 c.c. Volvo Amazon, tested on 6 June, 1958, out-performed the current larger-engined car on every one of the standing-start acceleration figures. Only on the standing quarter-mile was the current car quicker, by one-tenth of a second. Unfortunately, this confirms what was suspected at the time, that the 1,583 c.c. car provided direct from the factory at Gothenburg, and tested on the Continent, was in an above-normal state of tune. The figures achieved by the current car should be regarded as normal, and not considered in comparison with those of the earlier test car.

Engine and auxiliaries, indeed all the under-bonnet equipment, are more than usually accessible, the engine being surrounded by plenty of clear space. The finish beneath the bonnet is particularly good, the scarlet paint of the engine and its polished valve cover looking very smart. In these days of bent-wire screwdrivers and cheap, soft spanners it is a pleasure to find a tool kit of reasonably high quality, one that is comprehensive enough to carry out the majority of roadside repairs. Together with the jack, the tools are stowed in the space beside the spare wheel, which, in turn, stands vertically to the left of the luggage boot.

Without being flashy the Volvo combines a high degree of quality, good workmanship and much attention to detail,

Luggage space is reasonable, though of a very irregular shape. The boot lid is spring-loaded; tools are stowed between spare wheel and body side. The fuel filler will take the full flow from an electric pump

with clear-cut functional serviceability. Throughout the interior of the car the care taken to protect passengers in the event of a violent stop is gratifying, and the body shell gives the impression of being extremely robust. With no sacrifice in flexibility or smoothness, the engine gives an impressive performance with surprising economy. The basic "inventory" is comprehensive for a medium-priced family saloon, and the car provides comfortable, effortless travel for its four occupants.

Specification

ENGINE

Cylinders	...	4 in line
Bore	...	84·1mm (3·31in.)
Stroke	...	80mm (3·15in.)
Displacement	...	1,780 c.c. (108·5 cu. in.)
Valve gear	...	Overhead, pushrods and rockers
Compression ratio		8·4 to 1
Carburettor	...	Two S.U. Type HS6
Fuel pump	...	AC mechanical
Oil filter	...	Full flow, renewable element
Max. power	...	90 b.h.p. (gross) at 5,000 r.p.m.
Max. torque	...	105 lb. ft. at 3,500 r.p.m.

TRANSMISSION

Clutch	...	Borg and Beck s.d.p., 8·5in. dia.
Gearbox	...	Four-speed, all synchromesh Laycock-de Normanville overdrive on top
Overall ratios	...	O.D. top 3·5, top 4·56, third 6·3, second 9·1, first 14·3, reverse 13·34 to 1
Final drive	...	Hypoid bevel, 4·56 to 1

CHASSIS

Construction	...	Integral with steel body

SUSPENSION

Front	...	Independent with coil springs and wishbones. American Delco telescopic dampers; anti-roll bar
Rear	...	Rigid axle; coil springs, radius arms, Panhard rod. American Delco telescopic dampers
Steering	...	Cam and roller; 17in. dia. steering wheel

BRAKES

Type	...	Girling hydraulic; discs front, drums rear
Dimensions	...	F. 10·87in. dia. discs. R. 9in. dia. drums, 2in. wide shoes
Swept area	...	F. 226 sq. in.; R. 113 sq. in. Total 339 sq. in. (278 sq. in. per ton laden)

WHEELS

Type	...	Pressed steel, 5 studs, 4·0in. wide rim
Tyres	...	5·90-15in. Firestone tubeless

EQUIPMENT

Battery	...	12 volt, 38 amp.-hr.
Headlamps	...	45-40 watt
Reversing lamp	...	None
Electric fuses	...	4
Screen wipers	...	2, two-speed, self-parking
Screen washer	...	Standard, electrical pump type
Interior heater	...	Standard, fresh air type
Safety belts	...	Standard for both front seats
Interior trim	...	Plastic throughout
Floor coverings	...	Rubber
Starting handle	...	None
Jack	...	Screw pillar, winding handle
Jacking points	...	Four external sockets, under body sills
Other bodies	...	None

MAINTENANCE

Fuel tank	...	10 Imp. gallons (no reserve)
Cooling system	...	15 pints (inc. heater)
Engine sump	...	6·7 pints inc. filter. Change oil between 2,500 and 5,000 miles, depending on driving conditions. Change filter element every 6,000 miles
Gearbox and overdrive	...	1·25 pints SAE 80. Change oil every 12,500 miles
Final drive	...	2·25 pints SAE 80 hypoid. Change oil every 12,500 miles
Grease	...	8 points every 3,000 miles
Tyre pressures	...	F., 20 p.s.i. R. 23 p.s.i. (normal driving); F., 22 p.s.i. R. 25 p.s i. (fast driving)

Scale: 0·3in. to 1ft.

Cushions uncompressed.

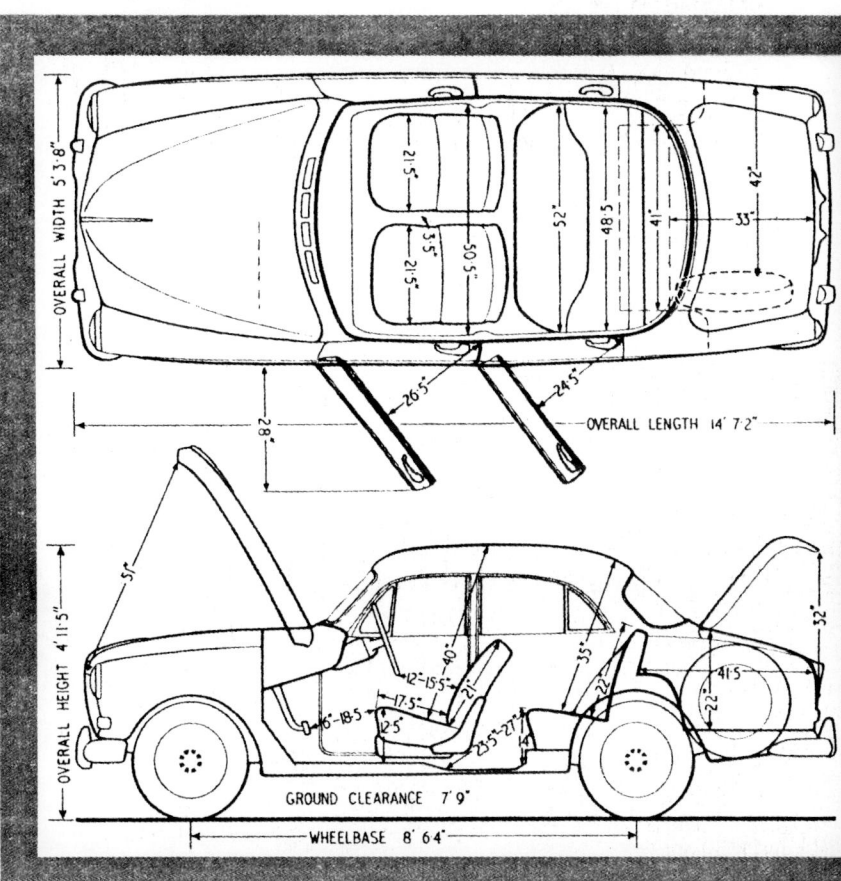

VOLVO B18-122S

in the early 1920's with two Swedes, Assar Gabrielsson and Gustaf Larson, who built 10 experimental cars in 1927. Then they raised enough capital to build a batch of 100, and borrowed an SKF Ball-Bearing Company factory on the island of Hisingen.

The factory had earlier been used by a daughter company of SKF, called "Volvo". Gabrielsson and Larson regarded this name as so suitable for cars, that they took it over.

Since then the concern has grown into the massive

THE SWEDISH VOLVO IS STILL RATHER an unfamiliar car in South Africa, as it has a comparatively short history here. The PV544 "Sport" model is probably the better known, but the 122S is rapidly gaining ground and becoming the more popular of the two Volvo models in the Republic.

The 122S has no claim to being particularly beautiful. On the contrary, it is a rather plain and unostentatious saloon. But, as we found on this test, it is a tremendously satisfying car to drive under all conditions. It feels, and is, outstandingly strong and competent. It is the sort of car that makes its driver feel it can do anything.

It does a shade under the magic 100 comfortably, accelerates like a true sports saloon, hugs its passengers in tailored seats, handles superbly, stops quickly, is very easy on fuel and is reliable mechanically. It is — to put if briefly — a very good car.

The name "Volvo" is interesting in itself. It is Latin, and means "I roll". (But not in the slang sense of the word — we imagine a 122S would be pretty difficult to turn over!)

According to the company's history, the car originated

Volvo Group of industries, producing 100,000 vehicles a year as well as such a variety of machinery as Bofors guns, road graders, jet aero engines, letterpress machinery, and marine and industrial engines.

twin carburettors have dry air cleaners with cheap, replaceable paper elements. These are efficient and easy to work with, but tend to be more noisy than oil-bath types.

There is improved independent front suspension, and disc brakes at front to increase stopping power. The electrical system has become 12-volt.

The interior of the car is well finished. The separate front seats are moulded and deep-sprung to provide great comfort, and the rear bench seat is similarly shaped to provide form-fitting seating for two big people. The 122S, though in appearance a fairly big car, is actually built to seat four in spacious comfort — even to a retractable centre armrest in the rear seat.

It would be awkward and impracticable to seat three adults in the rear for long periods. We tried it, and at least one of them has to sit forward on the edge of the seat. But with children, seating three is no problem.

As befits a car costing nearly R2,400, there is no skimping in the interior luxury. The quality of fittings, mats, head-lining (which is sound-absorbent) and instruments leaves no room for criticism.

The driving position is commanding and comfortable. The large steering wheel is mounted rather low and well back towards the driver, as both windscreen and dash are set well back. But the front seats are adjustable for both rake and height, and in the car under test were set in the highest position. The all-round glass area is not enormous by modern standards, but driver visibility

(Above) **The comfortable interior layout. Radio is an extra.**

(Right) **Although it looks bigger, the Volvo is designed as a spacious four-seater.**

(Left) **The 122S has neat and unpretentious lines.**

(Below left) **Cornering hard in rain. Note the right front wheel lifting.**

★

The Volvo emblem — a circle with an arrow at two o'clock — is universally used in biology as the male symbol, and this, too, is appropriate, as we would say offhand that the Volvo is very much a man's car.

Reaching South Africa by way of the Motor Assemblies plant at Durban, the 122S is not changed in appearance for 1962, but has some important major changes over previous models as well as a great many detail improvements.

To start with, it has the new B18 engine with a five-bearing crankshaft, which produces 90 brake-horsepower and drives through a 4·1 to 1 differential. The

is good — and too much glass can be a nuisance in a hot country.

The B18 motor is a tidy unit with good accessibility, and very cleanly finished. Included in the underbonnet are windscreen washers and a forced-draught heater/demister/ventilator unit. Doing the test on a day when the outside air temperature was 51 deg. F., we found the heating to be thoroughly efficient.

Working at full capacity, in fact, it would make the interior uncomfortably hot, but of course it can be adjusted on the dash to any required level of both heat

VOLVO (B18-122S)

SPECIFICATION

MAKE AND MODEL: VOLVO B18-122S 4-Door Saloon.

ENGINE: 4-cylinder, in-line, water-cooled, o.h.v., twin carburettors,

COMPRESSION RATIO: 8·5 to 1.

BORE AND STROKE: 3·313 x 3·15 in. (84·14 x 80·0 mm.)

CUBIC CAPACITY: 1·78 litres.

MAXIMUM HORSE-POWER: 90 b.h.p. at 5,000 r.p.m.

MAXIMUM TORQUE: 105 lb./ft. at 3,500 r.p.m.

ROAD SPEED IN DIRECT TOP GEAR AT 1,000 R.P.M.: 18·2 m.p.h.

PISTON SPEED AT MAXIMUM HORSE-POWER: 2,760 ft. per min.

BRAKES: Hydraulic three-cylinder discs at front, drums at rear. Total lining area: 147 sq. in.

SUSPENSION: (Front) Independent with coil springs and stabilizer. (Rear) Torque rods and coil springs.

TRANSMISSION: 4-speed manual, floor shift, all synchromeshed.

GEAR RATIOS: 1st: 3·13 to 1   Top: Direct
2nd: 1·99 to 1   Reverse: 3·25 to 1
3rd: 1·36 to 1

FINAL DRIVE RATIO: 4·1 to 1.   **TYRE SIZE:** 5·90 x 15.

LENGTH: 175 in.   **WIDTH:** 63·75 in.   **HEIGHT:** 59·25 in.

GROUND CLEARANCE (laden): 8·5 in.   **STEERING:** Cam and roller, ratio 15·5 to 1.   3¼ turns lock-to-lock. Turning circle: 32 ft.

FUEL TANK CAPACITY: 10 gal.   **BOOT CAPACITY:** 17·5 cu. ft.

LICENSING WEIGHT: 2,405 lb.   **WEIGHT AS TESTED:** 2,740 lb.

ANNUAL LICENCE: R16.   **PRICE:** R2,394 all round.

INTERIOR DIMENSIONS:
Width of front seat(s): 21·5 in.
Driver's seat squab to clutch pedal: (Max.) 40 in., (Min.) 35 in.
*Front seat headroom: 4 in.
Width of rear seat: 48 in.
Rear seat kneeroom: (Max.) 12 in., (Min.) 7 in.
*Rear seat headroom: 3 in.

(*Measurements taken with 6 ft. man seated, no hat.)

PERFORMANCE

ACCELERATION THROUGH GEARS:

M.P.H.	Sec.	M.P.H.	Sec.
0—30	4·0	0—60	14·7
0—40	6·9	0—70	20·5
0—50	10·5	0—80	28·3

ACCELERATION IN HIGHER RATIOS IN SECONDS:

M.P.H.	Top	Third
20—40	9·4	6·9
30—50	9·3	6·8
40—60	10·9	7·5
50—70	11·9	—
60—80	15·2	—

STANDING QUARTER MILE: 19·25 sec.

REASONABLE MAXIMUM SPEEDS IN GEARS:
1st: 28   2nd: 42   3rd: 67

MAXIMUM SPEED IN TOP: 95·3 m.p.h.

EMERGENCY STOPS (10 at 30-sec. intervals, from 50 m.p.h., in sec.): 3·8, 3·8, 3·6, 3·4, 3·8, 3·9, 3·8, 3·7, 3·8, 3·7. (Average 3·73 sec., no fade).

HANDBRAKE STOP (from 50 m.p.h., in sec.): 8·2.

FUEL CONSUMPTION AT STEADY SPEEDS IN TOP:

30 m.p.h.: 42·3	60 m.p.h.: 35·5
40 m.p.h.: 42·9	70 m.p.h.: 28·1
50 m.p.h.: 38·2	80 m.p.h.: 22·3

FUEL CONSUMPTION AT FULL THROTTLE: 15·8 m.p.g.

TOURING FUEL CONSUMPTION: 27·2 m.p.g.

SPEEDOMETER CORRECTION: (See Text):

Indicated:	30	40	50	60	70	80	90
Actual:	30	40	50	60	70	80	90

TEST CONDITIONS: Barometer reading 30·37; at sea level; air temperature 51 deg. F.; 0–5 m.p.h. down test strip; wet new tarmac surface; weather cold, occasional showers; 93-octane fuel.

CAR SUPPLIED BY: W. G. Thompson (Pty.) Ltd., Cape Town.

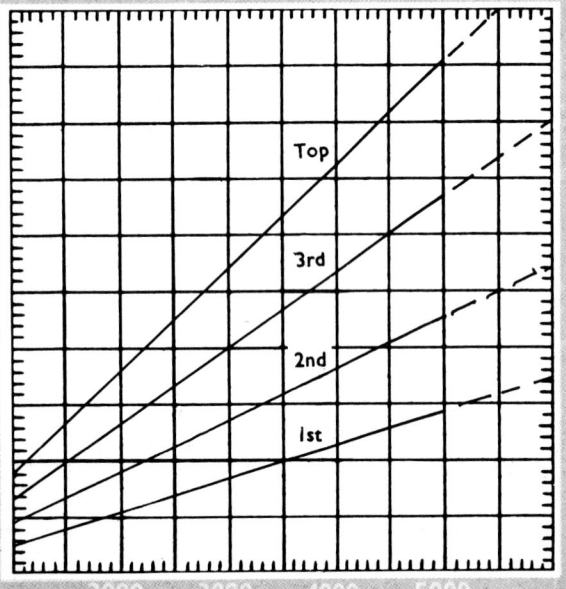

ENGINE SPEED

REVS. PER MINUTE

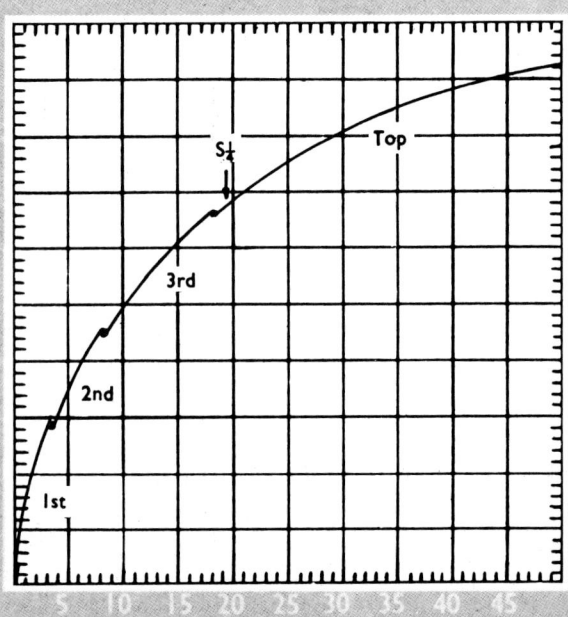

ACCELERATION

TIME IN SECONDS

and air intake. This, by the way, is typical of all the features of the Volvo — they all worked well.

The 90-brake-horsepower engine gives the car a very workmanlike performance in the sports saloon category, as will be seen from the accompanying performance table. The 90 horses are "big" ones, by international standards, and give very firm acceleration right up to top speed, as well as plenty of torque for about-town cruising.

Acceleration is accompanied by a booming noise — we put it down to exhaust resonance, though it might be augmented by induction roar — which reaches the interior, but not to any disturbing level. It is also heard at top speed, to a lesser extent.

But on the whole the interior noise level is low, and road noise is totally absent.

The B18 is as competent a "big-four" engine as one is likely to find anywhere, and we enjoyed its performance.

In the department of handling and roadholding, the 122S earns very high marks. It understeers safely and is very precisely responsive to the steering wheel. (This would have been helped by the special sports-type tyres fitted to the model tested.) Our cornering picture, if studied closely, gives an indication of its performance. The right front wheel is almost off the ground and the left front tyre dragging.

This was in a shower of rain (the wipers are working) yet the car did not break away at all.

The roll angle is small for such a severe turn.

At high speed the car handles well on all types of surfaces, and a slight tendency to wander is easily corrected.

The braking capabilities of the car are superb. On a wet, new tarmac surface it stopped from 50 in under 4 sec. every time, with no trace of fade. At the 10th emergency stop the pedal could be applied just as gently as for the first. The front discs answered the pedal immaculately and without locking. The rear drums tended to lock, but this could be attributed to the wet surface.

On a dry, good tarred roadway the stopping average would have been about 3 sec., which is excellent by any standards. The handbrake is a good auxiliary at 8·2 sec.

One of the Volvo's greatest virtues (proved by two class wins in the Mobilgas Economy Run during June) is its most moderate thirst.

The 1·8-litre engine. Circular fitting on the bulkhead is the heater/ventilator, which can be fan-driven at the flick of a switch.

We calculated the boot capacity as 17·5 cu. ft. Spare is mounted vertically.

For a car of this size and performance to return well over 40 m.p.g. at 40 m.p.h., and well over 20 at 80, is truly outstanding. Twin-carburettor efficiency results in the 40 m.p.h. figures being the best, and the 15·8 m.p.g. at full throttle (with the car doing well over 90 m.p.h.) is striking.

Driven hard on our touring run, with plenty of use of the gear-lever, the 122S returned an unusually good 27·2 m.p.g. On this basis, the car can be expected to cover about 300 miles of ordinary cruising on the full 10-gallon tank, while the Economy Run result shows that a careful driver could raise this figure to nearer 400 miles!

While on the subject of performance — we almost forgot to mention the gearbox. It is a beauty, synchromeshed right through and extremely smooth to operate. The floor-mounted gear-lever itself is a nicely finished, solid-feeling piece of machinery and most pleasant to use.

The speedometer (of the ribbon type) is one of those ambiguous types in which the ribbon tapers to a point. We have not yet met anyone who knows at which position the reading should be taken — whether at the point, or where the point tapers back into the ribbon.

On the 122S we took readings at both, and came to the conclusion that the correct reading is at the ribbon, and not the point. On th.s basis, the car's speedometer proved completely accurate right through the range from 30 to 100. But if the reading is taken at the point, the speedo registered 4 m.p.h. optimistic readings right through.

This would seem to be conclusive, but it would be much better if the correct reading could be taken at the point, which is easier to read.

The top speed of just over 95 recorded on the test coincides with the manufacturer's claim. In fact, the car is geared to do 91, but will reach very close to a true 100 under good conditions.

This Volvo B18-122S is not an inexpensive car, but the owner gets good value. At the same time it is very economical to run — even to falling just inside the R16 licence bracket! We enjoyed testing it and — as we said earlier — we were very impressed by its many capabilities.   ●

THE VOLVO 122

Not so long ago, the Volvo was almost unknown in this country. Indeed, when I first road-tested this car I had to go to Holland to find one. Now, the machine is remarkably popular, and one meets a fair number on any long journey. It is therefore worth pondering the reasons for the appeal of the Volvo.

The Volvo 122 is a completely conventional car. It is built in Sweden, but it incorporates the best components from other countries, including Britain. The four-cylinder over-square engine has been recently slightly enlarged to 1,780 c.c. and the car has rather more performance than its competitors. The roadholding is outstandingly good for a design with a rigid rear axle, particularly on wet roads. Perhaps the all-synchromesh four-speed gearbox sells a good many Volvos, and the superior finish of the body, coupled with many safety features, also attracts buyers. Above all, the car's reputation is the best form of advertisement.

The very rigid steel four-door saloon body of the Volvo forms its chassis. The front suspension is conventional with wishbones and helical springs. Behind, the axle is also on helical springs, with trailing arms and a Panhard rod for lateral location. The front brakes are Girling discs.

Absolutely normal in appearance, the engine has been developed to give a most impressive power output while remaining utterly reliable and having a very long life. It is smoother than it was and the characteristic "power roar" has been virtually eliminated. The gear-driven camshaft operates the valves through pushrods and rockers, while

carburation is by two horizontal SU instruments.

The performance of the car owes a good deal to the gearbox, which has well-spaced ratios and powerful synchromesh on all four speeds. The synchronized bottom gear is most useful, as it will exceed 30 m.p.h. The axle ratio is 4.1 to 1, or 4.56 to 1 when the Laycock-de Normanville overdrive is fitted. The test car was so equipped.

The Volvo gives the impression of being quite a big car and it is certainly a full five-seater. Being extremely well built, it is quite heavy, but the efficient engine makes light of its load. The lavish equipment includes safety belts for the front seats and copious crash padding, though the good roadholding and powerful brakes go far towards rendering such precautions unnecessary.

One sits quite high in the Volvo and there is a good field of vision. The clutch is smooth but does not tend to slip after rapid change of gear. There is remarkable freedom from wheelspin and axle tramp.

The maximum speed in the direct top gear is almost as high as that in over-drive, but, of course, the engine becomes "busy" at such high revolutions. A maximum speed of 94 m.p.h. is very satisfactory for so substantial a car with an engine of moderate size. Timed in one direction 96 m.p.h. was recorded, and rather more was occasionally achieved under favourable conditions. The acceleration is good without being startling but the figures are actually better than one would expect, thanks to the excellent gear change.

The steering tends to be heavy on sharp corners and for manoeuvring, but it is quite light at touring speeds. It gives a great feeling of control to the driver, the whole behaviour of the car being predictable. Slight understeer during the initial stages may be converted to oversteer during full-throttle cornering, but there is no tendency for the rear end to break away.

A fairly firm ride, with some sharp movements on inferior surfaces, gives an acceptable degree of comfort on all but the worst roads. This suspension ensures that the car does not roll excessively during fast cornering. The seats are comfortable for long journeys and the reduced noise level has also made the machine more attractive for such trips. The engine is still obviously a high-

---

SPECIFICATION AND PERFORMANCE DATA

Car Tested: Volvo 122 four-door saloon. Price, £1,294 including P.T. (overdrive extra).

Engine: Four-cylinder, 84.14 mm. x 80 mm. (1,780 c.c.). Pushrod-operated overhead valves. Compression ratio, 8.5 to 1; 90 b.h.p. at 5,000 r.p.m. Twin SU carburetters. Coil and distributor ignition.

Transmission: Single dry-plate clutch with hydraulic operation. Four-speed all-synchromesh gearbox with central control and Laycock-de Normanville overdrive. Ratios, 3.5 (o/d), 4.56, 6.2, 9.1 and 14.3 to 1. Divided propeller shaft. Hypoid axle.

Chassis: Combined steel body and chassis. Independent front suspension by wishbones, helical springs, and anti-roll bar. Cam and roller steering. Rear axle on trailing arms, Panhard

rod, and helical springs. Telescopic dampers all round. Girling disc brakes in front, drums rear. Bolt-on disc wheels fitted 5.90 x 15 ins. tyres.

Equipment: Twelve-volt lighting and starting. Speedometer, fuel and water temperature gauges, heating and demisting, flashing indicators, safety belts to front seats.

Dimensions: Wheelbase, 8 ft. 6½ ins. Track, 4 ft. 3½ ins. Overall length, 14 ft. 9 ins. Width, 5 ft. 3½ ins. Weight, 1 ton. 1½ cwt.

Performance: Maximum speed, 94 m.p.h. Speeds in gears: direct top, 93 m.p.h.; 3rd, 70 m.p.h.; 2nd, 48 m.p.h.; 1st, 32 m.p.h. Standing quarter-mile, 19.4 secs. Acceleration: 0-30 m.p.h., 3.6 secs; 0-50 m.p.h., 9.6 secs.; 0-60 m.p.h., 13.2 secs., 0-70 m.p.h., 19.2 secs.

Fuel Consumption: 23 to 26 m.p.g.

efficiency "four", but it is commendably smooth in spite of that. The transmission is pleasantly quiet and the divided propeller shaft avoids any vibration being felt from that component.

The brakes will stand up to the hardest driver. Like the other controls, the pedal is not outstandingly light in action, but firm pressure produces powerful, progressive and fade-free braking. The car does not tend to bow down during braking, nor does the rear axle become lively.

The fuel consumption is moderate, ranging from 23 to 26 m.p.g. I have met owners who claim 28 or even 30 m.p.g., but perhaps they drive with little less pressure on the accelerator than I do. It will be realized, therefore, that the Volvo is just as economical as other cars of its size, in spite of having a useful turn of speed.

For Scandinavian winters, a car must have really powerful heating. This make has always been well equipped in this direction, and the latest model has an even more potent heater, which would be able to deal with British winter conditions with the greatest of ease.

After a searching test, one respects the 122 as a really sound car with many virtues, and it is easy to understand the enthusiasm of Volvo owners. Although this is not a sports car, it appeals to the fast driver, while its many safety features endear it to the family man. Unlike some Continental cars, the Volvo holds its value well, so it can be regarded as a good investment.

A VOLVO TUNED
BY ROBERT BODLE

IMMEDIATELY after testing the Volvo 122 I was able to sample a similar car tuned by Robert Bodle, Ltd., of Dorchester Service Station, Oxon.

The Bodle-tuned car had an extensively modified cylinder head, the work being very beautifully done and everything finished to a high standard of polish. The compression ratio was raised to 10 to 1, the ports opened out, and the combustion chambers balanced. Special valve springs, permitting 6,800 r.p.m., were fitted. The work on the head would cost £39 10s. and the valve springs £3 19s. 6d. The inlet manifold was machined and balanced, but the standard carburetters were retained. The complete tuning operation was kept within Group II regulations.

The test car was heavier than standard, having a great deal of rally equipment. Nevertheless, its performance was better than that of the standard car. The 4.1-to-1 axle was fitted, with overdrive in addition. This permited delightfully easy cruising and a maximum speed (timed) of 102 m.p.h., while 100 m.p.h.

came up on the direct top. The increased engine revolutions allowed maxima of 37, 58 and 83 m.p.h. to be achieved on the gears without going to the absolute limit. A rev. counter was fitted, which proved that 6,000 r.p.m. could be easily exceeded with no fuss.

The standing quarter-mile was covered in 18.9 secs., the acceleration times being: 0-30 m.p.h., 3.5 secs.; 0-50 m.p.h., 8.3 secs.; and 0-60 m.p.h., 12 secs. These figures prove that the power output of the engine was usefully increased, but most unexpectedly the fuel economy was also better, 28 m.p.g. being recorded during hard driving.

This tuned Volvo was just as quiet as the standard car and had all its virtues. It demanded a little more gear-changing because it was higher geared, but the standard of flexibility was quite satisfactory. For those wishing to achieve a genuine 100 m.p.h. this quite moderate degree of tune should prove entirely satisfactory and have none of the drawbacks of a really "hot" conversion.

ACCELERATION GRAPH

MAX 94 M.P.H.

¼ MILE

VOLVO 122

M.P.H.

SECONDS

Appearance of the more powerful Volvo has not changed from the earlier 1600 cc model. Car is functionally attractive

VOLVO'S PRACTICAL nordic twins

From SLONIGER

German correspondent Sloniger recently tried two of the new Volvo sedans, found that although they look the same outside, their personalities are very different.

THE opportunity to compare two cars from the same line over the same roads and in the same weather conditions is rare and I jumped at an offer from Volvo, in Germany, to try the 121 and 122S cars with the 1.8 litre motors within one week.

Stepping from one to the other and back was highly enlightening. Before the Volvo fans — invariably seven-foot Swedes running six axe handles across the shoulders — come after me with a beam axle in each hand, one or two points bear explanation.

Both the acceleration times and the top speeds fell below my expectations and well under the boasts of the clan. For one thing the 121 had only 7500 miles on the clock when collected and the hotter 122S was just under the 5000 mark. "Only" is the operative word for a Volvo. They really aren't loose and quick to their potential without 10,000 showing. For another, this was a semi-winter test and both cars were fitted with snow tyres on the rear. These should cut between four and six mph off the top speed and do similar sneaky things to the standing starts. Of course, both cars were identically shod so the comparison remains valid.

For purposes of clarity I might explain, too, that 121 is the 75 (SAE) hp Amazon, four-door sedan with one carburettor, etc. The 122S is outwardly identical and fitted inside with exactly the same degree of comfort. The difference comes in the nose where the same engine is boosted to 90 (SAE) hp. Strangely enough Volvo quotes the same torque figures for both (104.6 ft/lb), but I will continue to nurse doubts. Volvo does admit that it comes in at 2800 rpm for the tamer car and at 3500 rpm in the sport-engined model.

Sport is the vital concept here, in separating the two cars. Changing to the 1780 cc engine (same as the P1800 coupe in size) didn't alter factory thinking.

Its standard four-door sedan is a tractable, willing five-seater with a nice turn of performance for its size. The 122S has the same creature comforts but it requires more attention to the gear lever. The engine is frankly intended for sporting types and is just plain fussier.

Before getting into the differences, let's amble through the similarities which cover just about all parts of the car.

The appearance is bound to come up sooner or later so I might as well admit the obvious — the Amazon range is not new but it is a design that has mellowed well. It looks as well balanced as it did some years ago and will some years hence. Compared to the frankly dated 544 range of two-door sedans it is positively handsome. Incidentally, there will be a two-door Amazon model on the export line very shortly. It is already to be seen in Sweden. Apart from the number of hatches it looks identical.

One drawback to the design age of the Amazons is a window area smaller than we have got used to in the past five years or so. The front pillars are a trifle thick and the driver can't see his other front guard very easily. Rear vision for reversing is similarly a matter of guesswork. On the other hand, those solid posts increase the feeling of absolute safety you enjoy in a Volvo. Panel fit and general finish is impeccable and everything opens or closes with that quality thud.

The luggage space is also somewhat less than current norms but the trunk will accept an extra amount of soft luggage. Four could tour a month with some restraint in packing and two could take an entire wardrobe. The lip is fairly low, making loading easy enough while the spare occupies one side, unprotected. At the other end the motor lid opens so high you have to jump for it and the compact four is about as accessible as such things can be. The 121 is naturally a hair better, with only one carburettor.

Access, front and rear, to the five seats is easy for the corpulent or aged and once you get settled com-

76

Profile shows generous door sizes, conventional but pleasant body lines. Test cars had snow tyres on rear.

Interior of both models is exactly the same. Those floor mats are extras, by the way. Note sturdy gearlever. Not visible are standard equipment seat belts.

fort is noteworthy. The front seats are individual with good fore-aft range and three back positions to suit drivers from Puritan ramrod spine to reclining Pasha. Unfortunately they don't go clear back, bedwise.

The driving position is very chair-like, an innovation these days. You have a sort of enthroned feeling which helps a little in minimising the lack of glass area. There is no real attempt to provide bucket seats, though the backs do curve slightly. Padding is firm in the manner that makes long runs comfortable and the standard Volvo seat belts—my choice for the best compromise possible—make true bucket seats almost frivolous. These are the combined diagonal-shoulder and single-lap straps with one attachment point. Barring a full harness, which few would bother to use, these are the best.

Once in place and belted you discover the thought put into control positions. The driver can reach every knob, lever and pedal with his strap on and without squirming or stretching. The handbrake is found outboard of his seat, with a neat ring over

Correspondent Sloniger tested Volvo in Germany. Here the car is parked next to an ornate fountain in Heidelberg.

Two SU carburettors make this the 122S. The 121 is otherwise the same but used a single Zenith carb.

the 122S, which needed it more, was clumsier. The trouble seemed to lie with the brake pedal. The 122S has front disc and no booster. You feel the heaviness of the pedal very soon and don't generally depress it as far, so that its relationship to the accelerator is not as good.

Since I have slipped into a comparison I might as well continue with the braking proper. Around town the 121 was more pleasant, with a light, progressive pedal that pulled the car up clean every time. When hustled down a winding mountain road, using the brakes exclusively, instead of gearing down, there was no fade whatsoever. On the other hand it was far easier to lock things up in the panic-stop test and harder to hold an absolutely straight line. The disc-braked 122S took much more effort in town (naturally didn't fade either) but pulled much truer in the 60 to 0 runs. Actually, it was hard to get enough weight on the pedal to lock the wheels and still the car had a better value on the Bowmonk dynometer and stopped shorter.

Back to driving. To start with, the cars were both relatively slow warmers, even with the radiator blind closed completely — by means of a reluctant chain that then swings and rattles around your feet. There is a great temptation to leave the blind down, keeping the chain short. Both motors jumped to life right away when choke was applied from cold and without a murmur after full speed runs. They just took their own time getting hot.

Fortunately the heater is such that even half-power at first is putting out warmth for demisting. Once it got going, full power would drive you out of the automobile in anything but an Arctic blizzard. There is a two-stage booster fan (noisy in high) and separate controls for air and dividing the heat between screen and feet.

Other evidence that Volvo designers think well is the wiper/washer control. One pull is low speed wipe, two is high ditto and three notches is wash. You only take one hand from the wheel for everything. The light knob next to this is equally easy and the blinker/turn indicator lever is right under

the button to prevent inadvertent release. Only rarely did it catch a trouser leg with the seat right back on its track.

The pedals are all pendants but spaced so that you can heel-toe as a general thing. I qualify this because the 121 was a natural for the practice while

VOLVO 121-122S

PERFORMANCE

	121	122S
0 to 40 mph	7.7 sec	7.2 sec
0 to 50 mph	10.4 sec	9.6 sec
0 to 60 mph	17.2 sec	14.9 sec
0 to 70 mph	24.2 sec	20.4 sec
Top speed	91.3 mph	94.8 mph
Speedometer error	4.0 percent slow	4.5 percent slow
Brakes: 0 to 60 to 0 %g	20.7 sec/75 percent	19.5 sec/88 percent
Fuel consumption	26.9 mpg	25.0 mpg

ENGINES:

Designation, B18A (B18D); four cylinders, in-line, 75 hp SAE at 4500 rpm (90/5000); compression, 8.5 to 1; bore/stroke, 84.1 by 80 mm; displacement, 1780 cc; maximum rpm, 4500 (5000); single Zenith 36 VN downdraft carburettor (twin SU HS6 horizontal); 12 volt electrics.

POWER TRAIN:

Four-speed gearbox, designation M40, with synchromesh on all forward ratios, central floor shift lever. Ratios: first, 3.13; second, 1.99; third, 1.36; top, 1.0; reverse, 3.25; rear axle, 4.1 to 1.

BRAKES AND STEERING:

Four-wheel hydraulic brakes with nine inch drums front and rear (disc brakes in front, 122S, with no booster). 147 in lining area for 121. Worm and roller steering, 16.5 to 1 ratio; turning circle, 31 ft 6 in.

DIMENSIONS:

Wheelbase, 102¼ in; track, 51¾ in, front/rear; length, 175 in; width, 63¾ in; height, 59¼ in; ground clearance, 6¾ in laden; kerb weight, 2560 lb; 34.1 lb/hp; 17.4 lb/in brake area; trunk capacity, 14.9 ft; fuel, 9.7-9 Imp gallons; water, 15 pts; crankcase, 5¾ pts on oil changes.

the steering wheel. It's almost too close since I got winkers once or twice when using the wheel briskly in driving gloves. The headlights dim with a floor switch that could be a little closer with the seat back. Add the cigar lighter and you have all necessary controls.

Amps and oil pressure are left to the all-or-nothing lights, I'm sorry to say, along with high beams and turn indicator warnings which make sense. There are dials for water temperature and fuel, both quite reasonably accurate. In addition to the total mileage counter there is a trip odometer with tenths wheel. The whole works has rheostat-controlled lighting.

To finish with the dash, the top is lightly padded and covered with a non-reflecting leatherette. There is *no* glove box, a major sin of omission in my book, and to compound the problem they threw in a small shelf on the passenger side that will hold gloves in place.

The steering is a little heavy for parking but comes into its own in fast corners where you have real feel with no kick-back worth mentioning and absolute precision in holding the car. The snow treads on the rear made maximum cornering impossible — the tail tended to hop a little — but the Amazon tracks well in cross winds and isn't bothered much by snowy ruts or ice. The floor shift is spring-loaded to the third and top side (cars also come with three-speed boxes, but I wouldn't know why). Once you get used to holding it over for second the box is simplicity to use and the synchro well-nigh unbeatable. Again the 121 with more miles had the more pleasant shift action. That needs breaking in, too.

This brings us to the realm where a Volvo is most at home — in motion. Taking the 75 hp car first, it had obvious torque in all the gears and was very willing to go over its rated 4500 rpm maximum at any time. (Incidentally, there is no handy place to mount the very necessary tachometer except in a nacelle above the dash. The test car had one in

front of the passenger, which was of academic interest for tests, but no use to the driver.)

Second gear proved a very handy item around town, pulling the car from near standstill to well over the legal limit with five aboard. There is a noticeable gap between first and second on race starts, but the rest of the steps are well matched. The 121 motor is reasonably quiet on the road, and the car will cruise comfortably and without fuss at 80 mph all day on relatively little premium grade fuel. It is quiet enough to make a whistle around the front wind wings annoying, by the bye.

The sport engine of 90 ponies is another beast. You need that synchro first gear in town because it just doesn't have the urge at low revs. As a matter of fact I would guess I shifted at least 35 percent more to keep the engine happy. Also there was a definite flat spot at the bottom end, making quick starts much trickier than with the 121. Thirdly, the engine roared and boomed at 80, although I must admit it was a nice husky tone.

Even admitting to a personal prejudice that always veers me to the hotter engine if offered, I would rate the 75 hp Volvo the better all-round buy — except for competition types, it goes without saying. Of course, you have to give up the discs and even with high pedal pressures there is nothing like too much brakes.

A definite factor here would be the optional Laycock overdrive for top gear, optional on the 122S (and sports car) but not on the 75 hp model. This was not fitted to our test car and would likely ease the cruising problem no small bit.

The small things like that itinerant speedometer ribbon and the door lock push buttons that were vaguer than necessary stand out on a Volvo because the rest of the car is so good. It's no surprise when one of these sedans wins a race, even out of its class. They are amazingly quick for 1.8-litres (and obviously happier with the extra 200 cc), robust beyond most modern habit, comfortable, and both fun to drive and solid enough on the road to fool your mother-in-law if she can't see the speedo.   #

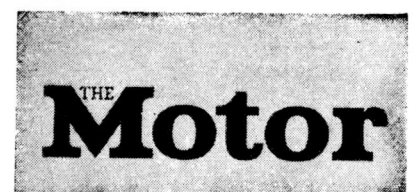

MAKE: Volvo. TYPE: 122

MAKERS: AB Volvo, Gothenburg, Sweden.

CONCESSIONNAIRES: Volvo Concessionaires Ltd., 28 *Albemarle Street, London, W.*1.

THE Motor ROAD TEST • No. 50/62

TEST DATA:

CONDITIONS: *Weather: Cool, misty, no wind. (Temperature 39°-44°F., Barometer 30·2 in. Hg.) Surface: Damp tarmacadam. Fuel: Premium grade pump petrol (98 Octane Rating by Research Method).*

INSTRUMENTS:

Speedometer at 30 m.p.h.	accurate
Speedometer at 60 m.p.h.	1½% slow
Speedometer at 90 m.p.h.	1½% slow
Distance Recorder	2½% fast

WEIGHT:

Kerb weight (unladen, but with oil, coolant and fuel for approximately 50 miles)	21 cwt.
Front/rear distribution of kerb weight	54/46
Weight laden as tested	24¾ cwt.

MAXIMUM SPEEDS:

Mean lap speed around banked circuit	94·8 m.p.h.
Best one-way ¼-mile time equals	96·8 m.p.h.

"Maximile" speed (Timed quarter mile after one mile accelerating from rest).

Mean of opposite runs	91·6 m.p.h.
Best one-way time equals	92·8 m.p.h.

Speed in gears

Max. speed in 3rd gear	80 m.p.h.
Max. speed in 2nd gear	57 m.p.h.
Max. speed in 1st gear	36 m.p.h.

FUEL CONSUMPTION

40 m.p.g.	at constant 30 m.p.h. on level
40 m.p.g.	at constant 40 m.p.h. on level
35½ m.p.g.	at constant 50 m.p.h. on level
32½ m.p.g.	at constant 60 m.p.h. on level
30 m.p g.	at constant 70 m.p.h. on level
25½ m.p.g.	at constant 80 m.p.h. on level
21½ m.p.g.	at constant 90 m.p.h. on level

Overall Fuel Consumption for 1,267 miles, 53·2 gallons, equals 23·8 m.p.g. (11·87 litres/100 km.)

Touring Fuel Consumption (m.p.g. at steady speed midway between 30 m.p.h. and maximum, less 5% allowance for acceleration) 30·2 m.p.g.
Fuel tank capacity (maker's figure) 10 gallons

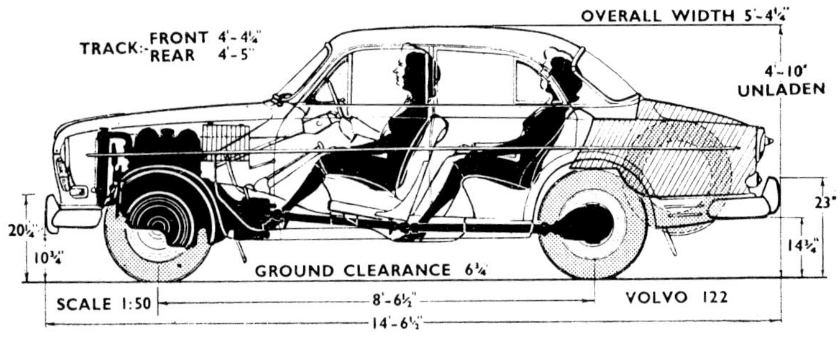
OVERALL WIDTH 5'-4¼"
TRACK:- FRONT 4'-4¼"
REAR 4'-5"
4'-10"
UNLADEN
23"
14¼"
20¼"
10¾"
GROUND CLEARANCE 6¾"
SCALE 1:50
8'-6½"
14'-6½"
VOLVO 122

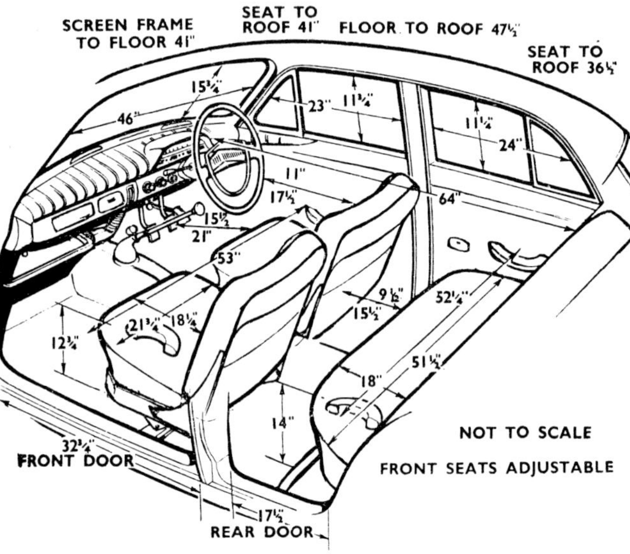
SCREEN FRAME TO FLOOR 41"
SEAT TO ROOF 41"
FLOOR TO ROOF 47½"
SEAT TO ROOF 36½"
15¾"
46"
23"
11¾"
11¼"
24"
11"
17½"
64"
15½"
21"
53"
9½"
52¼"
15½"
21¾"
18¼"
12¾"
51¼"
18"
14"
32¼"
FRONT DOOR
NOT TO SCALE
FRONT SEATS ADJUSTABLE
17½"
REAR DOOR

BRAKES from 30 m.p.h.

1·0 g retardation (equivalent to 30 ft. stopping distance)	with 125 lb. pedal pressure
0·85 g retardation (equivalent to 35 ft. stopping distance)	with 100 lb. pedal pressure
0·66 g retardation (equivalent to 45 ft. stopping distance)	with 75 lb. pedal pressure
0·48 g retardation (equivalent to 63 ft. stopping distance)	with 50 lb. pedal pressure
0·26 g retardation (equivalent to 115 ft. stopping distance)	with 25 lb. pedal pressure

HILL CLIMBING at sustained steady speeds

Max. gradient on top gear	1 in 10·6 (Tapley 210 lb./ton)
Max. gradient on 3rd gear	1 in 7·1 (Tapley 310 lb./ton)
Max. gradient on 2nd gear	1 in 4·7 (Tapley 470 lb./ton)

ACCELERATION TIMES from standstill

0-30 m.p.h.	4·1 sec.
0-40 m.p.h.	6·7 sec.
0-50 m.p.h.	9·8 sec.
0-60 m.p.h.	14·2 sec.
0-70 m.p.h.	19·4 sec.
0-80 m.p.h.	28·7 sec.
0-90 m.p.h.	42·4 sec.
Standing quarter mile	19·8 sec.

ACCELERATION TIMES on upper ratios

	Top gear	3rd gear
10-30 m.p.h.	10·5 sec.	7·0 sec.
20-40 m.p.h.	10·3 sec.	6·4 sec.
30-50 m.p.h.	9·4 sec.	6·5 sec.
40-60 m.p.h.	10 3 sec.	7·1 sec.
50-70 m.p.h.	12·0 sec.	9·4 sec.
60-80 m.p.h.	16·1 sec.	—
70-90 m.p.h.	23·2 sec.	—

STEERING

Turning circle between kerbs:

Left	31¼ ft.
Right	29¾ ft.
Turns of steering wheel from lock to lock	3¼

Specification

Engine

Cylinders	4
Bore	80·14 mm.
Stroke	80 mm.
Cubic capacity	1,780 c.c.
Piston area	31·3 sq. in.
Valves	Overhead (pushrods)
Compression ratio	8·5/1
Carburetters	Twin 1¾ in. S.U.
Fuel pump	AC mechanical
Ignition timing control	Centrifugal and vacuum
Oil filter	Full flow
Maximum power (net)	80 b.h.p.
at	5,000 r.p.m.
Piston speed at maximum b.h.p.	2,620 ft./min.

Transmission

Clutch	Borg and Beck 8½ in. s.d.p.
Top gear (s/m)	4·01
3rd gear (s/m)	5·58
2nd gear (s/m)	8·16
1st gear (s/m)	12·83
Reverse	13·32
Propeller shaft	Hardy Spicer open
Final drive	Hypoid bevel
Top gear m.p.h. at 1,000 r.p.m.	18·0
Top gear m.p.h. at 1,000 ft./min. piston speed	34·3

Chassis

Brakes	Girling, disc front, drum rear
Brake dimensions	10·88 in. discs drums 9 in. dia. × 2 in. wide
Friction areas	94 sq. in. of lining area operating on 345 sq. in. rubbed area of discs and drums

Suspension:

Front	Independent by transverse wishbones coil springs and anti-roll bar
Rear	Coil springs and live rear axle located by radius arms and Panhard rod

Shock absorbers: Front and rear	Telescopic
Steering gear	Gemmer cam and roller
Tyres	5·90-15 tubeless

VOLVO 122

A Swedish saloon that reflects its rugged rally reputation

THE Volvo is not a car to judge on brief acquaintance. Both its remarkable success in international rallies and the enthusiasm which it arouses amongst discriminating owners with a sporting bias reveal that this quite ordinary looking saloon with its conventional mechanical specification has unusual qualities to offer. These are best appreciated after a considerable spell of hard driving; amongst the members of our staff who tried this car, those who drove it farthest liked it best.

In Sweden fashion does not dictate frequent changes in appearance so that the body shell, which has not been altered appreciably for several years, no longer conforms with current European stlye. For this reason the driver's first impressions are not favourable; there is an old-fashioned air about the high waist and the high scuttle and bonnet which cut off the view of the nearside wing. The windscreen pillars are thick and the gear lever is longer and stouter than is now customary. It soon becomes clear, as we shall mention later, that the driving position has many compensating virtues and that there are very few gear changes in existence for which this one would be willingly exchanged.

Four years ago we tested the Volvo 122S, an almost identical car but with an engine capacity of 1,580 c.c. The alternative 1,780 c.c. B18 engine was introduced over a year ago and for a time the more powerful model was called the 122S B18 until this was simplified to just 122; other recent improvements include Girling disc brakes on the front wheels and a larger clutch. Optional extras which were not fitted to our test car include Laycock-de Normanville overdrive (in conjunction with lower final gearing) and a vacuum servo for the brakes; to complete the picture we should mention that there is also the cheaper Volvo 121 with a less powerful single-carburetter version of the B18 engine.

Remarkable Performance

THERE is no obvious reason why an apparently conventional and straightforward four-cylinder 1,780 c.c. engine should propel this car so fast and so flexibly. It has ordinary pushrod-operated valves in a cast iron cylinder head, a moderate compression ratio of 8·5 : 1, which causes a little pinking on mixture-grade fuel but is more than satisfied by Premium, and a crankshaft with five main bearings, a feature which is still unusual but which is rapidly becoming less so; two large S.U. carburetters supply the mixture. The maximum power output is high (80 b.h.p. at 5,000 r.p.m.), but even so the performance

figures proved unexpectedly good for a car which looks large, heavy and not very well streamlined.

Since it returned a mean maximum speed only a fraction short of 95 m.p.h. it may well be better streamlined than it looks. Certainly it is much lighter than it looks, turning the scales at only 21 cwt. unladen, ¼ cwt. less than the smaller P1800 sports coupé which, in spite of a less powerful engine, it can out-accelerate from rest to over 50 m.p.h. In fact, 0-50 m.p.h. takes only 9·8 sec. and the standing quarter-mile 10 sec. longer. It cruises happily and smoothly at speeds in the eighties with a lot of intake roar but hardly any wind noise at all if the windows and quarter lights are closed.

The engine starts easily from cold, sometimes without use of the choke at all, warms up without temperament and soon delivers full power, and will accelerate the car strongly and smoothly from under 10 m.p.h. in top gear. A radiator blind is a standard fitting which can be used to hasten the attainment of running temperature or to raise the thermostat control temperature in very cold weather. At low speeds and part throttle this engine seems to have none of the usual vibration periods and roughness and feels more like a six cylinder; at full throttle noise from the inadequately silenced carburetter intakes makes it more obtrusive and some vibration can be felt at high r.p.m.

In Brief

Price (as tested) £940 plus purchase tax £196 7s. 11d. equals £1,136 7s. 11d.	
Capacity	1,780 c.c.
Unladen kerb weight	21 cwt.
Acceleration:	
20-40 m.p.h. in top gear	10·3 sec.
0-50 m.p.h. through gears	9·8 sec.
Maximum top gear gradient	1 in 10·6
Maximum speed	94·8 m.p.h.
"Maximile" speed	91·6 m.p.h.
Touring fuel consumption	30·2 m.p.g.
Gearing: 18.0 m.p.h. in top gear at 1,000 r.p.m.	

The maximum speeds we recorded in the intermediate gears correspond to about 6,000-6,300 r.p.m. at which point violent valve crash sets a limit but, in fact, it is quite unnecessary to use such high speeds and all our other figures were recorded without exceeding about 5,700 r.p.m.

On the 1 in 3 test hill the Volvo would not restart in either first or reverse. In reverse this was largely due to axle tramp and wheelspin but in first the failure was a marginal one excusable in a gear in which the car will reach 36 m.p.h.

The all-synchromesh gearbox is delightful to use and all the gears, including first, are easy and light to engage and reasonably silent. The long lever, which has a short positive movement, is spring loaded to the right-hand side of the gate and a stranger may tend to go straight from first to top until he learns to apply a light counter pressure; a pleasing feature is the way in which changes across the gate are accomplished in one diagonal sweep instead of the usual three movements at right angles. In general, this transmission flatters the driver and the car seems to progress in a particularly smooth and well-cushioned fashion; if the rather heavy clutch is dropped in sharply at high r.p.m., as in standing start acceleration tests, there is a considerable bang as all the drive flexibility is taken up but low speed starts are perfectly smooth.

Driving Comfort

VOLVO thoroughness is typified by comprehensive provision for front seat adjustment. Fore and aft movement caters for the tall as well as the short and quick-action levers give three settings for the backrest angles. The seats can be raised or lowered bodily by putting the attachment bolts through alternative holes in the mounting brackets and screwed lugs under the front can be moved with a spanner to alter the inclination. With all these combinations, and seats which, if rather firm, are properly shaped for spinal curvature and for support under the knees, only the lazy need be uncomfortable.

For really fast motoring there is not enough squab curvature to give adequate lateral restraint and Volvo owners are obviously expected to specify and to use the optional seat harness for which mountings are built in; with the lap strap and diagonal belt done up tightly all the controls are still within easy reach. There are no dangerous projections to cause serious injuries in an accident, the sun visors are soft and flexible and there is a padded roll along the top edge of the facia which is covered with matt non-reflecting fabric.

An accelerator with a long travel and progressive action makes for easy control in heavy traffic but without the optional servo the foot brake needs above-average pressures. On a good surface the rear wheels lock first at a deceleration of less than 0·9g although quicker stops are still possible at the expense of control. As one would expect with brakes of such generous size there was no trace of fade or loss of power at any time. The handbrake, mounted between the driver's seat and the door, has a guard over the ratchet button so there is no danger of knocking it off when entering or leaving but it can be difficult to release if heavy gloves are worn. Similarly, an armrest on

VOLVO 122

The interior, with its moulded rubber floor covering is essentially practical. Note the guard over the handbrake release button and the broad flat spokes of the steering wheel which should reduce impact loading in an accident.

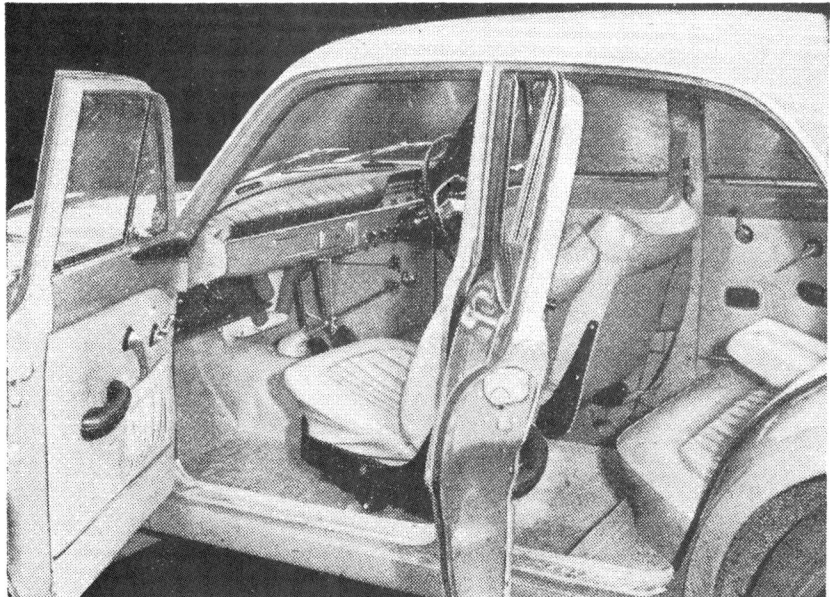

Left: A low floor level allows rear passengers to tuck their feet underneath the front seats and there is a folding central armrest.

Below: A large boot lid balanced by torsion bars gives easy loading, the spare wheel and tools remaining accessible even when a large amount of luggage is carried.

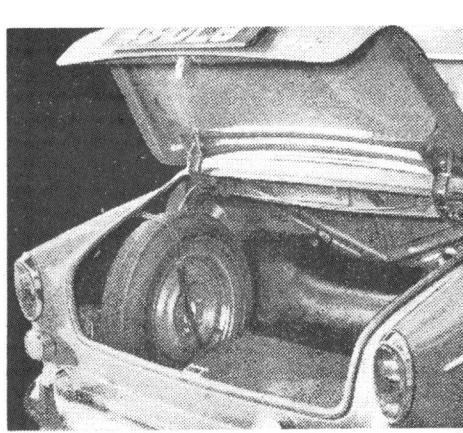

the door makes the lever awkward to reach if the driver is heavily clothed but this should seldom be necessary since the heater is powerful and its output at low speeds can be strongly boosted by a two-stage fan which is very quiet in the "slow" position but distinctly noisy at full flow.

Springing and Steering

COIL springs all round give firm suspension which earns more praise for its behaviour in extreme conditions than it does on ordinary roads. On really bad, unmade, potholed tracks the Volvo behaves better than most cars and seems to suffer very little. On ordinary roads there is quite a lot of vertical bounce and sometimes the car feels a little underdamped; the ride is appreciably better in the front seats than in the rear but both tend to improve with speed. In spite of the liberal use of rubber

Twin S.U. carburetters give the straightforward engine a sporting appearance which is more than confirmed by the performance.

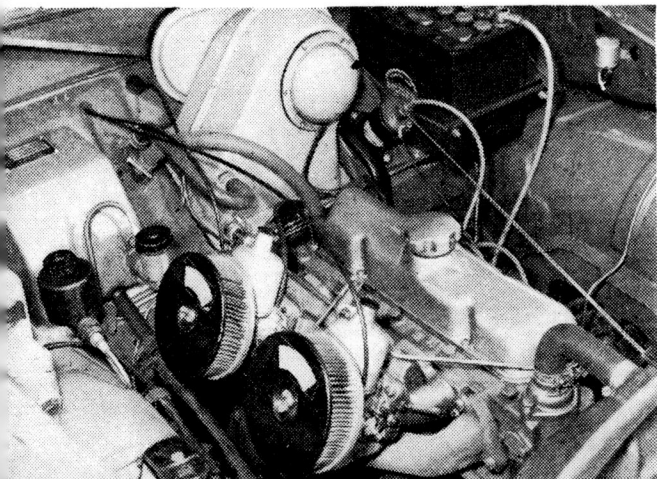

in the locating arms of the rear suspension, quite a lot of road noise is transmitted and, perhaps because of it, there is some rear axle tramp and shake at times and the car may change direction slightly, although this happens more on the straight than on corners.

Cornering roll is very moderate and the cam and roller steering has that smooth frictionless feeling that is characteristic of the best high efficiency mechanisms. With its positive action and freedom from play it feels much higher geared than it is and strong castor return action straightens the steering at just the right rate if the wheel is allowed to play through the fingers after a really sharp turn. It is certainly not heavy in any conditions but it is more freely reversible than is now usual so that side winds and camber changes are felt at the wheel and there is some kick-back on really bad bumps; this close feel of conditions at the front wheels gives rapid awareness of slippery road surfaces. In general the Volvo maintains a slight and consistent understeer in cornering but the car is throttle sensitive and the back can usually be brought round under power. At low speeds it is not difficult to pick up the inside rear wheel and spin it when accelerating out of a side turning.

Strictly speaking the 122 is a four-seater car, but if the central armrest in the back is folded away a fifth person can be tolerated for short journeys; the rear seat is comfortably upholstered and there is adequate leg room for most people unless the front seats are right back. Cut-outs for the rear wheel arches make the rear doors rather narrow at the bottom so that access to the back is less easy than it is to the front. All the doors can be locked from inside by pressing buttons on the window sills—if they are then closed from the outside it is possible to lock oneself out if the keys are left in the ignition.

The electrical system now has 12-volt operation; the headlights appeared to be good but they were adjusted too low for fast night driving to be possible. A headlamp flasher incorporated in the direction indicator is a valuable item of equipment. The moving red strip of the horizontal linear speedometer is quite clear in daylight but very difficult to see at night although the rest of the instrument panel is well illuminated. In general the controls and switches have been arranged with thought and care and we particularly liked the three-position pull switch giving slow and fast wiper speeds and then fast with continuous electrical washing sprays.

At a price of over £1,100 the potential market for the Volvo 122 must necessarily be confined to those who demand more than mere transport from their business or family saloon. Its notable qualities include sensible design and careful workmanship, good handling and pleasing controls, but its outstanding quality is performance; since we took full advantage of this, an overall fuel consumption of 23·8 m.p.g. is very creditable.

Left: The Volvo has particularly solid, well-made and well-finished bodywork although the car is much lighter than it looks. Flexible wheel flaps protect following cars from flying stones on Swedish gravel roads.

Coachwork and Equipment

Starting handle	.. .. No	Sun visors .. .. Two	Interior lights: One above windscreen operated by manual and courtesy switches and parcel shelf/map light.
Battery mounting	Under bonnet on near side	Instruments: Thermometer, speedometer, total and decimal trip mileage recorders, fuel gauge.	
Jack .. .. Screw pillar type			Interior heater .. .. Standard, fresh air heater and demister
Jacking points .. Two each side under body		Warning lights: Generator, main beam, direction indicators, oil pressure.	
Standard tool kit: Wheelnut and plug spanner and handle, adjustable spanner, pliers, screwdriver, Phillips screwdriver.		Locks: with ignition key .. Ignition only	Car radio .. .. Optional extras, Ekco 915 or Motorola transistor
		with other keys .. Doors and boot	Extras available: Webasto sunshine roof, brake servo, auxiliary lamps, reversing lights, wheel trim, wing mirrors, badge bar.
Exterior lights: 2 headlights, 2 sidelamps, 2 stop/tail lamps, 2 number plate lamps.		Glove lockers .. .. .. None	
Number of electrical fuses .. .. .. Four		Map pockets .. .. One in each front door	Upholstery material .. .. Leather cloth
Direction indicators .. Self-cancelling flashers		Parcel shelves .. .. One under facia and one below rear window	Floor covering .. .. Rubber
Windscreen wipers .. Electrical two-speed self parking		Ashtrays .. .. One on facia and two in rear	Exterior colours standardized .. .. .. Five
Windscreen washers .. .. .. Electrical		Cigar lighters .. .. .. One on facia	Alternative body styles .. .. None

Maintenance

Sump .. .. .. 6¼ pints, S.A.E. 20 (summer) S.A.E. 10 (winter)	Ignition timing .. .. .. 4° before t.d.c.	Tappet clearances (warm or cold) inlet ·018 in. exhaust ·018 in.
	Contact breaker gap .. .. ·016 - ·020 in	Front wheel toe-in .. .. .. 0-⅛ in.
Gearbox .. .. .. 1¼ pints, S.A.E. 90	Sparking plug type .. .. Bosch W 175 T	Camber angle .. .. 0-1° positive
Rear axle .. .. .. 2¼ pints, S.A.E. 90	Sparking plug gap .. .. .. ·028 in.	Castor angle .. .. 0-1° positive
Steering gear lubricant .. .. Hypoid gear oil		Steering swivel pin inclination .. .. 8°
Cooling system capacity .. 14 pints (3 drain taps)	Valve timing: inlet opens 32° b.t.d.c. and closes 72° a.b.d.c., exhaust opens 70° b.b.d.c. and closes 34° a.t.d.c.	Tyre pressures .. .. Front 20 lb. rear 23 lb
Chassis lubrication.. .. .. by grease gun every 3,000 miles to 8 points		Brake fluid .. .. Girling
		Battery type and capacity .. 12 V. 60 amp. hour

![Volvo Canadian car parked on gravel]

VOLVO CANADIAN

Prior to its Canadian introduction, CT&T had the pleasure of trying the Volvo Canadian on what was possibly the longest, most extensive road test conducted by a North American publication. For over seven thousand miles; from Toronto to Vancouver, to San Francisco and back to Toronto, the new car was given the opportunity to show that it would live up to its name. What name? First of all, the Volvo name has a reputation for quality and performance. Secondly, calling it the "Canadian" suggests a vehicle that will stand up to the rugged driving conditions of this country. Therefore "Volvo Canadian" was a name that had meaning the moment the car arrived on these shores.

Our road test was completed in less than two weeks, with stops at Vancouver, Seattle and San Francisco. It was done at higher than average speeds, in a trip that few normal (or sane) people would attempt.

The car had only 1300 miles on the odometer when we left Toronto, barely broken-in by Volvo standards. of the seats and interior spacing is equalled by the ride, smooth over broken roads yet firm on the high-

This became more obvious as the miles went by, and the cruising speed went up accordingly. As a result we had covered a thousand miles before bedding down for the night. Our test car was equipped with an optional Reutter-type reclining backrest, a most useful accessory on a trip of this distance, making it possible for the test crew to sleep in the car. Unfortunately, only the passenger's seat had this fitting and considerable twisting of the non-musical kind was necessary before two people could stretch out, so the nights were long and the sleep was short.

In normal use the Volvo seats are superbly comfortable. The front seats are separate, contoured, with three-position adjustable backrests. Horizontal movement will accommodate all types of individuals, as proven by our two-man crew with heights varying from 5 ft. 3 ins. to 6 ft. A person with exceptionally fat thighs might have some difficulty in fitting under the large steering wheel, which just clears the legs and rubs against them when braking. The rear seat is contoured at each end, with arm rests on both sides, recessed to provide more hip room, and a center folding armrest.

VOLVO CANADIAN

All passengers have ample leg and elbow room although the addition of a fifth person would naturally cramp the rear seat on a long trip. But let's be honest, no car can comfortably accommodate three adults in the back seat on a lengthy journey. Volvo is obviously aware of this and rather than make cramped space for six, has concentrated on maximum comfort for four.

Everywhere we went; gas stations, restaurants, truck stops, there were flattering remarks about the Volvo Canadian. We were finally moved to try and analyze the esthetics of the car because, while we liked it too, the unanimous agreement among the onlookers was surprising. Styling, of course, is a matter of taste but we finally concluded that Volvo's success was due to a combination of simplicity, a longish hood, large wheels and a rakish sweep from front to rear that states its solidarity with performance.

Driving the Volvo Canadian is pure pleasure. Anyone who enjoys driving in the grand sporting manner will have a ball at the wheel of this machine. Straight-line performance is equal to most 1.6 litre sports cars in stop-light acceleration or high speed passing. Fortunately Volvo's engineers have blessed the Canadian with disc brakes in the front and large drums in the rear, a combination guaranteed to haul it down like a huge iron hand. Driving hour after hour at high average speeds through the mountains failed to fade the brakes.

The Volvo's handling ability got the same crucial tests in the mountain areas. The CT&T test crew drove with considerable brio, diving deep into turns and drifting around the bends. In the faster corners the Volvo Canadian handles as though it were sired by a Grand Prix racing car. It is understeered slightly, but not once in seven thousand miles of rapid cornering did the rear end break loose. When a curve was taken too quickly the Canadian slipped sideways on all four wheels, requiring only adequate pavement (and perhaps a small prayer) to regain a forward grip. Slower corners were less its forte owing to the long wheelbase of the Canadian and the softness of the front suspension.

Perhaps the most startling facet of its handling

was discovered in the rain. Not one mile per hour of average speed was lost as the tires stuck in the wet like a ball of glue. The car received all-weather testing, too, going through rain, snow, hail, dust and thunderstorms. Travelling at 90 miles per hour in torrential rain produced a small leak which, oddly enough, failed to reappear in subsequent storms. The strong Swedish steel withstood collisions with at least five large birds that made losing flights from one side of the highway to the other.

A first acquaintance with the Volvo Canadian produces some trepidation at the size of the gear lever, which extends directly out of the gearbox at an angle to a point about two feet from the floor. But a few miles of driving makes it obvious that the gears can't argue when you change them with such a long and powerful lever. There is absolutely no mushiness nor doubt about which gear is being selected. The operation is stiff, though it loosens up after a few thousand miles, and this brings us to an interesting point about the Volvo Canadian. From a driving standpoint, it can be considered a "man's car". Women will enjoy riding in it, but some may find the stiff gearshift and somewhat heavy slow-speed steering a bit of a chore. A remote shift and smaller steering wheel could eliminate any feminine complaints.

The steering is quick, doesn't transmit bumps and at high speeds becomes lighter. Gear ratios are absolutely ideal. All gears are synchronized and speeds of 25 in first, 50 in second and 75 in third are attainable. In fourth gear the speedometer will indicate 100. On the other hand, the new B-18, five-bearing crank engine is very tractable, combining ample torque with its 90 horsepower. The Canadian can be driven at 30 miles per hour in any of the top three gears without strain.

Driving a new car over such a long distance can be expected to produce some faults, and a few repairs were anticipated. Nevertheless, we were pleased that only three loose bolts marred an otherwise perfect run. None were of a nature to prevent the car from continuing, and all three loosened-up before we reached Vancouver. The final four thousand miles were troublefree.

The average speed from San Francisco to Toronto, including fuel and meal stops, was 50.6 m.p.h. Our average for the entire road test, excluding stops, was 67 m.p.h. During this time the engine never missed a beat, delivering 26 miles per gallon. Cost for the 7,040 miles was $108.37 plus $1.06 for two quarts of oil.

Greatest problem facing the test crew was to find a point of criticism in the car. In desperation we finally settled for some negative comments on the dashboard design. Although the interior has a quality finish the dash detracts from the overall effect. Chief annoyance is the lack of a locking glove compartment. In its place is a large, collapsible shelf, a useful carry-all that should be retained. But the radio is placed where the glove compartment would normally be located, and the dials of the radio are not only hard to reach but obviate the padded dash by threatening to damage a passenger who flies forward in an accident. Neither did we like the horizontal strip speed-

ometer and would have much preferred a round dial.

Assets that must be mentioned are the large trunk, silent idling and generally silent riding with only a trace of wind noise at high speed; the heating system which will stand up to the toughest tests of any Canadian province. (Winnipeg will love it) the wide doors and easy engine accessibility.

We feel that the Volvo Canadian is the only reasonably-priced car to have successfully combined the best of the sports car and the family sedan. Its name has been admirably chosen, for it would be difficult to build a car better suited for Canadian conditions. As a closing thought, we'd like to see a special version available with the 100 h.p. P-1800 engine, adjustable shocks, dash similar to the coupe with tachometer included, and reclining backrests. It would be well worth an extra $200.00. But don't hold your breath waiting. Volvo have come up with a real winner and they'll be too busy keeping up with the demand.

DATA & SPECIFICATIONS

Engine:	4 cylinder in-line, water cooled
Bore:	3.313 in.
Stroke:	3.15 in.
Capacity:	1.78 litres
Compression ratio:	8.5:1
Max. torque:	105 lb/ft
Max. output:	90 hp (SAE) at 5000 rpm
Brakes:	Hydraulic discs in front, lo 7/8 in. x 1/2 in. Hydraulic V-type drums at rear
Transmission:	4 speeds forward, all synchro Ratios: 1st, 3.13:1; 2nd, 1.99:1; 3rd, 1.36:1; 4th, 1.1. Reverse, 3.25:1
Front Suspension:	Independent with coil springs and control arms carried in rubber bushings and ball joints Stabilizer.
Rear Suspension	Fixed longitudinal support arms, torque rods and track bar. Coil springs.
Steering:	Cam and roller type, 3¼ turns lock to lock
Wheelbase:	102½ in.
Overall length:	175 in.
Overall width:	63¾ in.
Overall height:	59¼ in.
Fuel tank capacity:	10 imperial gals.

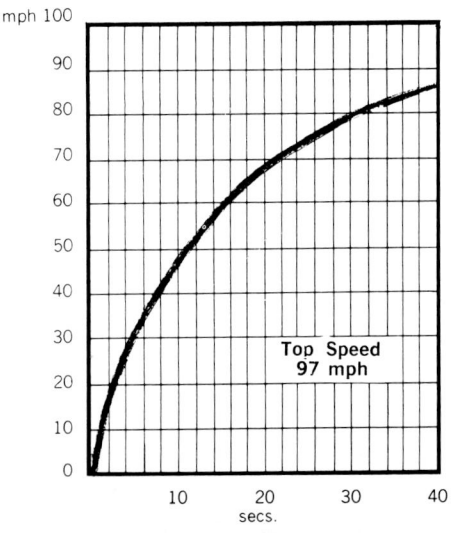

Top Speed 97 mph

to corroding or loose terminals; lights stay brighter longer, and the starter operates at peak for longer periods.

My test car had a bench seat in front—just wide enough for three, but not as form-fitting as the previous separate seats. However, these can still be specified—at no extra cost.

It also had a steering-column gearshift; again, the once-standard floor shift can be specified without raising the ante. I wished they had specified it for the test car.

Mind you, there's nothing seriously wrong with the column shift. It's fairly quick in action—but the spring that locates the lever in the plane of third and top is altogether too lusty. Changing across the box from third to second needs a hefty bit of wristwork—and you don't always locate the lever exactly for the final movement into second. Also, your knuckles rap the padded cowl round the instruments as you go into third, because the lever moves so close to the dash.

Little B18 flashes front and back are the only external clues to the bigger stable under the bonnet; otherwise, this new Volvo is unchanged from the original 122S model that came to Australia about 18 months ago.

Remarkable Performer

Painless performance is this car's forte.

Genuine top speed was 97.2 m.p.h.; 0-50 m.p.h. could be wrapped up in 10.1 seconds; cruise at any speed up to 90.

But my big surprise was fuel consumption. The factory claimed 23 m.p.g. for the original 122S; I got 23.1 on test. Here comes the B18, with bigger engine and more power —and I get 26.8 m.p.g.

Possibly the original test car — first Volvo to come out here—wasn't in perfect tune. But that wouldn't explain away the factory claim.

Another remarkable property of the engine is its flexibility. It will accelerate slowly but decorously from very low speed in top gear. Only a subdued ping or so will show its distaste.

Give it revs and it turns into a cannonball. Equally suited to demons or dowagers.

No change to the all-coil-springs

Bryan Hanrahan's test of new Volvo B18D, followed by trip report from David McKay. Both praised car lavishly, so we called this story . . .

POWER WITH GLORY

I WOULDN'T have been surprised to find that the paper-element air-cleaners on this new B18 Volvo had disappeared down the throats of its two big, fat SU carburettors after the first couple of performance runs.

When you get the revs just right, this otherwise amiable Swedish motorcar snorts like a racehorse — and takes off like one.

You see, engine size has been upped from just under 1.5 litres in the previous 122S engine to nearly 1.8 in the 122S B18—to give it its full, if not fulsome, title.

The four-cylinder, oversquare, o.h.v. unit now puts out 90 b.h.p. at 5000 r.p.m. on a compression of 8.5 to 1—15 more horses than before.

It is actually a detuned version of the 100 b.h.p. engine used in the Volvo P1800 sports coupe.

The other major mechanical change is disc brakes on the front wheels. I could never fault the old drum brakes, but the discs are even smoother and more powerful. Pedal pressure is still low, too.

Also useful is the swap to 12-volt ignition (instead of 6-volt). Higher voltage means greater electrical pressure and therefore less susceptible

NEW 1780c.c. engine gets its extra 200c.c. from enlarged bore (84.14mm. instead of 79.37). Compression is up to 8.5:1 (from 8.2) and twin SU carbies help boost output to 90 b.h.p. (SAE) at 5000 revs. Only external clues are B18 flashes on boot (top photo) and grille.

suspension is listed. Nevertheless, I am sure the car has had either spring rates or shocker reaction pressures made firmer.

At high speed on any sort of bitumen the ride was superb; on rough stuff it was positively uncomfortable. The back axle had a tendency to hop and power was lost through wheelspin.

More serious—a bump on a corner taken hard tended to throw the car off line. Altogether, riding and road-holding were not as impeccable as before.

The Volvo engine doesn't fuss and is not noisy at high revs—not even at the listed maximum of 5000. Top gear is always quiet, and there isn't much gearbox noise in the three indirect ratios, all beautifully synchronised.

So there's the power—now where's the glory?

Superb Finish, Equipment

Every square inch of paint, every fitting, every stitch in the upholstery is immaculate.

Standard equipment includes screen-washers, two-speed wipers, a built-in volcano for a heater-and-demister, boosted by a two-speed fan, a radiator blind, courtesy light working off all four doors, speedo with both trip and total mileage recorders, fuel gauge and water-temperature gauge, heavy safety padding on dash,

MAIN SPECIFICATIONS

ENGINE: 4-cylinder, o.h.v.; bore 84.14mm., stroke 80mm., capacity 1780c.c.; compression ratio 8.5:1; maximum b.h.p. 90 (S.A.E.) at 5000 r.p.m.; maximum torque 105lb./ft. at 3500 r.p.m.; twin SU carburettors, mechanical fuel pump; 12v. ignition.

TRANSMISSION: Single dry-plate clutch; 4-speed fully synchromeshed gearbox; ratios, 1st, 3.13; 2nd, 1.99; 3rd, 1.36; top, 1 to 1; reverse 3.25:1; hypoid-bevel final drive, 4.1:1 ratio.

SUSPENSION: Front independent, by coil springs, wishbones and stabiliser bar; solid axle at rear, carried on two support arms, with coil springs, torque rods and track bar; telescopic hydraulic shock-absorbers all round.

STEERING: Cam-and-roller; 3½ turns lock-to-lock, 31ft. 6in. turning circle.

BRAKES: Hydraulic, discs at front, drums at rear; mechanical handbrake.

WHEELS: Pressed-steel discs, with 5.90 by 15in. tyres.

CONSTRUCTION: Unitary.

DIMENSIONS: Wheelbase 8ft. 6½in.; track (front and rear) 4ft. 3½in.; length 14ft. 7½in., width 5ft. 3½in., height 4ft. 11¼in.; ground clearance 8in.

KERB WEIGHT: 23½cwt.

FUEL TANK: 10 gallons.

PERFORMANCE ON TEST

CONDITIONS: Fine, cool; no wind; dry bitumen; two occupants, premium fuel.

BEST SPEED: 98.3 m.p.h.

FLYING quarter-mile: 97.2 m.p.h.

STANDING quarter-mile: 20.1s.

MAXIMUM in indirect gears: 1st, 33 m.p.h.; 2nd, 54; 3rd, 76.

ACCELERATION from rest through gears: 0-30, 3.8s.; 0-40, 7.1s.; 0-50, 10.1s.; 0-60, 15.9s.; 0-70, 22.3s.; 0-80, 29.9s.; 0-90, 39.1s.

ACCELERATION in top (with third in brackets): 20-40, 5.1s. (8.3); 30-50, 11.4s. (7.5); 40-60, 11.8s. (7.9); 50-70, 12.6s. (9.1); 60-80, 15.2s.; 70-90, 21.3s.

BRAKING: 30ft. 2in. to stop from 30 m.p.h. in neutral.

FUEL CONSUMPTION: 26.8 m.p.g. overall for 220-mile test.

SPEEDO: Accurate throughout range.

PRICE: £1675 including tax

cigarette lighter, and variable-intensity instrument lighting.

If you wanted to enter an Alpine trial or drive across the Nullarbor, there's not one extra piece of equipment you'd need to get for a Volvo.

Steering is quick and light, clutch pedal travel short. Vision is good, except for rather thick windscreen pillars. Boot is a good size, with the spare upright at one side.

Upholstery is thick, rich-looking plastic. The rubber floor mats are a bit of a let-down—carpets would have been more in keeping with the rest of the car.

Engine is accessible, and straightforward in design. Crankshaft has five bearings instead of the usual three, to take care of any little larks you may want to indulge in.

Oil filtration is full-flow. The new-type paper-element air-cleaners call for replacement every 6000 miles.

Personally, I'd prefer the old twin seats and floor gearchange — but, as I said before, these can still be had as options, at no increase in price.

Anyhow, who am I to complain? Most people like bench seats and column levers, so everything they've done to the B18 is an improvement—except for the suspension settings.

At £1675 tax-paid, the Volvo 122S B18 is not exactly cheap—but anyone who wants a real motor-car won't mind parting with the money, once he's had a chance to sample this Swedish beauty.

VOLVO TEST

Continued from page 43

in. of friction surface. Operation is solidly positive. It didn't slip once under full throttle changes.

A divided propeller shaft takes the drive from the gearbox to a hypoid diff, keeping the floor transmission tunnel very low. And there was no sign of slack through the four universals. Most divided set-ups get loose and clunky as the drive is taken up, quite early in life.

Steering wheel has a medium ratio of 3½ turns lock-to-lock. Turning circle is an outstanding 32ft.

Suspension, Handling

Up front, suspension is independent by coil springs and double wishbones with a stabiliser bar; at the back, the solid axle is again carried on coil springs, with two control arms, two torque rods and a track bar. Telescopic shock-absorbers are fitted all round.

The front end is not hard, but the stabiliser ties the wheels together in perfect harmony.

Likewise, the back end has flexibility. The torque rods get the power through, while the track rod steadies the axle laterally.

You can let in the clutch solid at 3500 r.p.m. Very little traction is lost in wheelspin on smooth, hard surfaces; in loose rough stuff performance is magnificent.

The suspension allows a certain amount of body roll on corners — but a controlled amount, which leaves all wheels in firm contact.

There can be few better light saloon cars to corner. Slight understeer is constant right up to the point of a lazy sort of back-end breakaway. This gives the wheel a tautness in the hands which allows spot-on placing on line.

The excellent Swedish tyres fitted to the test car helped a lot. The casings were flexible but not baggy. They seemed to scrub into the road surface.

Although the handling is best sports-car standard, the ride is even and smooth on all normal going.

The brakes are self-centring, with two leading shoes at the front. Lining area is 165 sq. in. for the one-ton (kerb weight) car.

Pedal action is peculiar. Stroking the pedal to wash off speed is just a waste of time; it calls for a firm (but not heavy) pressure, which sets off one of the smoothest and quickest means of deceleration I've tried.

You have no chance of fading these brakes, or getting them flustered so they pull off line.

The Volvo, by the way, was not specially tuned for this test. Regent Motors' idea was to let out a car that would perform much as it will in the hands of a normal owner.

In competition tune, I'm told the car will do a genuine 95 m.p.h.

Finish is as good as you'll see on any production car. The unit-construction body is nothing if not solid. There wasn't so much as a mouse-rustle, rattle or squeak.

You come inevitably to the conclusion that this 122S has everything in the way of performance, handling, safety, comfort and precision engineering that either enthusiast or ordinary driver can wish for. I only wish I had £1700.

If they play their cards right, the importers (Regent Motors in Victoria, Antill Ranger in N.S.W.) should be kept pretty busy.

One more point: Volvo buyers in Sweden pay an insurance premium included in the purchase price. For five years this protects them against mechanical failure — and if at any time during the five years they write off the car, they get another one for free.

You are not likely to get this in Australia. Self-respecting insurance agents here would double their ulcer rates overnight. But Lloyds of London would certainly quote for it. Surely it would be worthwhile finding out if the quote would be impossibly high?

Firm's Background

Since the Volvo is a complete newcomer to this country, you will probably want to learn a little about its parentage.

The Volvo group of companies is the biggest engineering business in Sweden (Volvo is Latin for "I roll" —our word "revolve" is derived from this).

The original car company was formed in 1920, but its first two cars — a four-cylinder, 28 b.h.p. tourer and saloon — didn't appear until 1927 and 1928 respectively. Most of the components were produced by other firms and only assembled by Volvo.

In 1936 the first modern-style car —the PV36 Carioca—went into production. Less than 600 were made before war threatened and Volvo turned to military truck production.

As the firm expanded, it took over former associates — bearing, aircraft and heavy general engineering concerns.

Postwar full-scale car production resumed in 1947 with the PV444, a four-cylinder, 1½-litre four-seater saloon designed in 1944. By 1957 over 150,000 had been made, more than half of them being exported.

Last year the four-cylinder, 1.58-litre 121 and 122S saloons came off the line. All the postwar cars have won high reputations in motor sport, particularly in rallies.

Newest Volvo is the P1800 sports coupe; it is described elsewhere in this issue by our London correspondent, Doug Armstrong.

The P1800 is expected to be available in Australia by the end of this year as a companion model to the 122S. There are no plans to sell the two-door 121 model in Australia. ●

know your

VOLVO 122S

A fast saloon with fine handling and excellent finish

THE Volvo 122S makes no claims to being a sports saloon, but that is the description that leaps to mind when you have driven it even for only a short distance. Not only is it a lively performer, but it has been designed with driver comfort and efficiency very much in mind.

This has not been achieved at the cost of the passengers. As a 4/5-seater the standard of comfort is extremely high. The car, in fact, makes the best of both worlds.

As with many Continental cars in the medium price range, all the normal extras are fitted as standard, and there are many refinements which could well be adopted by some of the car manufacturers in this country. An instance of this is the front seats which give a firmly supported but relaxed ride. Both the backrest and the seat cushion are slightly concave, so holding the body in position even during fast cornering. The angles of both the backrest and the seat are adjustable, also the height of the seat above the floor. The rear seat is curved at both ends too in order to give lateral support and also to clear the rear wheel arches. It has a central armrest, and there are armrests on all the doors.

Well-placed controls

Anchorages for safety belts are provided all round, and the lap and diagonal belts fitted as standard proved extremely comfortable to wear with a simple fixing on the transmission tunnel. They are easy to get into, and after using them a couple of times they could be hitched up without looking. There is a "parking" stud on the door pillar

so they can be kept well out of the way when not being worn.

As might be expected with a firm that has been fitting seat belts for a long time, all the controls are within comfortable reach of the driver. The instrument cluster is under a cowl immediately behind the steering wheel. The speedometer is of the strip type, and under it are the temperature and fuel gauges and the trip and total mileage counters. These four dials are interspaced by the ignition, main beam indicator and oil pressure

warning lights. All are well placed.

The combined two-speed windscreen wiper and washer switch and the light switch are on the right of the steering wheel, with the choke (T-handled for easy identification in the dark), cigar lighter, two-speed heater booster and heater controls on the left. The radiator blind control and the switch for the glove shelf light which can also be used for map reading are on the under edge of the facia. The heater, designed for operation in cold climates, is very effec-

tive and is thermostatically controlled. The heater incorporates a fresh air ventilation system. The direction indicator switch under the steering wheel is also the headlight flasher.

Of the ancillary equipment only the window winders came in for criticism—they wind the wrong way for most people. Parcel space is adequate. There is a small glove tray on the nearside at the front, and a wide shelf under the rear window. This has been set low so that parcels put on it do not intrude into the window space and impair rearward vision. There are map pockets on both front doors. The floor covering is rubber.

Because of the adjustment on the front seats every one of the test team found the most ideal position for his size and driving style. All liked the handbrake with its round thumb hole, but most found the gate spring protecting reverse so strong that on occasion top was selected instead of second when changing up from first.

Mechanical noise is extremely low, and it

is a pity that this virtue is completely spoiled by the excessive road and wind noise at not too great speeds. On long fast journeys this proved to be disconcerting. Normal conversation is not possible.

The 1,780 c.c. four-cylinder engine, which has a five-bearing crankshaft and two horizontal S.U. carburetters, develops 90 b.h.p. gross at 5,000 r.p.m. It is a very accessible unit with plenty of working space under the bonnet, which is a boon to motorists who like to carry out their own maintenance.

A nice refinement which illustrates the thought that has gone into this car is the design of the coil which prevents a would-be thief shorting out the starter switch. The lead from the switch to the coil is in an armoured cable, and the switch terminal on the coil is concealed.

Lively performance

The Volvo is a vehicle in which the driver immediately feels he is at home. The controls fall naturally to hand, and not one of the test team could suggest an improvement in the siting of any of the controls.

Acceleration is brisk 0-30 m.p.h. in 4.4 seconds, with a mean maximum speed of 95.2 m.p.h., and a top one-way speed over the flying quarter-mile of 97.7 m.p.h. It is this lively performance which gives the Volvo its definitely sporting characteristics.

It does, however, lack the harder ride of a normal sports saloon, and on bad road surfaces the independent coil spring suspension and telescopic dampers do their job effectively. On corners taken healthily there is some roll, but the car holds its line well. There is a slight trace of understeer.

Steering is light and precise, and even in parking requires very little effort. The turning circle is remarkably good.

The brakes well match the performance, and pull the car up in a dead straight line. The pedal pressure, however, is fairly high.

Apart from the too strong spring on the gear lever, the fully synchronised four-speed box has a finger tip change which can be extremely quick. The synchromesh is unbeatable.

Clutch pedal pressure is light, and with a servo on the brakes the Volvo would be a car that could be driven by women without the slightest fatigue.

Safety angles

The combination of performance and comfort has been achieved happily, but the performance has to be paid for in petrol consumption. At an average of 60 m.p.h. on a wide but winding road it sucked in petrol at the rate of 23.5 miles a gallon, but when the average dropped to 40 m.p.h. the petrol consumption also fell to a commendable 34.6 miles a gallon. It should be pointed out that the excellent roadholding made this second test almost a constant speed consumption assessment.

Boot space is good. The spare wheel is upright on the nearside, and there is space behind it to accommodate the running tools, but there is a fairly high threshold which makes loading a little difficult.

The finish of the car is excellent, no doubt because the vehicle was designed to stand out of doors in a climate more severe than ours. But probably its most endearing points are its comfort and performance and the obvious meticulous attention to detail which has made it a true driver's car.

In all this the safety angle has not been overlooked. The top scuttle is heavily padded, sun visors are flexible, and even the parcel shelf is pliable. And, of course, there are the safety belts.

The short verdict is a delightful and most roadworthy motor-car.

Picture Points

the VOLVO 122S at a glance

USABLE GEAR SPEEDS
1st 25 m.p.h. (max. 29 m.p.h.)
2nd 43 m.p.h. (max. 45 m.p.h.)
3rd 65 m.p.h. (max. 70 m.p.h.)
Top 95 m.p.h.

ACCELERATION FROM REST
0-30 m.p.h.	4.4 sec.
0-40 m.p.h.	6.8 sec.
0-50 m.p.h.	10.0 sec.
0-60 m.p.h.	15.4 sec.
0-70 m.p.h.	22.8 sec.

ACCELERATION ON THE MOVE
	Through Gears	Top
20-40 m.p.h.	6.6 sec.	9.8 sec.
30-50 m.p.h.	7.0 sec.	10.0 sec.
40-60 m.p.h.	7.8 sec.	10.0 sec.
50-70 m.p.h.	10.6 sec.	11.0 sec.

FLYING ¼-MILE
Mean, 95.2 m.p.h.
Best, 97.79 m.p.h.

SPEEDOMETER
At 30 m.p.h., accurate.
At 60 m.p.h., accurate.

FUEL CONSUMPTION
23.48 m.p.g. at 60 m.p.h. road average.
34.59 m.p.g. at 40 m.p.h. road average.

ENGINE. Four-cylinder, 1,780 c.c., o.h.v. Compression ratio 8.5 : 1. Power output 90 b.h.p. gross at 5,000 r.p.m. Full-flow oil filter. Five-bearing crankshaft.

COOLING. Thermostat-controlled circulation with by-pass valve.

CARBURATION. Twin horizontal HS6 S.U. carburetters. Mechanical fuel pump. Tank capacity 10 gallons.

GEARBOX. Four-speed, fully-synchronised. 1st speed ratio 12.8 : 1, 2nd 8.2 : 1, 3rd 5.6 : 1, top 4.1 : 1. Floor change.

STEERING. Cam and roller. Turning circle 32 ft. 3¼ turns of steering wheel from lock to lock.

SUSPENSION. Front: independent with coil springs and wishbones. Telescopic dampers. Roll bar. Rear: coil springs, radius arms and Panhard rod.

BRAKES. Front: Girling discs 10⅞ in. × ½ in. Rear: Drums of 9-in. diameter, 2-in. wide shoes.

TYRES. Tubeless 5.90 × 15 in.

DIMENSIONS. Wheelbase, 8 ft. 6½ in.; overall length, 14 ft. 7 in.; overall width, 5 ft. 3¾ in.; overall height, 4 ft. 11¼ in.; kerb weight, 2,405 lb.

PRICE. Total £1,293 10s. 3d. (including £353 10s. 3d. purchase tax).

CONCESSIONAIRES. Volvo Concessionaires Ltd., 28 Albermarle Street, London, W.1.

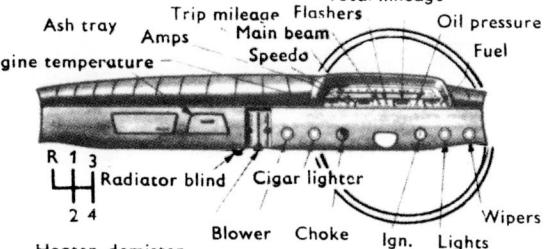

Manufacturer : AB Volvo, Göteborg, Sweden
Concessionaires : Volvo Concessionaires Ltd., 28 Albemarle Street, London, W.1.

Test Conditions

Weather .... Dry, light wind, 10-15 m.p.h.
Temperature...................65 deg. F. (18 deg. C.)
Barometer...29·2in Hg.
Dry tarmac and concrete surfaces.

Weight

Kerb weight (with oil, water and half-full fuel tank)
    23·1cwt (2,586lb-1,173kg)
Front-rear distribution, per cent F, 53·5; R, 46·5
Laden as tested .... 26·1cwt (2,922lb-1,325kg)

Turning Circles

Between kerbs .... L, 32ft 4in; R, 32ft 7in
Between walls.............. L, 34ft 3in; R, 34ft 6in
Turns of steering wheel lock to lock 3·5

Performance Data

Top gear m.p.h. per 1,000 r.p.m. .... 16·2
Mean piston speed at max. power ... 2,480ft/min
Engin revs. at mean max. speed ...... 5,430 r.p.m.
B.h.p. per ton laden .... 52·5 (net)

FUEL AND OIL CONSUMPTION

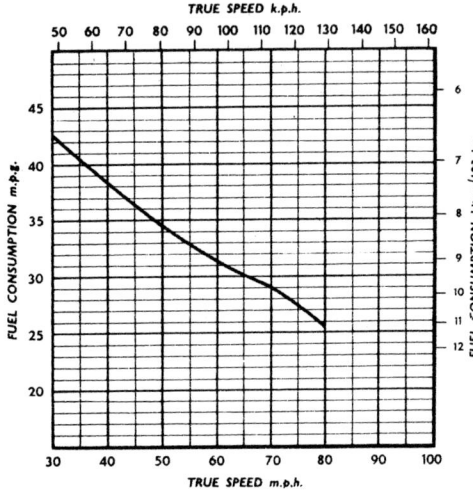

FUEL ....Premium Grade
                                (97 octane RM)
Test Distance ....1,309 miles
Overall Consumption ....29·5 m.p.g.
                                (9·57 litres/100 km.)
Normal Range ....28-36 m.p.g.
                                (10·08-7·85 litres/100 km.)
OIL: SAE 30 ......Consumption 10,000 m.p.g.

HILL CLIMBING AT STEADY SPEEDS

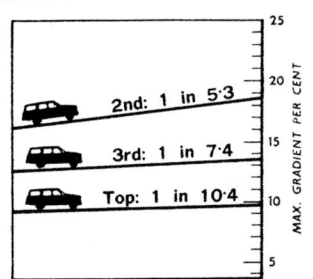

2nd: 1 in 5·3
3rd: 1 in 7·4
Top: 1 in 10·4

GEAR	Top	3rd	2nd
TAPLEY READING (lb per ton)	215	300	410
Speed Range (m.p.h.)	36-41	27-31	18-22

MAXIMUM SPEEDS AND ACCELERATION TIMES (mean)

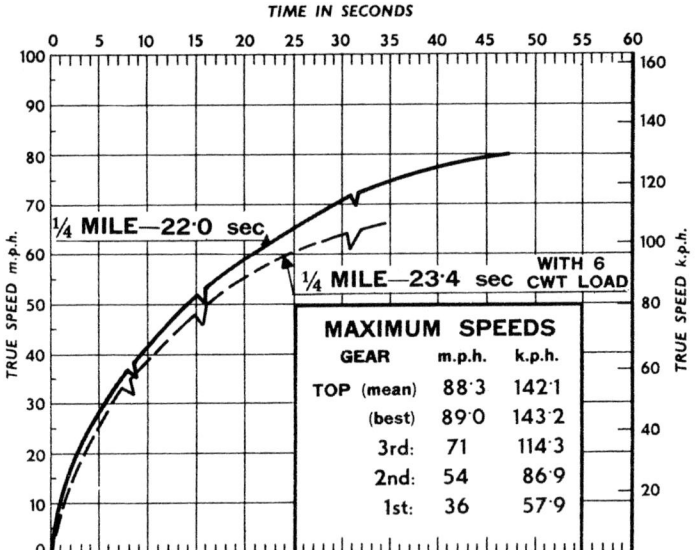

¼ MILE—22·0 sec
¼ MILE—23·4 sec WITH 6 CWT LOAD

MAXIMUM SPEEDS		
GEAR	**m.p.h.**	**k.p.h.**
TOP (mean)	88·3	142·1
(best)	89·0	143·2
3rd:	71	114·3
2nd:	54	86·9
1st:	36	57·9

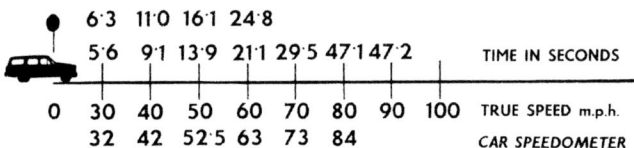

	6·3	11·0	16·1	24·8				
	5·6	9·1	13·9	21·1	29·5	47·1 47·2		TIME IN SECONDS
0	30	40	50	60	70	80	90 100	TRUE SPEED m.p.h.
	32	42	52·5	63	73	84		CAR SPEEDOMETER

Speed range, gear ratios and time in seconds

m.p.h.	Top 4·56 to 1	3rd 6·2 to 1	2nd 9·1 to 1	1st 14·3 to 1
10—30	—	8·4	5·2	4·3 (6·1)
20—40	11·3 (13·5)	8·0 (13·5)	5·6 (6·7)	—
30—50	10·9 (13·4)	8·2 (9·6)	7·3 (8·7)	—
40—60	11·9 (16·2)	9·9 (12·6)	—	—
50—70	15·3	15·5	—	—
60—80	22·5	—	—	—

(Figures in brackets with 6cwt load)

BRAKES (from 30 m.p.h) in neutral)	Pedal load	Retardation	Equiv. distance
	25lb	0·15g	203ft
	50lb	0·45g	67·5ft
	70lb	0·90g	37·3ft
	Handbrake	0·32g	94ft

CLUTCH   Pedal load and travel—48lb and 4·25in.

Volvo 121 Estate Car 1,780 c.c.

SWEDEN is a land of high engineering standards, as well as being a member of the European Free Trade Association. This means that its vehicles can be imported by Britain at just over half the rate of duty levied on the products of members of the Common Market. It is not surprising, therefore, that the rugged, well-built Volvo family of cars, offering unusually high performance from prosaic looking four-cylinder pushrod engines, has gained for itself an enviable reputation in a short time.

The Volvo 121 Estate Car is the latest model of the range to be imported into the United Kingdom, although it has been in production for some time in its native country, where the agricultural community ensures a ready market for this type of vehicle. It is a chunky, practical car, well finished in plain fashion, with a 6ft long loading space when the seats are folded down. It comes completely equipped with heater, seat belts, electric screen-washer, radiator blind and a host of minor refinements usually regarded as extras, and belies its appearance by offering 88 m.p.h. performance coupled with safe handling. These go a long way towards justifying its relatively high price of £1,227 in the U.K. market.

This is the first Volvo tested by *Autocar* to be fitted with the single-carburettor 1,780 c.c. engine introduced in mid-1961. In this form it gives 68 b.h.p. net at 4,500 r.p.m. compared with the 80 b.h.p. (net) at 5,000 r.p.m. of the R18D unit installed in the 122 saloon, the subject of our last Volvo road test (*Autocar*, 4 May 1962). However, the less powerful engine, with its 101 lb. ft. of torque at 2,800 r.p.m.

(very little less than the 105 lb. ft. at 3,500 r.p.m. of the B18D unit), is more suitable than the higher performance unit for a vehicle intended for hard slogging work in the hands of drivers who are not normally addicted to gear-changing. Pulling this 24-cwt car it averaged 28 m.p.g. for the whole of the 1,300-mile test, and consumed 1·25 pints of oil.

There is occasional need to use the choke for starting after a night in the open, but the control can be pushed in immediately the engine fires. A couple of miles are needed to bring the engine up to running temperature, although this period can be shortened by closing the radiator blind fully. With the blind in the half raised position, the engine fan caused the blind to "buzz" against the gills of the radiator core, the resulting noise penetrating into the car at speeds above 50 m.p.h.

The Volvo needs a mile or so to build up to top speed from 80 m.p.h. and this figure is best regarded as the usable maximum cruising gait; 70 m.p.h. is reached quickly and maintained easily, but because of the low overall gearing

PRICES		£	s	d
Estate Car		1,015	0	0
Purchase tax		212	0	5
	Total (in G.B.)	1,227	0	5

Left: Both halves of the rear door have sturdy locks, but are heavy in use; the lower half can be closed and the upper left fully open, or ajar, held by the small stay on the left-hand pillar. Right: With the rear seats folded, the base forms a buffer against heavy articles sliding forward under braking. The ring handle on the floor turns to lock the seat squab firmly in the raised position

Volvo 121 Estate Car . . .

and some slight fussiness of the engine at more than 70 m.p.h., one tended—unless in a hurry—to drive somewhat slower than this. By keeping to this modest limit and changing up early, fuel consumption figures of around 32 m.p.h. are obtainable without unduly affecting point-to-point speed averages.

Heavy duty rear springs are fitted and the car handles and rides better with rear seat passengers or a load on board, because of the improvement in rear wheel adhesion. With the driver only the ride is firm without being hard, the car feeling taut. The general handling characteristic, laden or unladen, is one of modest understeer; with the car lightly laden the oversprung rear wheels would patter out when cornering fast and, occasionally, a mild but quite controllable slide could develop. The steering is light and reasonably high geared with sufficient sensitivity to give the feel of the road without feedback. Three and a half turns are required from lock to lock, giving a turning circle of 34ft 6in.—tight for such a big car.

Coming from a country with a high proportion of unmade

roads, naturally the car gives a good ride over rough going. On bad *pavé*, incidentally, a surface which does not exist in Sweden, it keeps its direction well at speeds up to 40 m.p.h., but above this there are signs of "float". This was confirmed later on the same surface in the wet, when the car needed the full width of the track at 35 m.p.h. and 30 m.p.h. was a safe speed. However, this lack of control could be as much a function of the tyres as of the suspension, a point which was noted when the Volvo 122 was tested. On an artifical washboard surface the front suspension irons out the surface at about 30 m.p.h., but we experienced a persistent though not severe vibration from the rear suspension right up to the maximum speed of 60 m.p.h. at which this section was tried. Ordinary rough roads do not induce roar but rough gravelled surfaces or cobbles transmit a certain amount of noise to the interior.

On the car tested, the clutch was notably sweet and it continued to take up the drive smoothly after repeated standing start tests. It was equally tolerant of full throttle gear changes and had light pedal action. The transmission itself was free from backlash and gear noise, but the gear change was not very good. The lever is quite long and stiff,

The driving compartment is notably free from sharp protrusions and there is a full-width, firmly padded facia roll

Although clearly built for hard work, with high ground clearance and 15in. wheels for good traction, the Volvo 121 Estate Car is a handsome vehicle

and is heavily spring-loaded into the third and fourth gear plane. It needs a strong push towards first and second and a still harder push towards reverse. The outcome is a heavy not too precise change. Synchromesh is provided for all ratios.

The braking system is a curious international cocktail of components, the master cylinder being German Schaeffer, the wheel cylinders British Girling, the shoes American Wagner and the drums Volvo. Despite the care taken in selecting parts, the result is a system which is progressive only at moderate pedal loads, retardation building up rapidly as loads increase, until the brakes grab quite sharply. Best braking figures were obtained at 70lb pressure; pushing harder on the pedal simply caused the brakes to lock. On a dry surface the forward weight transfer prevents the front wheels from locking fully, but the lightly laden back wheels then lock and slide quite readily. With the test load on board, frequent stops from moderate speeds caused the pedal to "go soft," the first indication of fade; the same effect was noted when driving hard with one up. Despite the relatively short movement of the handbrake lever this held the car securely on a 1-in-3 gradient.

Although the separate front seats are well shaped, with curved backs, the impression is that one rides on the car rather than in it. This results partly from the high driving position and partly from the fact that the seat cushions, though wide, are built up on a relatively narrow base, so that there is a tendency for the whole top of the cushion to roll on the main seat springs with a heavy driver or

Throughout the test the engine stayed spotlessly clean and free from oil leaks. The main auxiliaries are all easily accessible, but the oil filler cap, on the rocker box cover, is partly obstructed by the air filter and is stiff to remove

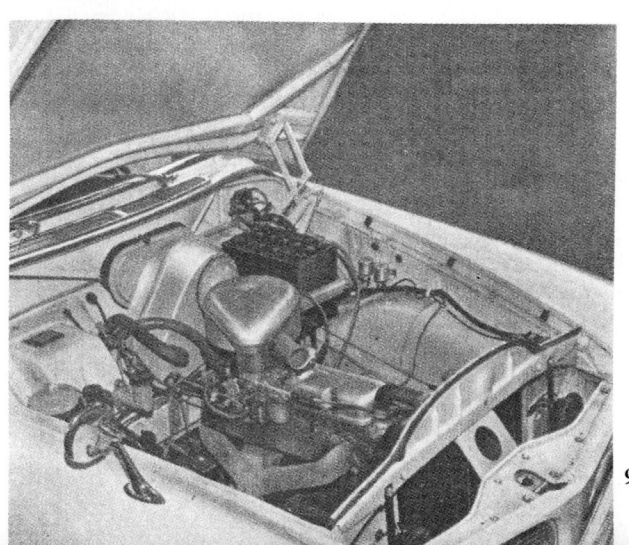

passenger. Nevertheless, the high seat gives a good view of the road forward and to the rear through the big window; both front wings are visible without craning. The windscreen seems narrow and the screen pillars rather thick compared with some of the latest cars, reminding one that this basic model has been on the go for some years.

Both front seats slide fore and aft over a 6in. range and the rake of the squabs is adjustable through about 15 deg by means of a small lever located at the outside edge of each seat; there are three positions of rake. Further adjustments for the height and angle of the whole seat can be made with a spanner, adjustable front anchorages and a three-position rear anchorage being provided to give about 1·5in. of height variation. Trim is in good quality plastic with firm upholstery, which is shaped to give raised "pillows" at the sides of the seats, effectively preventing the passengers from sliding about during fast cornering; there is adequate support in the small of the back and long distances can be covered without fatigue.

Attention to Safety

Layout of the driving compartment confirms the impression of solid practicability found throughout the car. Safety has been given serious study; the edge of the scuttle has a firmly padded, black, plastic-covered roll along its edge, which would give good protection in a collision. All the switches have flat knobs and the steering wheel has two broad spokes. Soft, padded sun vizors, a plastic-framed driving mirror, and a firm but collapsible parcels shelf with a padded edge, complete the list of safety features.

Volvo were the first company to fit safety belts as standard equipment; the diagonal and lap straps fitted to these cars hold the passengers firmly in their seats. All the driving controls are so placed that one can easily reach them when the seat belt is in use.

In an accident a heavy load in an estate car might move forward and do a deal of damage. Volvo have gone to some trouble to prevent this possibility by providing a really sturdy double bolt, actuated by a ring handle in the seat back, to hold the seat in the upright position. The bolts engage with strong die cast catch plates bolted to the side of the body, and the arrangement locks capable of restraining loads of several hundredweights. With the seats folded, the pressed-steel base of the cushion forms a buffer across the full width of the car.

The ribbon-type speedometer with trip recorder and all the gauges are contained in a single, horizontal panel in front of the driver, hooded by a lip in the scuttle padding to prevent reflections in the screen. Two needle pointers below the speedometer graduation indicate water temperature and fuel level. However, neither gives any quantitative indication, the thermometer being labelled simply "hot"

Volvo 121 Estate Car . . .

and "cold," with two graduations between; the petrol gauge has five divisions between full and empty. The lowest one of these is blocked in white to give warning of impending fuel shortage. On the test car it was something of a false prophet, three-and-a-half gallons remaining in the tank when the needle entered the white sector. Warning lights indicate generator failure, headlamp main beam, oil pressure drop and operation of the direction signals.

The minor controls work smoothly, but the choke control, next to the steering column and to the left of it, is hidden by the spokes of the wheel and is likely to be forgotten, since there is no warning hiss from the carburettor when it is in use. A two-speed screenwiper switch has three positions, slow, fast and fast with the washer working. Unfortunately the lighting switch, which is also of the pullout type, is located next to it; it was easy to switch off the lights when making a blind push in the dark to turn off the washer or wipers.

In the prevailing warm weather the main requirement from the heating and ventilating system was fresh air, which came in reasonable quantities when all three keys of the heater control were depressed fully. A cool, early morning gave the opportunity to try the heater from a cold start. With the radiator blind up, warm air started to come through after about three-quarters of a mile of running. With the heater fan going and the full capacity of the heater directed on to the screen, a torrid blast of warm air, which must be sufficient to defrost the screen quickly in the worst conditions, is directed upwards by the carefully ducted screen slots.

Under the bonnet, the engine room is neatly laid out, with lots of clear space; all the auxiliaries are easily accessible and the engine retained its pristine red finish unsullied by oil smears throughout the test. The jack, spare wheel and tools are stowed beneath a trap door at the front of the loading platform, where they are safe and dry, but the spare wheel is rather awkward to extract.

Service attention is required at a minimum of 3,000-mile intervals for nine greasing points; engine oil changes are necessary at the same intervals. Reliability should remain high, for the car obviously is assembled to stay together under hard conditions of driving and terrain; although it was driven hard by different people, there was no instance of any slackening off of adjustments, nor did rattles or loose fittings develop.

Outwardly an ordinary quantity-produced car, the Volvo has the quality that comes from careful selective assembly and great attention to detail. Combined with good performance and handling, its assets make it an attractive vehicle for the sporting-minded family man, and for the country dweller. For what it can do, the engine is surprisingly economical and there is no need of extras with this Volvo.

Specification

Scale: 0·3in. to 1ft.

Cushions uncompressed.

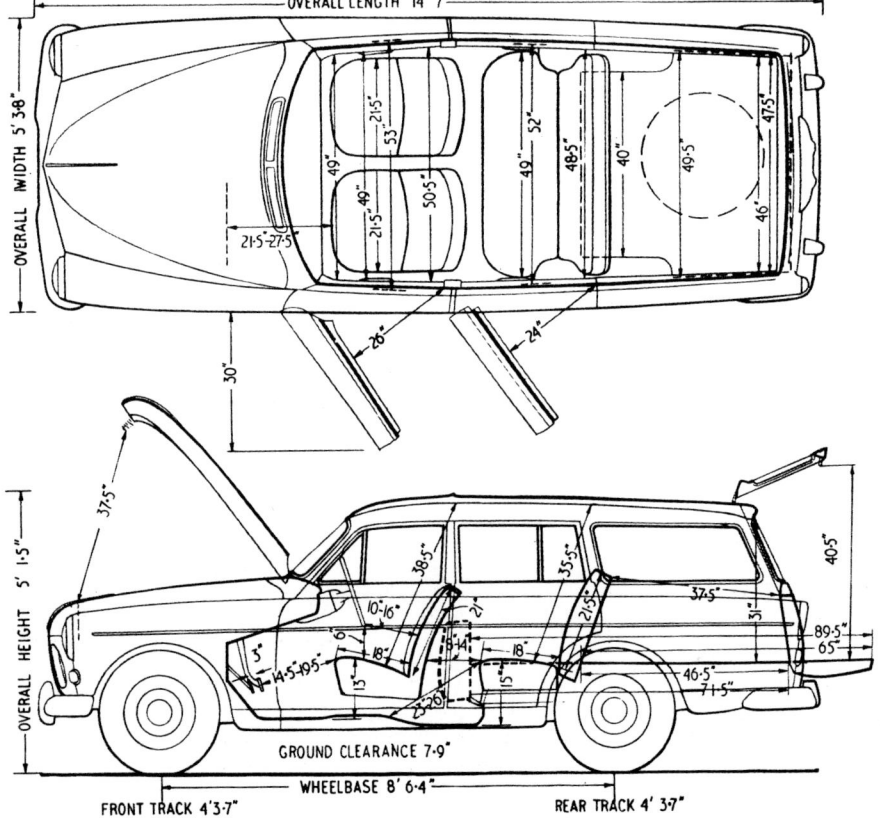

ENGINE

Cylinders ...	...	4 in line
Bore ...	...	84·1mm (3·31in.)
Stroke ...	...	80mm (3·15in.)
Displacement ...	...	1,780 c.c. (108·5 cu. in.)
Valve gear ...	...	Overhead, pushrods and rockers
Compression ratio	8·5-to-1	
Carburettor ...	...	Single Zenith downdraught, Type 36 VN
Fuel pump ...	...	AC mechanical
Oil filter ...	...	Full flow, renewable element
Max. power ...	...	68 b.h.p. (net) at 4,500 r.p.m.
Max. torque ...	...	101 lb. ft. at 2,800 r.p.m.

TRANSMISSION

Clutch ...	...	Borg and Beck single dry plate, 8·5in. dia.
Gearbox ...	...	Four-speed, all synchromesh
Overall ratios ...	...	Top 4·56, 3rd 6·2
Second ...	...	2nd 9·1, 1st 14·3, Reverse 14·8 to 1
Final drive ...	...	Hypoid 4·56 to 1

CHASSIS

Construction	...	Integral with steel body

SUSPENSION

Front ...	...	Independent with coil springs and wishbones. Delco telescopic dampers, anti-roll bar
Rear ...	...	Live axle; coil springs, radius arms, Delco telescopic dampers
Steering ...	...	Cam and roller; wheel dia., 17in.

BRAKES

Type ...	...	Hydraulic drums F. and R.
Dimensions ...	...	F. 10in. dia., 2in. wide shoes. R. 9in. dia., 2in. wide shoes.
Swept area ...	...	F. 125·5 sq. in., R. 113·5 sq. in. Total: 239 sq. in. (57·5 sq. in. per ton laden)

WHEELS

Type ...	...	Pressed steel, 5 studs, 4·5in. wide rim
Tyres ...	...	Firestone 6·40—15in.

EQUIPMENT

Battery ...	...	12-volt 60-amp. hr.
Headlamps ...	...	Bosch 45-50-watt
Reversing lamps ...	...	2
Electric fuses ...	...	4
Screen wipers ...	...	2, two-speed, self-parking
Screen washer ...	...	Standard, Bosch gear pump type
Interior heater ...	...	Standard fresh air type
Safety belts ...	...	Standard, lap and diagonal type
Interior trim ...	...	P.v.c.
Floor covering ...	...	Rubber in front and rear foot-wells, carpet in loading space
Starting handle ...	...	None
Jack ...	...	Screw pillar
Jacking points ...	...	Four external sockets, under body sills
Other bodies ...	...	Saloon

MAINTENANCE

Fuel tank ...	...	10 Imp. gallons (no reserve)
Cooling system ...	...	16 pints (inc. heater)
Engine sump ...	...	7 pints SAE 30. Change oil every 3,000 miles; Change filter element every 6,000 miles
Gearbox and over-drive ...	...	1·5 pints SAE 30. Change oil every 12,000 miles
Final drive ...	...	2·25 pints SAE 90 Hypoid. Change oil every 12,000 miles
Grease ...	...	9 points every 3,000 miles
Tyre pressures ...	...	F. 20; R. 24 p.s.i. (normal driving). F. 22; R. 25 p.s.i. (fast driving). F. 20; R. 30 p.s.i. (full load)

Volvo 122 S and Peugeot 404 Station Wagons

The introduction of two new station wagons within a few months of each other, both developed from rally-winning sedans, provides us with a rare opportunity for a comparison road test. European station wagons quite frequently are small trucks with windows; these two, however, measure up to their sedan origins in steering, braking, and even handling, if not quite in overall performance.

The Volvo 122-S and Peugeot 404 wagons have a lot in common. Both are stylish and modern, well-equipped and unusually versatile for their compact size. They have unit-construction four-door bodies that share a large number of pressings with the sedans from which they derive. The power plants and drive trains of both are the same as in the sedans (except for lower-geared final drive ratios). Both have avoided the use of truck springs at the back—the axles are positively located by radius rods and a Panhard rod on each, with coil springs (single ones on the Volvo behind the axle housing, and double ones on the Peugeot).

The greatest single difference between the two cars is performance. The Volvo has much more power and is superior on both acceleration and top speed.

In braking, the advantage again lies with Volvo; the 122-S has discs on the front wheels, while the 404 has drums all around. Both have four-speed transmissions, but Volvo's quick if seemingly unwieldy floor shift has better synchromesh. Peugeot has a column shift with a first gear that is only nominally synchronized, and a pattern that is precise but not very quick until fully run in. Peugeot, however, has the edge in steering, with a wonderfully light and precise rack-and-pinion gear, while Volvo uses a worm-and-roller gear which gives slower response.

Beyond these differences, personal preference governs the choice. The Volvo seats are harder, narrower, and have a three-position lever for backrest adjustment. Peugeot uses wide and very soft seats which seem to give more side support, and backrest adjustment goes in closely spaced notches. We have to admit we were comfortable in both.

The 404 steering column is installed Indy-style, at an angle which places the wheel so that the driver can straighten his left arm, but still has to bend the right one. This sounds more severe than it is, for in actual driving the effect is not unpleasant at all. The pedals are not offset, but the accelerator is set so far back that the right leg has a sharper bend at the knee than the left one. This is quite inconvenient until you find a good place to rest your left foot, and could be a continuing small discomfort.

No compromise has been made in the seating of the Volvo, although the seat sides do tend to collapse under heavy cornering loads. The steering wheel is squarely in front of the driver and all pedals are about equidistant. There is also better provision for resting the left foot on the toeboard.

Looking out from the driver's seat presents different views in spite of the reasonably similar styling of the cars. In the Peugeot, the short, sloping hood directs one's eyes down at the road; the Volvo's hood is longer, and stands higher than the fenders. On the whole, the Peugeot offers a better view, as the Volvo windows are small and 1950-ish.

Both cars have one-piece, gently curved windshields, with a sharper rake on the Volvo than the Peugeot. Front doors on the 404 have a small cutout reminiscent of some GM cars with "panoramic" windshields, but in this case entry and exit are aided rather than hampered. Front door vent panes have been dispensed with on the 404; the 122-S still has them. In direct contrast, the rear side windows on the French car slide open while on the Swedish wagon they are fixed. Both cars, incidentally, use Wilmot-Breeden door latches, but the Peugeot has the more expensive "zero-torque" version that lets the doors close easily and effortlessly.

Both companies have gone to great lengths in their efforts to eliminate vibration and resonance in the unitized bodies, and both have achieved remarkably good results. Vibration is a condition of existence with four-cylinder engines, but none of it reaches the passengers or the driver (except through the gear lever on the Volvo).

The Peugeot is slightly roomier—especially in the space behind the rear seat. This isn't particularly strange, as the wheelbase is longer than in the 404 sedan, while Volvo builds its wagon on the sedan wheelbase. From back of rear seat to tailgate is 47 inches in the 122-S versus 57 inches in the 404. With the rear seats folded, the luggage space is 72 inches long in the Volvo and 81 inches in the Peugeot. The French car is wider (56 inches against 50) but slightly lower inside (33½ inches against 34).

Tailgate designs differ greatly. The whole tailgate swings up on the Peugeot, possibly as a safety precaution, as French owners are likely to load down any surface that comes close to level and will withstand loading, regardless of its effect on weight distribution. The Swedes are less sanguine when extra-wheelbase loads are involved, and Volvo has provided a horizontally split two-piece tailgate. On the 122-S, those who have bulky loads can find support for great length, as the flap below the belt line extends the rear floor some 18 inches. This top flap is all window and swings upward.

Volvo has gone very scientifically into corrosion and rust-protection at the manufacturing stage. The cars now carry less undercoating, but have better sealing (after an early history of water leaks in some of the sedans), ventilation, and drainage of all spots exposed to humidity. This work has also resulted in a small weight saving.

With a normal load, the 122-S wagon is indistinguishable from the sedan under most driving conditions; if there is a difference it is a bit more harshness in the wagon. The Volvo has some understeer, increasing as speed rises, reasonable body roll, and a margin for braking all the way through a turn, as is sometimes necessary in traffic. The Peugeot definitely feels longer and heavier than the 404 sedan but is very little less maneuverable. It understeers all the way but responds excellently to a firm hand at the wheel. Michelin X tires are fitted as standard equipment, giving an advantage in cornering and longer tire-life over the Good-

WAGONS

year-equipped Volvo. The French preference for braced-tread tires has in fact gone so far that the Volvo importers in Paris fit all cars with tires of this type before delivery. From experience with Volvo sedans we know how much braced-tread tires improve their handling, and perhaps the cost-cutting boffins at the factory will one day give in to the course of progress, as they did in the case of the P-1800.

In the engine department, the attitude is quite different. The highly conventional pushrod overhead-valve Volvo engine has been given the benefit of a Magnafluxed five-bearing crankshaft, fully machined combustion chambers, and extremely critical inspection of such externally manufactured parts as valves (Farnborough), bearing shells (Vandervell), pistons (Mahle), and rings (Perfect Circle). Certainly the Volvo engine is one of the best engines of its type and displacement anywhere, and the whole engine compartment is a masterpiece of attention to detail.

For 1964, Peugeot is also going to a five-bearing

More modern aspect of Peugeot's styling is exaggerated by the wide-angle lens. Result is better visibility, closer view of the road.

The Peugeot one-piece tailgate prevents its use as a loading platform, and gives a higher sill to clear when loading heavy objects.

crankshaft (which the 404 needs about as much as it needs a metallic multi-plate clutch). What it does need is a larger carburetor throat and a freer gas flow. This engine has an ingenious valve gear, with inclined valves (at a 48° included angle), giving an almost hemispherical combustion chamber—but it just doesn't breathe well enough to really approach its full potential. This design originated as a policy of understress, as Frenchmen have only two positions of the throttle foot: On or Off. But the time has come for Peugeot to reconsider policy; acceleration, rather than top speed, is becoming more and more important in traffic, and in this country good acceleration is unquestionably a safety factor.

Cast iron blocks are the basis of both engines, but Volvo installs theirs vertically while Peugeot's is slanted 45° and has an aluminum head (Volvo sticks with cast iron). Both are known for their silence, and it is curious that the Volvo has no sound-dampening material under the hood when the Peugeot, which has a smaller expanse of hood metal, uses it all over.

The two factories are fairly comparable in size: Volvo makes about 120,000 cars annually (plus some 25,000 trucks and buses), and Peugeot puts out about 180,000 cars. But while Peugeot makes almost every part itself, except for such special items as tires and electrical equipment, Volvo buys many components from various suppliers, right up to the rear axle assembly (Spicer).

The sedans that form the basis of these two wagons are known for their endurance and good trade-in value. Certainly the wagons themselves would seem to be of the same well-built, well-finished, serviceable, and robust nature. And a buyer would be hard put to find *any* other station wagons on the market that were anywhere near as responsive and fun to drive.   **C/D**

A Volvo from a Peugeot, and a Peugeot from a Volvo; it's a choice between size vs. speed, then a matter of personal taste.

Volvo's a compacter compact, but the Peugeot looks like this year's car. Extra wheelbase allows full size rear door on the Peugeot.

VOLVO 122 S STATION WAGON

Price as tested: $2,895 P.O.E. N.Y.
Importer: Volvo Distributing, Inc.
452 Hudson Terrace
Englewood Cliffs, New Jersey

ACCELERATION:

Zero to	Seconds
30	3.7
40	6.3
50	10.2
60	14.6
70	20.7
80	30.6
Standing ¼-mile	20.4

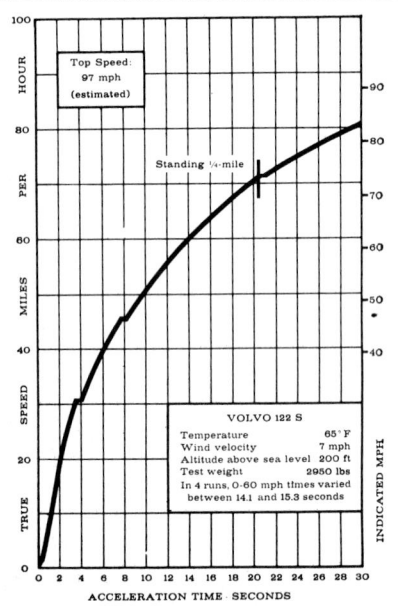

VOLVO 122 S

Temperature 65° F
Wind velocity 7 mph
Altitude above sea level 200 ft
Test weight 2950 lbs
In 4 runs, 0-60 mph times varied between 14.1 and 15.3 seconds

Top Speed: 97 mph (estimated)

Standing ¼-mile

ENGINE: Water-cooled in-line four, cast iron block, five main bearings.

Bore x stroke . . . . . 3.31 x 3.15 in, 84.14 x 80 mm
Displacement . . . . 108.6 cu in, 1,780 cc
Compression ratio . . . . 8.5 to one
Carburetion . . . . . Twin SU semi-downdraft H-6
Valve gear: Pushrod-operated in-line vertical overhead valves
Power (SAE) . . . . 90 bhp @ 5,500 rpm
Torque . . . . 105 lb-ft @ 3,500 rpm
Specific power output . . 0.83 bhp per cu in, 50.4 bhp per liter
Usable range of engine speeds . . 1,250–6,000 rpm
Electrical system . . 12-Volt, 60 Amp-Hr battery, 30-Amp generator
Fuel recommended . . . . Premium
Mileage . . . . 20-30 mpg
Range on 12-gallon tank . . . . 240-360 miles

DRIVE TRAIN:

Clutch: Borg & Beck 8.5-in single dry plate.
Transmission: 4-speed all-synchromesh gearbox

Gear	Ratio	Over-all	Mph/1000 rpm	Max mph
Rev	3.25	14.82	−5.0	−30.0
1st	3.13	14.34	5.1	30.5
2nd	1.99	9.90	7.5	45.0
3rd	1.36	6.21	11.8	70.8
4th	1.00	4.56	16.2	97.0

Final drive ratio . . . . 4.56 to one

CHASSIS:

Unit construction, all-steel structure.
Wheelbase . . . . 102.5 in
Track . . . . 51.8 in
Length . . 177 in   Width . . 64 in   Height . . 60 in
Ground clearance . . . . 7 in
Dry weight . . . . 2,440 lbs
Curb weight . . . . 2,625 lbs
Test weight . . . . 2,950 lbs
Weight distribution front/rear 51/49
Pounds per bhp (test weight) 34.90
Suspension: F: Ind., wishbones and coil springs, anti-roll bar. R: Rigid axle, radius rods and torque arms, Panhard rod, coil springs.
Brakes: Girling 10⅞-in discs front, 9-in drums rear, 350 sq in swept area.
Steering . . . . Worm and roller
Turns, lock to lock 3.25
Turning circle . . . . 33 ft
Tires . . . . 6.40 x 15

PEUGEOT 404 STATION WAGON

Price as tested: $2,795 (East Coast and Gulf Ports), $2,875 (West Coast).
Importer: Peugeot, Inc.
97-45 Queens Boulevard
Rego Park 74, New York.

ACCELERATION:

Zero to	Seconds
30	6.1
40	10.2
50	14.6
60	20.5
70	30.5
Standing ¼-mile	22.4

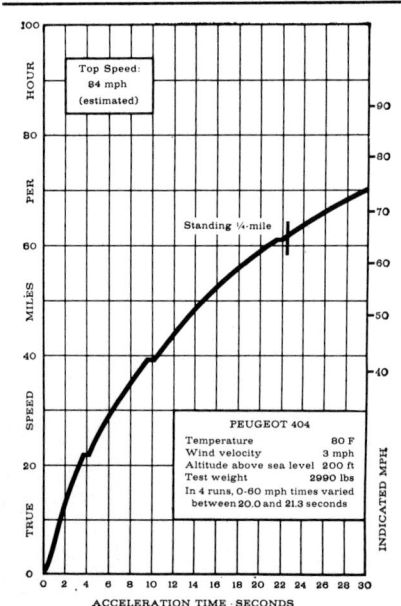

PEUGEOT 404

Temperature 80 F
Wind velocity 3 mph
Altitude above sea level 200 ft
Test weight 2990 lbs
In 4 runs, 0-60 mph times varied between 20.0 and 21.3 seconds

Top Speed: 84 mph (estimated)

Standing ¼-mile

ENGINE: Water-cooled in-line four, cast iron block, 5 main bearings.

Bore x stroke . . . . 3.31 x 2.87 in, 84 x 73 mm
Displacement . . . . 98.7 cu in, 1,618 cc
Compression ratio . . . . 7.3 to one
Carburetion . . . . Single Solex 32 PBICA
Valve gear: Pushrod-operated inclined overhead valves
Power (SAE) . . . . 72 bhp @ 5,400 rpm
Torque . . . . 94 lb-ft @ 2,250 rpm
Specific power output . . . 0.73 bhp per cu in, 44.5 bhp per liter
Usable range of engine speeds . . . 900-5,600 rpm
Electrical system . . 12 Volt, 55 Amp-Hr battery, 300 Watt generator
Fuel recommended . . . . Regular
Mileage . . . . 20-30 mpg
Range on 13.2-gallon tank . . . . 265-400 miles

DRIVE TRAIN:

Clutch: Dentel 8.5-in single dry plate.
Transmission: 4-speed, all-synchro gearbox

Gear	Ratio	Over-all	Mph/1000 rpm	Max mph
Rev	4.33	20.60	−3.6	−20.2
1st	4.00	19.00	3.9	21.9
2nd	2.24	10.62	7.0	39.2
3rd	1.44	6.88	10.8	60.5
4th	1.00	4.76	15.6	84.0

Final drive ratio . . . . 4.76 to one

CHASSIS:

Unit-construction, all-steel structure.
Wheelbase . . . . 112 in
Track . . . . F 53 in, R 51.5 in
Length . . 180 in   Width . . 64.25 in   Height . . 55 in
Ground clearance . . . . 6 in
Dry weight . . . . 2,580 lbs
Curb weight . . . . 2,710 lbs
Test weight . . . . 2,990 lbs
Weight distribution front/rear 51.5/48.5
Pounds per bhp (test weight) 40.4
Suspension: F: Ind., McPherson coil spring strut with lower wishbone, anti-roll bar. R: Rigid axle, torque tube, dual coil springs, Panhard rod and radius rods.
Brakes: 11-in drums front and rear, 230 sq in swept area
Steering . . . . Rack and pinion
Turns, lock to lock 4
Turning circle . . . . 35 ft
Tires . . . . 165 x 380

VOLVO CANADIAN AUTOMATIC

☐ This seems to be the year for European cars to shift into the automatic transmission market . . . and Volvo is one of the early birds. The company has selected its rugged, popular Canadian (122S) four-door sedan to launch this venture. We have tested this car before but feel it is worth reviewing as an automatic, particularly because it indeed makes a gentleman of this hardy competitor. The transmission used in the Volvo is the Borg Warner, model 35. Basically a hydraulically controlled unit with three forward speeds and reverse, it is driven by a torque converter. Somehow we miss the traditional Volvo floor shift but the automatic does provide a quick, smooth run through various gear ratios. It is also designed to lock in first and second gears, which makes it very handy for hard accelerating and braking. We found a slight tendency to over rev the engine, but this can be corrected as the feel of the transmission becomes more familiar. The addition of the automatic undoubtedly will help enhance Volvo's sales picture.

coachwork

Simplicity of design and a superb, durable finish are two outstanding features of the Volvo Canadian. The long hood and set-back rear hump combine to give it a sporting appearance while preserving its conventional sedan qualitites. The grille and bumper attachments are substantial but conservative. Whitewalls and mud guards are standard equipment. In brief, its styling is well suited to its character: tough but polished, with plenty of muscle and a minimum of frills.

101

interior

Volvo's interior appointments are a good match for its exterior styling. And everything is marked by excellent workmanship and materials. Up front, bucket seats with curved backrests give ample support for both city driving and distance motoring. The backrests are easily set in any of three different positions and there is a generous adjustment of horizontal movement, to accommodate drivers and passengers of assorted sizes and shapes. The rear seat is contoured for comfort and spaciousness. Upholstery is of vinyl plastic. The dashboard is shielded by safety padding with a smooth, rounded-off underside. Map pockets are attached to both front doors and the rear seat is banked by side and centre-folding arm rests, as well as two ash trays. Three-point safety belts are standard equipment on the front seats. Rear seat safety belt attaching points are built in.

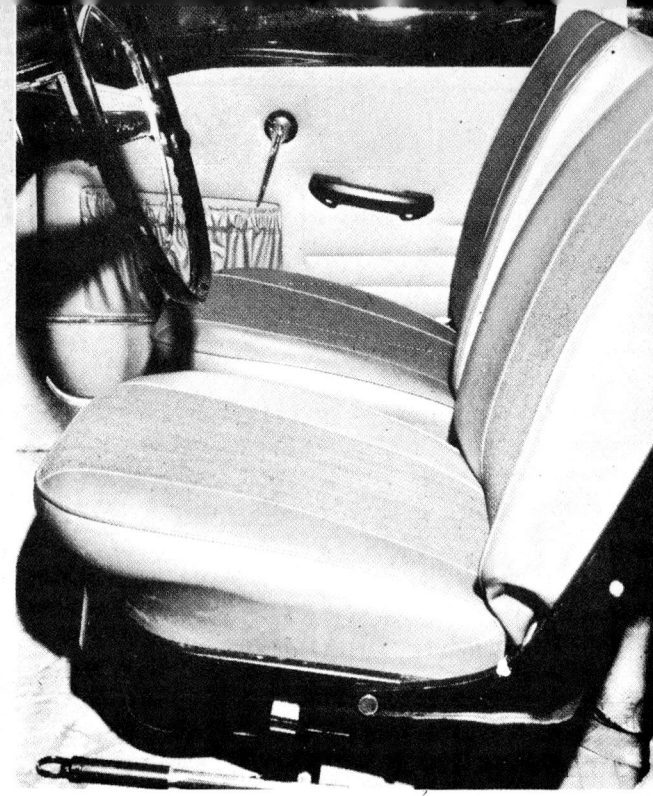

instruments

Volvo instruments are compactly-grouped for easy viewing through the top half of the steering wheel but the horizontal strip speedometer leaves something to be desired. It is difficult to get an accurate reading of speed without intent peer. We would prefer another arrangement, such as a dial or arc. All controls, including those for heating and ventilation, are within easy reach of the driver and the steering wheel is dash mounted, allowing additional leg room. Defroster nozzles blow warm or cold air over a large windshield surface and are well suited to combat Canadian winter driving conditions. Wipers are of the electric, two-speed variety. The automatic gear selector is stick-shaped, steering column mounted.

engine

The tried and tested B18-D engine which powers Volvo's 122S vehicles (two- and four-door sedans and four-door station wagon) is a rugged unit indeed . . . pound for pound one of the best on the market. Featuring a fairly high compression ratio, individual induction ports and fully machined combustion chambers, it is engineered to provide high output, strong torque and low fuel consumption. Four cylinders provide 90 hp (at 5,000 rpm) with performance facilitated by an "eight-port" cylinder head and two SU carburetors. The engine is surprisingly quiet for the workhorse that it is, helped no doubt by five over-sized lead-bronze crankshaft bearings. An added attraction, and a welcome bonus for winter driving, is a radiator blind controlled, via a chain, from the driver's seat.

trunk

Surprisingly roomy for a compact sedan, Volvo's luggage compartment is adequate for almost any occasion . . . certainly large enough to take four or five average sized suitcases. Another joy is that the spare can be lifted out without first removing the luggage. The trunk lid opens and closes easily and effortlessly. It is lined with rubber matting throughout.

handling

The Canadian is a versatile piece of goods on almost any terrain, taking the bumps with ease and the flat runs with agility. It incorporates the best of sports car strength with sedan softness, being firm but cosy. It corners on an even keel, hugs curves and loves to hold the pavement. Sexy little beast. On particularly fast corners it shows a tendency to understeer slightly, but not with the rear end threatening to break loose. Front wheel suspension is independent, with coil springs and rubber mounted control arms and ball points. Rear wheel suspension is keyed to a rigid axle. Brakes are disc in front, drums at rear, the combination giving immediate, solid gripping power, with a minimum of fade. Steering is somewhat sluggish in slow speed traffic but becomes lighter and quicker as speeds increase. Bumps and chatters don't vibrate through to the steering wheel either. In all, the Canadian is a dandy to handle and a joy to ride in, a tight blend of some of the finer features of sports car snap and family sedan comfort.

performance

The automatic transmission cannot be expected to provide the get-up-and-go characteristics of its stick shift counterpart, especially in short spurts. There is no doubt, however, that it gets full, and smooth, performance from Volvo's hard charging B18 engine. The progression through various gear ratios is accomplished with fluid power and just a hint of hesitation along the way. The performance is particularly good at highway speeds and the Canadian will cruise almost nonchalantly at 80 to 85 mph. Acceleration is slingshot quick, perhaps even a shade faster than the conventional gear shift leap ahead. Volvo's all-welded single unit body construction provides not only safety and strength but a quiet, comfortable ride. It is, in a nutshell, a well-balanced car . . . tough, handy in traffic, fine for weekend outings. Addition of the automatic transmission should win more support for Volvo Canadians from the distaff side of the family.

summary

With or without the automatic, we're sold on the Volvo. It is difficult to pick holes in this machine, except for some picayune matters such as dashboard readings, lack of locked glove compartment, a bit heavy handling at slow speed. Aside from its splendid performance and neat workmanship, it has many other outstanding recommendations, such as a tough hide (six coats of paint), numerous no-charge extras (whitewalls, seat belts, backup lights) and economy of operation (25 miles and more per gallon). Moreover, while it is a good belter, it also has the feel of safety about it . . . in particular, a road-holding and braking quality that inspires confidence in driver and passenger alike. The price is a bit above average for compacts but, in our opinion, this is merely the cost of excellence. The powerful starter motor and efficient heating system are ideal aids to help you through the wintery blasts.

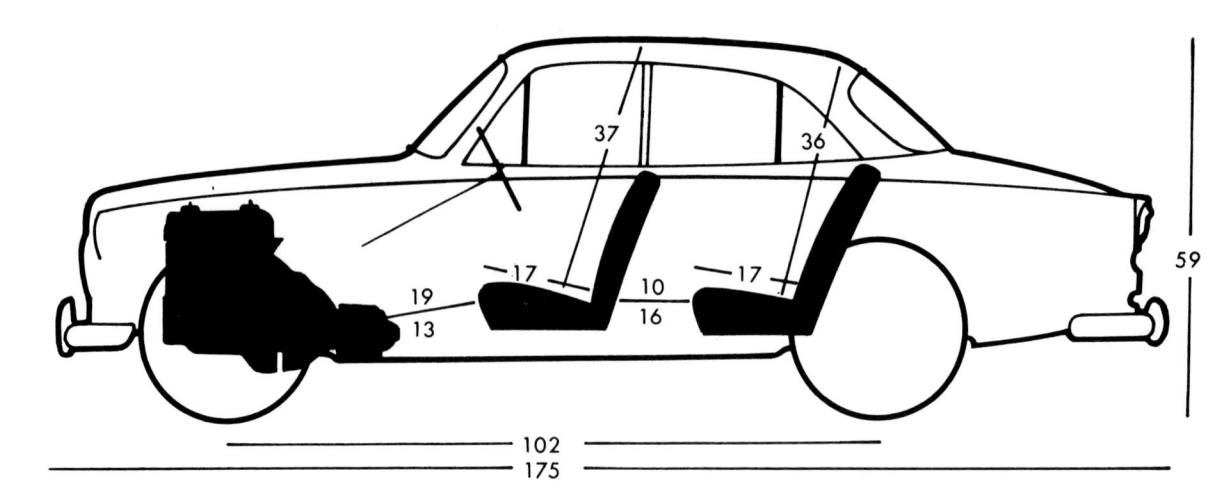

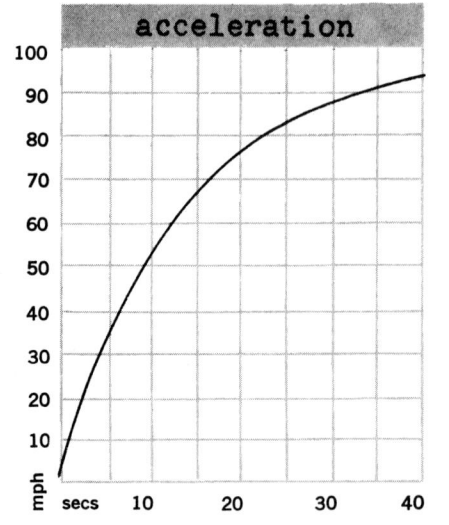

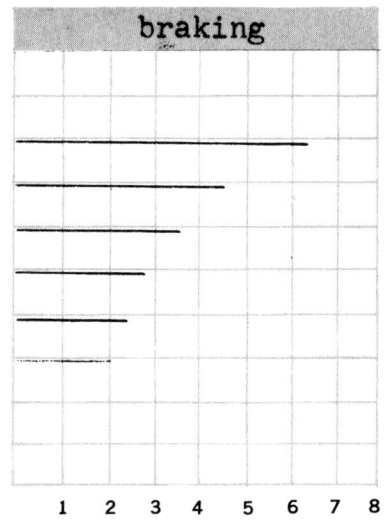

performance

ACCELERATION

0-30— 4.0 seconds
0-40— 5.7 seconds
0-50— 7.5 seconds
0-60—10.9 seconds
0-70—16.0 seconds
0-80—23.2 seconds

PASSING SPEEDS

30-50— 4.5 seconds
40-60— 6.0 seconds
50-70— 8.9 seconds
60-80—11.1 seconds

VOLVO CANADIAN AUTOMATIC

**TEST CAR COURTESY
VOLVO (CANADA) LTD.**

ENGINE—
Location: front.
No. of Cylinders: 4.
Head Type: OHV
Compression Ratio: 8.5:1.
Carburetors: 2 SU's.
Cooling: water .
Bore: 3.313''
Stroke: 3.15''
Displacement: 1.78 litres (109 cu. in.)
BHP: 90 (SAE) @ 5,000 rpm.
Torque: 105 lb. ft. @ 3,500 rpm.
TRANSMISSION—
No. Forward Speeds: 3.
Gear Ratios: 1st: 2.39:1; 2nd: 1.45:1;
3rd: 1:1.
Axle Ratio: 4.1:1.
DIMENSIONS—
Wheelbase: 102½''
Track f and r: 51¾''
Length: 175''
Width: 63¾''
Height: 59¼''
Hip Room (front): 52¾''

Fuel Capacity: 9.6 imperial gallons.
Weight, Curb: 2,400 pounds.
Weight Distribution
front: 52%.
rear: 48%.
Tire Size: 6.00x15'' L.
STEERING—
Type: cam and roller.
Turns, Lock to Lock: 3¼.
Turning Circle: 31.5 feet.
SUSPENSION—
Front: independent, coil and stabilizer.
Rear: fixed, longitudinal support arms, coil springs, torque rods and track bar.
BRAKES—
Front: discs.
Rear: drums.
STANDARD EQUIP.—3-point safety belts, adjustable back rest bucket seats, electric wipers, washer, lighter, whitewalls, backup lights.
PRICE AS TESTED— $3,300.00 (approx.)

VOLVO
122-S AUTOMATIC

Now Volvo offers a car
for senior citizens too. But if we
ever live to retire . . . .

We don't think that the lack of an optional automatic transmission has been a restrictive factor in Volvo's market penetration. However, Volvo has long felt a desire to offer an automatic, whether the demand was real or not, a trap that Renault and Fiat fell into about 12 years ago. Now it's Volvo's turn.

The choice fell on the Type 35 Borg-Warner in preference to developing a passenger-car version of the Volvomatic (used in buses) and after long tests of the Hobbs MechaMatic. When British Ford indicated a preference for Borg-Warner over Hobbs, Volvo took the same position. Engineers from Borg-Warner Limited in England worked closely with AB Volvo to match the transmission to the engine characteristics, and the results are notably better than those obtained with the same transmission in Hillmans and Fords. Of course, the 90-bhp twin-carburetor 1.8-liter Volvo engine may be less fastidious about transmissions than the less-powerful British-made units, but it still works a lot better with a standard 4-speed gearbox. Also, the

efficiency of the Type 35 Borg-Warner torque converter is substantially less than that of the Daimler-Benz hydraulic coupling as used on the 1.9-liter sedan. We are disappointed that Volvo didn't choose to ignore hydraulic transmissions altogether and concentrate on the MechaMatic, if indeed any alternative to their own excellent four-speed all-synchro gearbox is needed.

The lively 122-S loses not only some of its performance by using the Borg-Warner automatic, but also its sporting personality to an alarming extent. The engine fortunately has enough power to both slush the transmission fluid around *and* drive the car, but the effervescent performance which comes effortlessly with the stick-shift calls for concentrated planning (and prayer) with the automatic. The automatic upshifts at part throttle makes it difficult to keep the revs up in preparation for a turn or a hill. One gets the impression that the transmission takes a stranglehold on the engine, to be broken only by a frank and permanent kickdown beyond the full throttle position.

The transmission is set to upshift from 1st to 2nd between 5 and 37 mph according to throttle opening, and from 2nd to 3rd between 10 and 63 mph. Downshifts from 3rd to 2nd can be accomplished at any speed up to 56 mph by kickdown, and to first gear up to 30 mph. The selector is on a column-mounted quadrant.

As usual with other PRNDL-equipped cars, D range will give first-gear starts while L range serves as a hold switch for first. Selecting Low range at speed will get second or first, according to speed and throttle opening.

On our acceleration runs, 0-60 mph times varied between 17 and 18 seconds, and the starting quarter-mile was covered in about 21 seconds. When we last tested a stick-shift 122-S, we averaged 14.6 seconds on the 0-60 mph runs and did the standing quarter-mile in 19.9 seconds with a terminal speed of 70 mph. We did not test the top speed of the automatic Volvo, but we know from experience that the Borg-Warner Type 35 takes 2-3 mph off the top end on cars of this size, as well as getting there considerably later.

In spite of it all, we are bound to admit that the Volvo will stay with the "popular-priced" domestic compacts away from the traffic lights and even beat them down to the next intersection.

There may be people who care nothing about the performance of a Volvo but buy one solely for the quality of its design, its materials and workmanship. For such a man, who might also be happy with the more expensive Rover 3-liter, the 122-S automatic is a fine automobile.

One gets the same well-thought-out interior as in the standard car, with lots of room for four persons, and a multitude of safety features including extensive padding and three-point seat harness. Getting rid of the gear lever did not increase useful space, however. The firmly-upholstered, easily adjustable front seats remain separately mounted, which is, in fact, the only sensible thing with a total interior width of 50.5 in.

Mechanically, the Volvo stands out for the great attention to detail that is evident in all its aspects but especially under the hood. This will not be as much of a blessing to those who love to tinker as they tend to expect, however, as the Volvo needs very little checking and adjusting. Even the two semi-downdraft SU carburetors just seem to stay in tune forever.

All-in-all, the automatic 122-S is as much a no-nonsense car as its manually shifted counterpart. The transmission of their choice has proved completely reliable, and it will no doubt win many friends who think a certain loss of performance is a small price to pay for two-pedal control. **C/D**

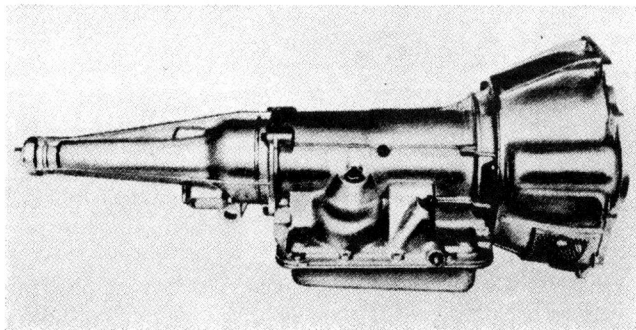

Borg-Warner's type 35 is light, compact, inexpensive, and does not require any maintenance. But it just kills the Volvo's performance.

AUTOSPORT Road Tests a

RUDDSPEED VOLVO

Sussex ex-racing driver Ken Rudd's aim has always been to take good cars and make them even better. **PATRICK McNALLY** has tested Rudd's 108 b.h.p. version of the Volvo B18 and has obtained a maximum speed of 107.2 m.p.h. and a standing quarter-mile time of 17.95 secs.—similar to, or better than, many 2-litre sports cars . . .

THE Ruddspeed Volvo is aimed directly at the man who wants sports car performance and handling but must have a five-seater for domestic reasons. There are few cars currently marketed around the thousand-pound bracket which will permit sporting instincts to be exercised yet still provide a comfortable, reliable, family carriage.

Rudd's version of the B18 is one of the few and fills the gap admirably. Capable of making exceedingly rapid progress through the twists as well as down the straights, it is capable of satisfying the G.P. streak without the usual comment from office and family alike.

Ken Rudd's aim has always been to take good cars and make them even better—with the Ruddspeed Volvo he has certainly succeeded.

The car varies from standard in many important aspects, and most modification is done in the suspension and engine departments.

The rugged four-cylinder push-rod power unit, with its sturdy five-bearing crankshaft, is fitted with a Rudd camshaft and a polished and gas-flowed head, polished manifolding and different valve springs.

The new cam gives increased lift and overlap while the stronger valve springs make higher r.p.m. possible. The polished head puts the compression ratio up to 10.5 to 1. Twin SU HD4s are retained, but are re-needled to suit the more efficient manifolding and hotter cam.

The exhaust system is completely new, with a fabricated four-branch manifold leading into a straight-through silencer and exits via twin tail pipes. These modifications raise the power output to approx. 108 b.h.p. at 6,000 r.p.m., a figure which obviously varies slightly from engine to engine.

The gearbox and clutch remain unaltered, the overdrive version with its low axle offering exceedingly close ratios.

The suspension receives a great deal of attention so as to be able to cope with the extra power. Both the front and rear coil springs are cut down and reset, this lowering the car by some 1½ inches. Koni dampers replace the standard ones, and Pirelli Cinturatos complete the picture.

The interior, with its reclining seats, remains unchanged with the exception of Rudd's own sound-deadening kit which has the effect of reducing the increased engine noise to standard. At a casual glance a Ruddspeed Volvo looks no different—the knowledgeable might notice the lowered suspension, which gives it a very purposeful look.

On the road the first thing that strikes the driver is the almost uncanny road-holding—even in the wet the car retains almost neutral handling characteristics.

This is a little unnerving to start with, for one expects the back to break away at any moment and it just doesn't move an inch. It is only a real excess of speed and stern driving which move the tail out, and then it goes, believe

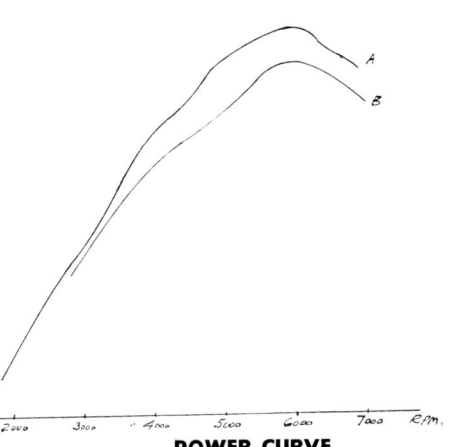

POWER CURVE

A. 11:1 Compression Ratio. 42 mm. Webers
B. 10.5:1 Compression Ratio. HS6 SUs

it or not, progressively and predictably.

With perhaps the exception of the Janspeed Mini-Cooper S I drove last year I have never driven a car which cornered so quickly with so little fuss. One does get the impression that the Pirellis play a big part in this dynamic cornering act—it would be interesting to try the car with more conventional tyres and record the differences. As regards the lack of dramatics when cornering, this can be put down to the absence of body roll and pitch, both of which have been obviated almost completely.

A criticism that can be made with regard to suspension is that it is possible to induce axle tramp when cornering hard employing full throttle; and also when busying oneself with full-blooded take-offs. I found the latter fault could easily be overcome by a little discretion. Ken Rudd tells me the reason for this is simple: the cars are set up to give a really smooth ride; if the shock absorbers were set harder, all tramp would disappear, but the ride would suffer a little. So, in fact, the Konis are at a compromise setting and can be altered to a customer's individual requirements. Driven on the road by Mr. Average neither situation should ever be encountered.

The increased power makes itself really felt, and coupled as it is with an excellent four-speed all-synchromesh gearbox, the acceleration times are simi-

lar to, or better than, many 2-litre sports cars. The maximum is upped to 107.2 m.p.h., and although the last seven m.p.h. need long straights before they are ever realized, 100 m.p.h. is often encountered in everyday motoring. At this speed the car is stable and little affected by adverse road conditions. Too many cars feel like ships in a storm, even below their maximum, as soon as a cross-wind comes up.

The acceleration times reflect the gear ratios to a great extent. Thirty is reached in 3.65 secs., 50 m.p.h. takes 7.4 and 60 m.p.h., 11 secs. dead. The most impressive time, however, is to 80 m.p.h., which is 17.8 secs., almost identical with a quarter-mile time of 17.95.

All this increased performance has been extracted without any apparent loss of flexibility. The car can be driven around all day in top if one feels so inclined, for the engine doesn't give you gear-change mania often associated with modified cars.

The interior noise level is surprisingly low, and though the exhaust note has a touch of the tweaked about it, inside the

car one might well be in a standard carriage—the sound-deadening does its job that well.

Any disadvantages of this potent performance and leech-like roadholding were not noticed The fuel consumption, never particularly good on the standard car, is hardly affected and we recorded 25 m.p.g. on several occasions, although it did drop to 20 m.p.g. when the corners were many and the car was being exercised to the full. Oil consumption appeared negligible, about a pint per thousand.

The Volvo guarantee doesn't cover some of the modifications, but Ken Rudd does, so the horror of driving a new car uncovered by the guarantee is no longer there.

I have always had a special liking for the Volvo Saloon (perhaps that is why Ken asked me to do the test) and I must say this was the best so far. Well, perhaps not quite the best, for I have just had the opportunity of trying Rudd's new Volvo. It has remote gear-change, a very necessary rev. counter, a hotter cam, and Webers.

ON WEBERS (above). Paddy McNally also tried another, more modified car running on Webers. Notice the fabricated inlet manifold and the efficient-looking cold air box. ON SUs maximum power was 108 b.h.p. at 6,000 r.p.m. (left). Picture shows the fabricated manifold and the extremely functional layout of all the under-bonnet accessories.

I took it down the test strip outside the Rudd factory at Ford, near Arundel, Sussex, and I must say it was a real flyer. The 0-60 time was in the region of 9 secs. and the quarter-mile was reputed to be down below 16 secs.! It obviously can't be quite as flexible as the SU version, but during my short drive no vices were noted.

PERFORMANCE DATA

Acceleration: 0-30 m.p.h., 3.65 secs.; 0-50 m.p.h., 7.4 secs.; 0-60 m.p.h., 11.0 secs.; 0-80 m.p.h., 17.8 secs.
Standing quarter-mile: 17.95 secs.
Maximum speed: 107.2 m.p.h.
Price: £1,245.

VOLVO 122-S

A rugged, proved performer
that even the toughest critics
find difficulty in faulting

Automotive journalism can be a disillusioning business. After years of dewy-eyed enthusiasm, a newcomer arrives in the *Car and Driver* office, only to discover that most of what he believed to be gospel about cars is patent nonsense.

With barrages of new information and qualified opinion pummeling him from all sides, a fair number of his sacred cows are destroyed and he suddenly realizes that *truly* good automobiles are as rare as black pearls. One of the few vehicles that survives under this ruthless criticism, without getting torn apart for being over-priced, poorly-made, stupidly-designed or outdated, is the Volvo 122-S. In this sense the Volvo belongs to a very select group. Automotive journalists are hyper-critical, yet you'll travel a fair distance before you find a professional who won't agree that the pride of Göteborg, Sweden isn't one of the best cars in the world and one of the biggest bargains in history.

The Volvo 122-S is not the prettiest car known to man, nor is it the fastest. But it may be the strongest. "Car of the Year" awards and "The World's Seven Best-built Cars" notwithstanding, the Volvo is possibly the toughest vehicle anywhere this side of the Aberdeen Proving Grounds and there is a growing legion of happy owners in the United States who will be glad to verify the point. This ruggedness is backed up by an alert, aggressive sales and service organization that rivals the Volkswagen setup for efficiency.

It has been storied that various Detroit manufacturers have spent large sums of money to make car doors slam shut with the solid, reassuring sound of quality. It is doubtful whether it involves anything intentional, but shutting a Volvo door sounds like about eight-and-one-half million dollars. This is symbolic of the entire Volvo body, which utilizes an all-welded unit construction of immense rigidity. The body metal is phosphated, giving it a slightly etched surface that enables the paint to cling more effectively. Anti-corrosive oil and undercoating are used liberally throughout the assembly process. Extra effort like this means a definite increase in the Volvo's resistance to the elements, especially to moisture and salt.

Volvo's B-18 4-cylinder engine may be the closest thing to an unbreakable production powerplant ever developed. It is a straightforward in-line, overhead valve, slightly oversquare layout that, like the rest of the vehicle, has undergone years of painstaking refinement. Aware that bottom-end strength is the key to really long engine life, Volvo's 1800cc engine has an exquisitely rigid five-main bearing crankshaft and

enough total bearing surface for a powerplant three times its size. For example, the new, five main-bearing BMC 1800cc engine has a total bearing surface area (including main and rod bearings) of 24.2 sq. in., while the same displacement Volvo has 42.8 sq. in! At the other end of the scale, the very strong Chevrolet 327 has 30.23 sq. in. and the old Chevy 409 has 41.02 sq. in.—both less than the Volvo!

Fitted with a pair of SU carburetors and operating with a compression ratio of 8.5:1, the B-18 is delivered in the 122-S with what seems to be a conservatively-rated 90 hp. The engine is highly flexible, easy to start and reasonably silent for a pushrod four-cylinder. It is one of the few engines that can be revved to valve float and beyond without damage. Brave souls have found that the B-18's valve action begins to get confused at about 6300 rpm, but will smooth out again at approximately 6500 rpm. That this can be done without immediate danger of bursting the works is a testimonial to the engine's strength.

The 122-S we tested was the four-door model, which has been imported since 1959. The similar two-door model was introduced here in 1963 and both remain essentially unchanged for 1965. The excellent four-speed all-synchro transmission with the long shift lever, and the beefy front disc brakes remain, as does the general feeling of soundness and quality of previous years.

The big changes come in the form of different wheels, with larger vents, a slightly larger pair of front grilles, and most important, super-adjustable seats. There is a diminishing, but still vocal group of so-called automotive pundits which maintains that a seat must be as firm as an oak board to be comfortable. Volvo apparently subscribed to this theory and their seats tended to be rather brutal on the back and shoulders during long trips. Additionally, the vinyl covering, though as durable as rhinocerous hide, did not breathe and caused nasty cases of prickly heat and other maladies resulting from Torrid Zone posterior temperatures. Both problems have been cured on the new models. The seat covering is now textured so that some ventilation exists and the frames have been fitted with no less than seven adjustment points so that anyone but an ape or a midget can fit behind the perfectly-positioned steering wheel. A screwdriver is needed to do the job, but one nevertheless can adjust the bulge in the seat for small-of-the-back support and that marks some sort of "first" in the science of driver comfort. Our taller staff members still complained about a lack of shoulder support on the new seats, but they should be satisfactory for people of average height. Volvo has also added new heater ducting to the rear seat—and that's a constructive step, though even the old setup could turn the entire interior into a Bessemer converter at will and we wonder why Volvo felt it needed improvement.

The Volvo's initial reputation was made from its giant-killer performance and that characteristic remains today. The car will accelerate to 60 mph in 15 seconds, has *usable* speeds to 90 and will carry four passengers and luggage in solid comfort at 75 mph for hours on end. It will corner with any sedan of its size and weight and will probably out-brake most of the competition.

When we consider that this car can be purchased with a fair number of options for less than $3000 and at that price will outperform most and outlast anything that can be considered remotely competitive, you better believe that you are getting one helluva automobile for one helluva bargain. Unfortunately, there are precious few makes that share that distinction. **C/D**

VOLVO 122-S

Importer: Volvo Inc.
Rockleigh.
New Jersey

Price as tested: $2630 POE East Coast

ACCELERATION

Zero to	Seconds
30 mph	4.3
40 mph	7.0
50 mph	10.2
60 mph	14.9
70 mph	20.2
80 mph	28.0
90 mph	45.0
Standing ¼-mile	70 mph in 20.2

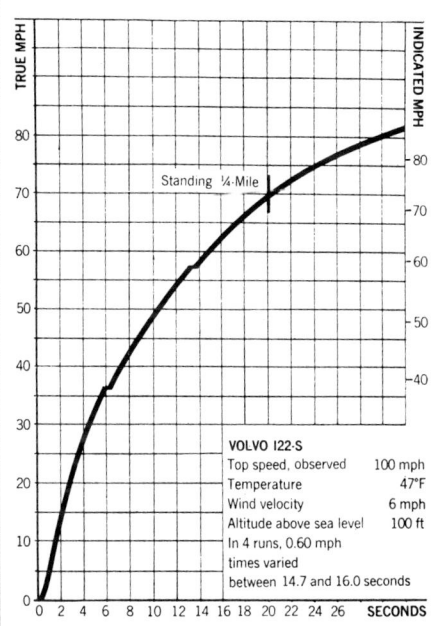

VOLVO 122-S

Top speed, observed 100 mph
Temperature 47°F
Wind velocity 6 mph
Altitude above sea level 100 ft
In 4 runs, 0.60 mph
times varied
between 14.7 and 16.0 seconds

Standing ¼-Mile

ENGINE

Water-cooled four-in-line, cast iron block, 5 main bearings
Bore x stroke . . . 3.31 x 3.15 in, 84.14 x 80 mm
Displacement . . . . 108.6 cu in, 1780 cc
Compression ratio . . . . 8.5 to one
Carburetion . . . . 2 SU type HS-6
Valve gear . . . Pushrod-operated overhead valves
Power (SAE) . . . . 90 bhp @ 5000 rpm
Torque . . . . 105 lbs-ft @ 3500 rpm
Specific power output . . . 0.83 bhp per cu in, 50.5 bhp per liter
Usable range of engine speeds. 1000-6000 rpm
Electrical system . . 12-volt, 60 amp-hr battery, 360 W generator
Fuel recommended . . . . Premium
Mileage . . . . 24-34 mpg
Range on 12-gallon tank . . . . . . 290-310 miles

DRIVE TRAIN

Clutch . . . . 8.5-inch single dry plate
Transmission . . . . 4-speed all-synchro gearbox

Gear	Ratio	Over-all	mph/1000 rpm	Max mph
Rev	3.25	13.15	−5.9	−35.5
1st	3.13	12.80	6.2	37.2
2nd	1.99	8.16	9.7	58.0
3rd	1.36	5.58	13.9	84
4th	1.00	4.11	19.2	100

Final drive ratio . . . . 4.11 to one

CHASSIS

Unit construction, all-steel body
Wheelbase . . . . 102.5 in
Track . . . . F 51.5 R 51.5 in
Length . . . . 175 in
Width . . . . 64 in
Height . . . . 59 in
Ground clearance . . . . 7.5 in
Dry weight . . . . 2310 lbs
Curb weight . . . . 2380 lbs
Test weight . . . . 2665 lbs
Weight distribution front/rear . . . . . . 53/47 %
Pounds per bhp (test weight) 29.7
Suspension: F Ind., unequal-length wishbones and coil springs, stabilizer bar
R Rigid axle, radius arms and torque rods, coil springs, panhard rod
Brakes . . . . Girling 10.85-in discs front, 9-in drums rear, 339 sq in swept area
Steering . . . . Cam and roller
Turns, lock to lock 3.33
Turning circle . . . . 31.5 ft
Tires 5.90-15
Revs per mile 830

CHECK LIST

ENGINE
Starting . . . . Good
Response . . . . Fair
Noise . . . . Fair
Vibration . . . . Good

DRIVE TRAIN
Clutch action . . . . Good
Transmission linkage . . . . Very good
Synchromesh action . . . . Excellent
Power-to-ground transmission . . . . . Good

BRAKES
Response . . . . Excellent
Pedal pressure . . . . Good
Fade resistance . . . . Excellent
Smoothness . . . . Good
Directional stability . . . . Excellent

STEERING
Response . . . . Good
Accuracy . . . . Good
Feedback . . . . Very good
Road feel . . . . Very good

SUSPENSION
Harshness control . . . . Very good
Roll stiffness . . . . Good
Tracking . . . . Very good
Pitch control . . . . Very good
Shock damping . . . . Fair

CONTROLS
Location . . . . Excellent
Relationship . . . . Very good
Small controls . . . . Very good

INTERIOR
Visibility . . . . Good
Instrumentation . . . . Poor
Lighting . . . . Good
Entry/exit . . . . Very good
Front seating comfort . . . . Very good
Front seating room . . . . Excellent
Rear seating comfort . . . . Very good
Rear seating room . . . . Very good
Storage space . . . . Excellent
Wind noise . . . . Good
Road noise . . . . Good

WEATHER PROTECTION
Heater . . . . Excellent
Defroster . . . . Very good
Ventilation . . . . Good
Weather sealing . . . . Excellent
Windshield wiper action . . . . Good

QUALITY CONTROL
Materials, exterior . . . . Excellent
Materials, interior . . . . Very good
Exterior finish . . . . Excellent
Interior finish . . . . Excellent
Hardware and trim . . . . Excellent

GENERAL
Service accessibility . . . . Excellent
Luggage space . . . . Very good
Bumper protection . . . . Very good
Exterior lighting . . . . Very good
Resistance to crosswinds . . . . Good

Ruddspeed Volvo Estate Car

Apart from the Pirelli Cinturato tyres and very slightly lowered suspension, the Ruddspeed Volvo Estate Car looks normal

ESTATE cars are too often regarded as beasts of burden, used for carting around crates and boxes, ferrying children to school, shopping from the market or moving friends' furniture from place to place. A man with an estate car is never without friends, but a man with a tuned estate car will have even more friends. Recently we have been driving two—the first being the Ruddspeed Volvo estate car.

The standard Volvo estate car is a solid piece of Swedish engineering capable of carrying a lot of luggage and people at reasonably high speeds. It has the single carburettor version of the 1,778 c.c. B18 engine and a lower back axle ratio to give reasonable acceleration when a heavy load is being carried.

Ruddspeed bring the standard engine up to P1800 specification by fitting their own modified cylinder head and S-type camshaft. Twin S.U. carburettors are added, along with a special free-flow exhaust manifold and silencing system. To cope with the extra speed, especially when carrying a load, the front drum brakes are replaced by Girling discs as on the 122S saloons. Koni dampers are also fitted to stiffen the ride and give better handling at speed. As a final touch, the whole car is given the Ruddspeed Silent Ride sound-proofing treatment.

As soon as one drives in the car one can tell that all is not standard under the bonnet. The twin carburettors have no air-cleaners—just bell-mouth ram pipes which create a healthy roar even at quite low engine revs.

The estate car's low overall ratios, combined with the Pirelli Cinturato tyres slightly smaller diameter, give it real sports-car get-away. The standing quarter mile took 18·6sec, while 40 was reached in 6·0sec, 50 in 8·1sec and 60 in

12·2sec. The test car suffered from slight clutch slip, so the times might have been fractionally better with it biting firmly.

Part of the "conversion" includes a stick-on strip of metal which gives direct r.p.m. readings in third and top gears from the speedometer. However, the smaller-radius tyres affect the speedometer so much as to make it almost usless—98 m.p.h. true showing as 120 m.p.h. In addition, the mileage recorder was 17·8 per cent optimistic.

Our best one-way speed was 101·5 m.p.h. and the mean maximum exactly 100 m.p.h. This gives an indication of the low axle ratio used, showing that the engine was running at very near its maximum power output.

The test mileage was 500 miles, most of which was made up of very hard driving and taking the performance figures. The overall fuel consumption worked out at 20 m.p.g., but spot checks during more gentle driving put this up to around the 24 m.p.g. mark. With the 10-to-1 compression ratio, super premium fuel has to be used; one pint of oil was added at the very end of the test to bring the level back to full again.

Handling is very much improved, the car feeling firm on corners and barely rolling. When the load included four passengers, two big suitcases and a large packing case full of books, the ride still remained good and the extra performance made it very easy to keep up with unladen saloon cars of similar capacity. With 118 b.h.p. to order, the Ruddspeed Volvo Estate car is always an interesting way of travelling.

Twin S.U. carburettors and a vaccum servo for the brakes are the external signs of Ruddspeed's conversion

Performance Data

Maximum speeds in gears:

		m.p.h.	k.p.h.
Top (mean)	.. ..	100·0 (88·3)	161·0 (142·1)
(best)	.. ..	101·5 (89)	163·2 (143·2)
3rd	.. ..	82 (71)	132 (114)
2nd	.. ..	55 (54)	89 (87)
1st	.. ..	36 (36)	58 (58)

Standing quarter-mile 18·6 sec. (22·0 sec.)

From rest through gears to:

				m.p.h.
30 m.p.h.	.. ..	3·6 sec.	(5·6 sec)
40 „	.. ..	6·0 „	(9·1 „)
50 „	.. ..	8·1 „	(13·9 „)
60 „	.. ..	12·2 „	(21·1 „)
70 „	.. ..	16·5 „	(29·5 „)
80 „	.. ..	23·6 „	(47·1 „)
90 „	.. ..	33·0 „	(— „)

PRICE: Ruddspeed Volvo Estate Car: £1,399 0s 0d.
Ruddspeed Ltd., The Aerodrome, Ford, Arundel, Sussex.

Figures in brackets are for the Volvo Estate car tested in Autocar of 6 September 1963.

Acceleration Times: *Speed range, gear ratios and time in seconds:*

m.p.h.	Top	Third	Second	First
10—30	— (—)	7·3 (8·4)	4·1 (5·2)	3·1 (4·3)
20—40	9·7 (11·3)	6·4 (8·0)	4·0 (5·6)	
30—50	9·0 (10·9)	6·0 (8·2)	4·8 (7·3)	
40—60	9·2 (11·9)	6·4 (9·9)		
50—70	10·6 (15·3)	8·0 (15·5)		
60—80	10·9 (22·5)	13·3 (—)		
70—90	15·3 (—)			

Overall fuel consumption for 500 miles: 20·0 m.p.g.; 14·1 litres/100 kms. (29·5 m.p.g.; 9·6 litres/100 km.)

VOLVO 122-S COMPETITION

In full racing trim, the
122-S is the most accurate sort of
guided missile

A racing Volvo? Yessir, and how! Everything about it just shouts about the joy of being a racing car. It has that mean and purposeful look, it smells like a racing car, it sounds like two or three racing cars, and boy, it goes like a racing car!

We have tested race-tuned versions of other cars, mainly British sports cars, but none has been tuned to the pitch of this Volvo. The transformation from a sporty sedan is so complete that one almost begins to doubt that any of its mechanical components can have the remotest connection with the standard parts.

With a stripped body, it's stark in the extreme, and a couple of minutes at the wheel of this thing will bring out the Walter Mitty in the best-adjusted man alive. Now we know what Collins and Hawthorn and Musso felt like when driving the Grand Prix Ferraris back from the Reims circuit to the garage through traffic on open roads. It's pretty exhilarating. The engine doesn't want to idle and has no inhibitions about letting you know it. Power comes in at about 2500 rpm and peak torque is reached between 3500 and 3700 rpm. It reaches peak power at about 6000 rpm but revs up to over 7000 with an ecstatic scream. But the car isn't just a hot-rod. The chassis has been worked over as religiously as the engine, with phenomenal results.

It rides very flatly and the springs are fairly stiff, with hard damping. Body roll is non-existent, and wheel travel is severely restricted. Yet, incredibly, its ride comfort is superior to some standard sports cars, and our wives had no fear of going on long trips.

It wants to go straight. The front wheels are set to resist any deviation from a straight path, and self-centering action is very strong. The cornering characteristics are strongly neutral, and its behavior is always uncannily predictable. For all its smooth and unspectacular way of going very fast on fast bends, it's quicker on sharp turns to hang the tail out in good dirt-track fashion and steer more with the throttle than with the wheel. It also makes the driver feel mighty heroic.

The racing successes of the Volvo PV-544 are legendary, but the racing career of the 122-S is more interesting, because this model began its life with a less obvious competition potential and is, in fact, still under development. From tentative beginnings in European rallies it has become a leading contender in production touring car races all over the world. This is all the more remarkable as the engine started out as a highly conventional non-sporting unit. It was originally a 40-bhp 1.4-liter, and its output was almost doubled before it was enlarged to 1.6-liters, when the production engine put out 85 bhp. The current 1.8-liter gives 108 bhp in its highest-performance version as supplied in the P-1800-S. As raced in the most competitive areas of Europe and America, the 1.8-liter puts out about 120 bhp. And they are strong, healthy horses!

In the preparation of the Volvo engine, all the usual stuff was done—opening up the ports and polishing them, matching the manifolds, balancing piston-and-con-rod assemblies, and so on. But race preparing Volvo engines is easier than some other power plants—the combusion chambers, for instance, are fully machined as installed in all their production cars. No further work was needed there. Factory experience indicates three highly rewarding fields for the engine tuner: breathing, compression, and balance. Improved breathing and higher compression will raise the thermal efficiency, and better balance will improve reliability and durability. In addition, displacement can be increased by boring out the block to 0.040-in oversize (using oversize pistons with a 0.003 to 0.0035-in skirt-to-wall clearance). This results in raising the displacement from 1780 to 1860 cc.

The car has a lightened flywheel and a standard P-1800 clutch. The complete crankshaft with clutch assembly is balanced, and it runs in the standard P-1800 bearing shells. For drastically raised compression ratios, Volvo can supply a special head gasket. Standard P-1800 valve timing is satisfactory for rally work, but for racing a special camshaft is needed. Volvo has developed an optional camshaft with 0.203-in lift and greater overlap:

Intake opens	32° BTC
Intake closes	72° ABC
Exhaust opens	70° BBC
Exhaust closes	34° ATC

The valve springs are special, and are installed without the rubber gaskets and steel shims used on the production engines. Valve clearances are increased to 0.020-0.022 in. The P-1800 engine has an oil/water heat-exchanger as standard, and this can be fitted to all current Volvo car engines. The racing car has proper gauges for both oil pressure and oil temperature, while the production car makes do with a simple warning light for the oil pressure. Until last year, Volvo's racing cars had the standard SU HS-6 carburetors with TZ needles and a new type of spring-loaded fuel needle, but in 1963, Webers twin-choke sidedraft units were fitted for a dramatic gain in power. Total engine weight is 352 pounds.

Volvo uses Spicer axles, and consequently a wide range of final drive ratios is available. The car can be geared for any circuit, and we tested it with the 4.88-to one ratio used for Lime Rock, Thompson and Marl-

Steering wheel spokes are drilled for lightness; instrument panel is fully reworked, with great VDO dials for everything.

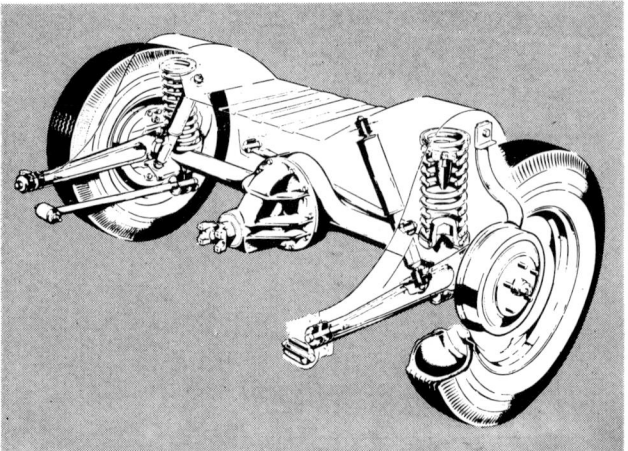

Volvo has one of the best attachment systems for a rigid axle, with radius arms, torque rods, panhard rod, and coil springs.

Special fuel tank with large-diameter filler neck and a long breather pipe severely limits use of the trunk for luggage.

Beautiful bell-mouthed Weber carburetors gape as eagerly as a cuckoo in a thrush's nest, but actual consumption is modest.

boro. On faster circuits, the standard 4.56 ratio is preferred, and in Europe the 122-S is often raced with the optional 4.10 to one ratio. Bob Perry of Volvo GmbH in Frankfurt claims that their cars with 4.10 final drives will reach 6500 rpm in top gear (with 165 x 15 tires), equivalent to 112 mph. Goodyear Blue Streaks are preferred for American race tracks, but the car was fitted with Dunlop SPs for street use during our test. The wheels are 5½-in Dodge, which offer a better support for racing tires that the 4½-in rims of the P-1800.

Chassis modifications went in two directions: lowering the center of gravity, and stiffening the suspension. The front coil springs were shortened by two inches; three inches were cut off the rear leafs. The standard Delco shock absorbers were replaced by Konis, with the fronts set at maximum hardness and the rear ones at medium. With the standard ¾-in front anti-roll bar, cornering at racing speeds will inevitably lift the inside rear wheel, so the racing car has a 1⅛-in diameter bar, which effectively keeps the rear wheels down and totally eliminates body roll. The front roll center is at ground level, and the basic understeering tendency is counteracted by giving the front wheels two degrees of negative camber.

Volvo's 122-S began life with 9-in brake drums all around, with two-leading shoes front. In 1958 the front brakes were enlarged to 10-in diameter and the duo-servo system was adopted. By that time, however, the standard cars were so fast that they really deserved disc brakes. Girling and Lockheed both developed disc systems for the front wheels of the P-1800, and Volvo chose the Girling one when the car was ready for production. The same system was adapted for the 122-S and introduced concurrently with the 1.8-liter engine in 1961. The only change undertaken in the brake system on the racing car is the fitting of cerametallic front pads and rear linings. Pedal pressures are high, but the car can also be brought to a standstill in amazingly short distances. And the brakes never get tired, never overheat, and do not need frequent adjustment.

Lightening the car did not involve putting the drivers on a diet, but stripping the car has been very thorough. The entire rear seat, upholstery and panelling, was thrown out, and the bumpers were removed. Other big weight factors were the heater unit, the floor mats and the window winding mechanisms. Only one windshield wiper is retained, one sun visor is kept, and plexiglass was substituted for the rear window and rear side windows. The standard front seats are replaced by lightweight bucket seats fitted on the standard rails. The result is a curb weight of 2110 lbs. against 2665 for the standard model.

For actual racing, Volvo uses a special exhaust pipe with a mouth on the side of the car, in front of the rear wheels; but for street use the car had a muffler and a single tail pipe, looking very standard.

Nobody in their right minds would want this kind of car for everyday use. However, it contains several modifications that can successfully be undertaken on "civilian" Volvos by handling-conscious and performance-oriented owners. Some of the most pleasant and memorable privately owned Volvos of our experience have had standard engines but were greatly improved by the use of braced-tread tires, Koni shock absorbers, and stronger anti-rolls bars. Those who feel they need more power can easily get it by milling the head (up to 0.080 in) and installing the special camshaft and valve springs. Volvo co-operates by carrying the racing options in stock, but they are not available as factory-installed options on a new car. We hope the factory will be interested in exploring this field, for better customer satisfaction as well as higher profits. **c/d**

VOLVO 122-S COMPETITION

Importer: Volvo Import, Inc.,
452 Hudson Terrace, Englewood Cliffs,
New Jersey

ACCELERATION

Zero to	Seconds
30 mph	2.7
40 mph	4.4
50 mph	6.5
60 mph	9.6
70 mph	13.0
80 mph	17.5
90 mph	22.9
100 mph	31.2
Standing ¼ mile	79 mph in 17.1

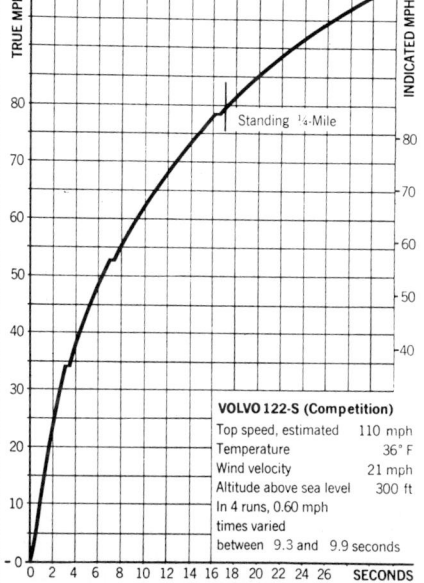

VOLVO 122-S (Competition)

Top speed, estimated	110 mph
Temperature	36° F
Wind velocity	21 mph
Altitude above sea level	300 ft
In 4 runs, 0.60 mph times varied between 9.3 and 9.9 seconds	

ENGINE

Water-cooled in-line four, cast iron block, 5 main bearings
Bore x stroke ..... 3.35 x 3.15 in 85 x 80 mm
Displacement .... 113.3 cu in, 1860 cc
Compression ratio .... 11.0 to one
Carburetion .. Two Weber side-draft 42 DCOE/8
Valve gear .. Pushrod-operated overhead valves
Power (SAE) .... 120 bhp @ 6000 rpm
Torque .... 127 lb-ft @ 3600 rpm
Specific power output ..... 1.06 bhp per cu in, 64.8 bhp per liter
Usable range of engine speeds .. 2000–7000 rpm
Electrical system ... 12-volt, 60 amp-hr battery
Fuel recommended .... Super premium
Mileage .... 20–30 mpg
Range on 25-gallon tank ...... 500–750 miles

DRIVE TRAIN

Clutch .... Borg & Beck 8.5-inch single dry plate
Transmission .... 4-speed all-synchro

Gear	Ratio	Over-all	mph/1000 rpm	Max mph
Rev	3.25	15.87	-4.72	-33.0
1st	3.13	15.16	4.95	34.5
2nd	1.99	9.75	7.69	53.7
3rd	1.36	6.63	11.30	79.0
4th	1.00	4.88	15.36	105.0

Final drive ratio .... 4.88 to one

CHASSIS

Unit-construction, all-steel body.
Wheelbase .... 102.6 in
Track .... F 51.75 R 51.75 in
Length .... 170 in
Width .... 64 in
Height .... 54 in
Ground clearance .... 3½ in
Curb weight .... 2110 lbs
Test weight .... 2330 lbs
Weight distribution front/rear .. 54.0/46.0%
Pounds per bhp (test weight) 19.45
Suspension: F Ind., unequal-length wishbones and coil springs, anti-roll bar.
R Rigid axle, trailing arms and torque rods, panhard rod, coil springs.
Brakes .... Girling 10⅞-in discs F, 9-in drums R, 350 sq in swept area
Steering ...... Cam and roller (15.5 to one ratio)
Turns lock to lock 3¼
Turning circle .... 32 ft
Tires .... Dunlop SP 165 x 15
Revs per mile 820

CHECK LIST

ENGINE
Starting .... Good
Response .... Excellent
Noise .... Terrible
Vibration .... Good

DRIVE TRAIN
Clutch action .... Good
Transmission linkage .... Excellent
Synchromesh action .... Excellent
Power-to-ground transmission .. Excellent

BRAKES
Responsiveness .... Good
Pedal pressure .... Poor
Fade resistance .... Excellent
Smoothness .... Average
Directional stability .... Excellent

STEERING
Responsiveness .... Good
Accuracy .... Good
Feedback .... Average
Road Feel .... Good

SUSPENSION
Harshness control .... Poor
Roll stiffness .... Excellent
Tracking .... Excellent
Pitch control .... Good
Shock damping .... Fair

CONTROLS
Location .... Fair
Relationship .... Fair
Small controls .... Fair

INTERIOR
Visibility .... Fair
Instrumentation .... Good
Lighting .... Good
Entry/exit .... Fair
Front seating comfort ...... Average
Front seating room .... Fair
Rear seating comfort .... —
Rear seating room .... —
Storage space .... Poor
Wind noise .... Fair
Road noise .... Average

WEATHER PROTECTION
Heater .... —
Defroster .... —
Ventilation .... —
Weather sealing .... Excellent
Windshield wiper action .... Poor

QUALITY CONTROL
Materials, exterior .... Good
Materials, interior .... Good
Exterior finish .... Average
Interior finish .... Poor
Hardware and trim .... Average

GENERAL
Service accessibility .... Excellent
Luggage space .... Poor
Bumper protection .... —
Exterior lighting .... Good

VOLVO SWEEPS SHELL 4000

Above: Rugged, picturesque terrain marked B.C. part of rally.
Below: Car heads down first special speed stage in Cascades.

1964 event termed tough test

☐ Volvo won all the marbles and the rally organizers gained international praise in the 1964 running of the Shell 4000. Competitors and observers alike agreed it was the toughest, truest test of cars and drivers in the four-year history of the event, elevating it to realistic world championship calibre. Forty-four of the 62 starters completed the course.

Led by Klaus Ross and John Bird, who lost a scant four points through 4,044 miles and six days of gruelling road warfare, Volvo works cars swept the four major prizes of first over-all, first in class, manufacturers' team and Coupe des

Dames. Three works TR4's edged the six-car Ford contingent to take second place in the manufacturers' category

Maurice Carter and Ian Worth, in a dealer-sponsored Chevy II finished second over-all — after leading for the first two days of the rally — with a loss of 11 points. George Merson and Brent Davies in a Falcon Sprint captured third over-all, and first in their class, for the Ford entry.

International rallyist Olivier Gendebien, who placed fourth in a Volvo, echoed the sentiments of most drivers when he termed this year's Shell 4000 an enormous improvement over previous competitions. It was a challenge truly worthy of its world championship classification the Belgian driver said.

Well-chosen roads, difficult to navigate under any conditions, and the addition of five special high-speed stages (closed to public traffic) made the event a tough test for driver ability and car reliability.

The cars rolled out of Vancouver Saturday night, April 18, at two minute intervals, moved through Hope and Princeton in B.C. and headed straight into the first special speed stage . . . in the Cascades. Thirty-three miles of twisting mountain gravel road took its toll of cars, with Terry Sumner, in a Valiant, and Norm Namerow, in a Volvo, both flipping and Gene Henderson's Falcon Sprint left perched precariously on the edge of a cliff. Esko Keinanen, of Finland, was forced to drop out with clutch trouble in his Valiant.

Cutting through the Rockies and into the Foothills, the first part of the six-section program ended in Calgary with Scott Harvey/Bob Millman in a Valiant sharing first place with the Carter/Worth team, neither having lost a point during the 20-hour drive.

The next section ran to Saskatoon, with the second speed stage, arranged at Camp Wainwright, Sask., taking last year's winners — Dick Doyen and Clay Gibbs — out of action when their Chevy II

Above: Gendebien at Calgary Hillclimb in Volvo. Event ended first day of the rally.

Below: Batori/Valsamis MG 1100 gets taste of prairie mud on Yorkton, Sask., farm field.

Above: Speed test at Riding Mountain Park in Manitoba.

Left: Crowd greets driver on arrival at Wawa, Ont., check.

VOLVO SWEEPS SHELL 4000

rolled. Harvey/Mollman also flipped, lost their first place position but continued on. The Swedish team of Bo Lungfeldt/ Fergus Sager retired after rolling their Falcon at Wainwright.

Maurice Carter came through clean for the second consecutive day and remained in the lead as the cars reached Saskatoon. Ross/Bird moved into a second spot tie with John Merriman and Paul Manson in a Chevelle, both having lost but a single point.

Snaking across the prairies, the third day of the rally ended in Winnipeg, with the Ross/ Bird and Merriman/Manson crews sharing first place, both having come through clean from Saskatoon. A snow-packed road in the Yorkton, Sask., area stalled a number of cars and caused considerable point losses. The day's run also eliminated one of the leading crews, Lou Lalonde and John Jones, when their Chevy II was damaged in a collision on a rural Manitoba road.

Drowsiness became the major obstacle in the section from Winnipeg to Sault Ste. Marie, travelled mainly over Trans-Canada highway pavement. The Ford team lost another crew when the Ann Hall/Jean Steagall Falcon rolled over near Dryden, Ont.

Merriman/Manson came through the day without any point loss, to preserve their first-place position. Ross/Bird lost a single point and dropped to second, with Carter/Worth third, minus six points.

The section from Sault Ste. Marie to Toronto was described by many competitors as the toughest of the rally, with muddy, rutty, slippery roads in the Muskoka district testing drivers and cars alike. One 20-mile stretch of mud, south of Parry Sound, caused more point losses than the previous

Above. A Triumph rolls along shore of Lake Superior heading for Sault Ste. Marie during fourth day of Shell 4000.

Right: Muddy roads take toll south of Parry Sound as three cars slide into ditch travelling through slippery S-turn.

Below: Ross/Bird dance around curve at Mosport during special speed stage. Drivers completed five laps of circuit.

Left: Diana Carter, left, with navigator Gillian Field, topped Coupe des Dames.

Below: Volvo victors Klaus Ross, left, and John Bird at finish.

OFFICIAL FINAL RESULTS
1964 SHELL 4000

	Driver/Navigator	Car	Point loss
1.	Ross/Bird	Volvo	4
2.	Carter/Worth	Chevy II	12
3.	Merson/Davies	Falcon	13
4.	Gendebien/Kerry	Volvo	19
5.	Curran/Carney	Peugeot	20
6.	Bunch/Edwards	Skoda	25
7.	MacLennan/Dempsey	Falcon	43
8.	Merriman/Manson	Chevelle	44
9.	Houser/Remington	Valiant	50
10.	Wenzel/Proctor	VW 1500	53
11.	Grant/Katilla	Chevy II	66
12.	Hochreuter/Lachner	VW 1500	67
13.	Harvey/Mollman	Valiant	70
14.	Bobek/Luce	Skoda	71
15.	Felton/Riddick	Mini-Cooper	83
16.	Rasmussen/Coombe	TR-4	88
17.	Jennings/Homsey	TR-4	95
18.	Graham/Acteson	Volvo	104
19.	Namerow/Bick	Volvo	107
20.	Jellett/Anderka	Valiant	124
21.	Thuner/Fidler	TR-4	130
22.	Henderson/Bickham	Falcon	140
23.	McLean/Gallop	Falcon	168
24.	Carter/Field	Volvo	173
25.	Mazuch/Dodsworth	Skoda	182
26.	Errington/Callon	Volvo	190
27.	MacGregor/Bailey	Morris 1100	193
28.	Morgan/Hartley	Falcon	197
29.	Pepper/Jackson	Volvo	202
30.	Lindquist/Koelmel	Valiant	228
31.	Hayes/Wilson	Falcon	250
32.	Bartels/Teubler	VW 1500	284
33.	Rainville/Catto	Volvo	305
34.	Gerry/Gibbons	Volvo	364
35.	Andreasen/Smith	VW	422
36.	Maters/Louden	Renault	637
37.	McQuirk/McQuirk	Bentley	667
38.	Deno/Pittock	VW	875
39.	Sumner/Meyden	Valiant	831
40.	Marchildon/Jackman	Mini-Cooper	947
41.	Ramsey/Elliott	Renault	1220
42.	Peter/Andrews	VW	1359
43.	Murphy/Calvin	Studebaker	1381
44.	Martin/Griffin	Triumph	1721

3,000 miles. During the day, the top 20 positions changed hands.

Ross/Bird, jumped into a commanding first place lead. Merriman/Manson tumbled from leadership ranks when they became mud-stuck for over 40 minutes. Carter/Worth took over second place and Harvey/Mollman edged into third.

The final day's competition shook up the standings again as unexpectedly challenging roads caught some drivers unaware. Harvey/Mollman dropped from third to 14th place, following an accident and the Grant McLean/Doug Gallop Falcon tumbled from fourth to 23rd place when a front suspension member snapped. The special stage at Mosport was a delight for most drivers but didn't result in any major point losses.

Action at end of Calgary hill climb.

PRICE: Secondhand £575; New—Basic £930, with tax £1,319

Petrol consumption	22-25 m.p.g.	*Date first registered*	16 March 1961
Oil consumption	250 m.p.pint	*Mileometer reading*	42,402

WHEN buying a used car, it is sound policy to choose a make which has built up a good reputation for durability and trouble-free service, yet although the Volvo comes very definitely into this class, buyers tend to shy away from second-hand foreign cars, and prices are surprisingly attractive. The buyer thus gets the best of both worlds—a car that lasts well, and one which is particularly good value.

Both these aspects are confirmed by the condition and price of this four-year old example. In fact, when we arrived in Manchester late in the evening to collect the car from John Wallwork Ltd., it seemed absolutely spotless by the showroom lights, and we could scarcely believe that it was old enough for the series. Daylight inspection confirmed its excellent condition, but revealed a few permissible signs of use. Uniformly drab appearance of the fabric roof lining is the biggest give-away, and there are a few shallow creases on the p.v.c. upholstery. Otherwise, the interior is extremely clean, including the carpets in two shades of mottled green. There is no reason why the facia and side trim of a car ever should become scratched and marked, but it is still quite a rare pleasure to find them as good as in this car.

Light green is the exterior colour and this, too, is exceptionally well-preserved. A few tiny blemishes at edges and joints have been retouched. Here and there a few specks of rust are seen on the chrome, though most of the brightwork shines well and again is good for its age. Volvos are made to stand up to the excesses of salt used on the winter roads of Sweden, and this 122S showed no ill-effects of surface rust after similar conditions in foggy, icy weather on motorways swimming in brine.

1961 was the year in which the B18 was introduced with 1,780 c.c. power unit. This is the basic 122S, with 1,582 c.c. four-cylinder engine, yet its performance is creditable for a five-seater of such sturdy build; as well as the lively acceleration, covering a standing quarter-mile in 20 sec, and reaching 70 m.p.h. from rest in 21·6sec, it cruises at 85 m.p.h. The engine is quiet on small throttle openings, its note changing to a purposeful, rather obtrusive snarl under hard acceleration. Starting is reliable, but by no means "first time," and with the 6-volt electrical system (discontinued on the next model) the starter is sluggish. One can even detect a momentary delay before the headlamp filaments light up. These points are characteristic, and do not suggest a faulty battery. The engine pulls vigorously almost as soon as it fires, but is slow to warm up. A new thermostat probably is needed, as unless the radiator blind (a standard fitting) is used, the running temperature is too low.

At first there was a tendency to clutch judder, which disappeared during the test. A rather springy floor-mounted lever, angled sharply to the rear, controls the four-speed gearbox; the indirect gears are quiet, and the synchromesh is still effective.

A firm and well-damped ride, with excellent insulation of wheel noise, continues much as with the car when new, and gives a feeling of great strength and rigidity. The steering is light and free from any tremors to suggest out-of-balance wheels, but there is a trace of free play. This car precedes the change to front disc brakes, and the response from the drums is poor at first, increasing abruptly with heavier pressure on the pedal, so that it is not too easy to stop from low speeds without a jolt. A pull-up lever to the right of the driving seat controls the handbrake, and holds the car securely on a steep gradient.

A fairly new set of Pirelli Cintura tyres is fitted to all wheels except the spare, which has a rather tired Firestone. A jack is stowed in the boot, beside the spare, but we were unable to find any other tools—not even a wheelbrace. A handbook is in one of the door pockets.

Everything on the car is working, with the exception of the dip switch, which tends to stick and turn the headlamps off altogether—calling for prompt use of the headlamp flasher. The speedometer reads 7 fast from 60 m.p.h. onwards, and flicks wildly when the car is moving at a crawl in traffic. The final drive is inaudible most of the time, but at high speeds there is some annoying resonance between engine and transmission noises.

Both mechanical and body conditions of this Volvo are far better than the average for the age and mileage; and the price asked for it sounds very fair.

There is still a radio aerial with the car, but the set has been removed and the facia hole neatly covered with the appropriate Volvo plate. Initial equipment included safety belts, adjustable seat backrests, an efficient windscreen washer, and a heater of suitable efficiency for the car's native climate

PERFORMANCE CHECK

(Figures in brackets are those of the mechanically similar Volvo Amazon, Road Tested 6 June, 1958)

0 to 30 m.p.h.	4·0 sec (3·7)	0 to 80 m.p.h.	32·4 sec (28·9)
0 to 40 m.p.h.	6·6 sec (-)	0 to 90 m.p.h.	54·0 sec (41·3)
0 to 50 m.p.h.	9·7 sec (9·7)	Standing quarter-mile	20·0 sec (19·9)
0 to 60 m.p.h.	14·8 sec (14·0)	20 to 40 m.p.h. (top gear)	9·2 sec (9·5)
0 to 70 m.p.h.	21·6 sec (20·0)	30 to 50 m.p.h. (top gear)	9·2 sec (9·7)

Car for sale at: John Wallwork Ltd., Sackville Street, Manchester, 1. Telephone: Manchester CENtral 8011.

Over rough, sandy, Mojave Desert trails

Floyd Clymer Road Tests the Volvo 122S

In the 1,045-mile road test of the Volvo 4-door sedan we gave the car the toughest tests of every conceivable nature, with altitudes ranging from sea level to over 5,000 ft. in the mountains, and over back roads and trails in the Lake Arrowhead "Rim-of-the-World" Highway district, through the winding San Gorgonio Mountain roads, then across to Lancaster, and on the trails and back roads of the hot and windy Mojave Desert.

The 4-cylinder, overhead-valve, five-bearing crankshaft engine of 109 cu. in. displacement, with compression ratio of 8.5:1, is of Volvo design, which they manufacture in their own engine factory at Skovde, 200 miles from their main factory at Gothenburg, Sweden. SAE horsepower is 90 at 5,000 rpms, with maximum torque of 105 lb. ft. at 3,500 rpms. (108 hp in the P-1800 Sports model.) Bore is 3.113", stroke is 3.15".

Through the mountains, around winding roads, over the trails of the desert and in some test runs on the Willow Springs road racing course near Rosamond, I found many interesting features to report on Volvo. While actually the car is smaller than our compacts and slightly larger than most imports, it has the feeling and road-holding qualities of a larger and higher priced car. Volvo out-accelerates most imported cars of similar displacement and some of our own compacts; the 4-cylinder engine at all speed ranges above 50 mph is as smooth

as many V8's or 6's. With a rigid five-bearing crankshaft, the engine is about as vibration-free as any car I have ever driven. Every component seems to be stronger than needed or "oversize" and therefore engineered for extra long life. At extreme high speeds there is practically no hood flutter or front fender tip vibration. The hood is wider, longer and more massive from the driver's seat than most foreign cars and some U.S. compacts. It does not handle, or even feel, like a small car. Wheelbase is 102 1/2"

The roadability and cornering characteristics are more like those of a high priced sports car and, in some respects, a racing car. There is practically no roll on the corners and it holds the road as though it were running on rails. You really "aim" the car in the desired direction rather than steer it. It recovers beautifully from self-imposed corner skids. The cam and roller design steering gear (only 3 1/4 turns of the steering wheel from lock to lock) makes it a "fun car" to drive. There is a real feeling of

safety, even when driving at high speeds or around winding curves. The unique Volvo 3-point seat belts are standard equipment and they are excellent. The over-the-shoulder harness and across-the-lap belt adds to the feeling of safety in driving this car.

At idling speed Volvo has about the quietest engine that I have ever found in any car. With throttle closed, sometimes you wonder if the engine is operating. At high speeds there is very little body rumble and the wind noise is quite low.

Front suspension is independent, with rubber-mounted control arms and ball joints, coil springs and excellent stabilizers. Rear suspension is good, with a rigid rear axle mounted on two longitudinal, rubber-mounted support arms and two longitudinal rubber-mounted torque arms. Coil springs are also used. The rear axle is hypoid-type with a 4.1:1 gear ratio, a good choice.

The $8^1/_2$-inch, single dry-disk clutch is hydraulically controlled and as smooth as any I have ever used.

The transmission is 4-speed, fully synchronized, with a long shift lever which has a good solid shift feeling and a rather long movement of the lever. I like this better than the short, stubby, short-action shift levers found on some stick-shift cars. The shift lever is unique in that it is spring-loaded and, when in a neutral position, it is held to the right side so that the movement from third to fourth gear, and vice versa, can be made very quickly. However, to engage first and second it is necessary to exert hand pressure on the knob and push the lever across the neutral position to the left firmly to get into first or second gears and to shift between them. This tension on the lever makes it apt to slip over in the neutral slot into fourth gear if one is unaccustomed to its operation; thus to the new driver, the occasional Volvo driver and some others this shift arrangement seems more complicated than it should be. Fortunately, there is an attractive shift pattern indicator at the top of the shift lever knob.

Hydraulic disc brakes of the 3-cylinder type are self-adjusting and protected by splash guards for the front wheels. The rear brakes are of V-type drum design, hydraulic and self-centering. This combination of disc and drum is highly satisfactory, and the pedal operates with light pressure. In quick stops and while descending winding mountain curves, I found less brake fade than in most cars, either foreign or domestic.

The hand brake operates mechanically on the rear wheels and is unique in that it is located between the driver's seat and the left door, thus left-hand operated. An excellent feature is a loop over the top of the release button, which makes it almost impossible for it to be released by accident or by children — on most cars this button is unprotected.

Windshield, wide and rear vision is good and there are many fine interior features for the driver and passengers to enjoy. The doors are as solid as on any custom-built job I have ever driven. The body is tight, joints are well-fitted. No dust leaked in and in high-pressure water tests, no water leaks developed. The all-welded single-unit body construction is ideal for strength, safety and low body noise or rumble. Three turns of the crank lowers or raises the windows. The doors can be locked without use of the key, by depressing an inside door button, which I like — unlike General Motors cars, the outside handle button does not have to be depressed to lock the door. Also, the door handle push-buttons operate easily and are located higher on the door than on most cars — very convenient.

The interior is sound-proofed and the sun visors have the softest padding that I have ever noticed on any car. The dome light is located in the center just above the rear view mirror and it can be manually operated as a map or reading light, and turns on automatically when the front doors are opened. The rear view mirror is a trifle small — it should be larger.

In many respects Volvo has followed American practices insofar as the instrument panel and controls are concerned. The ignition is operated by the right hand and the key is larger than the door key, which makes for quick and easy selection of the needed key even in the dark. Volvo has even copied

There are many winding, dusty and steep mountain trails for ones who like adventure, such as shown in the photo.

the idiot lights used on most American cars, which is not to their credit. Most drivers prefer gauges, as I do. However, there are temperature and fuel gauges. General and oil signals are warning lights. The German-built VDO speedometer has a ribbon-type horizontal indicator (which American cars feature some years and abandon in others). It fluctuates very little.

Unlike many foreign cars, the instruments in the Volvo are so Americanized that a person almost feels he is driving an American-built car. There are no "mickey-mouse" odd ball operations or tricks to sound the horn, operate the lights or ventilating and heating system. The direction indicator is on the left of the steering column where it should be — which some foreign makers have yet to learn. The steering wheel is large and the driving position extremely comfortable. The horn is operated by a typical U.S.-style horn rim. The choke is manual, which I happen to prefer to the automatic choke; and this is especially useful in cold weather regions and in warming up the engine. The heating and ventilating system is operated by three small vertical hand levers, and a 2-speed fan is controlled by a push-pull knob. The heater is excellent — in fact, one of the finest I have ever used. Volvo realizes the importance of an efficient heater, as the car is built in a very cold country. If any heater can keep the Swedes warm, it can satisfy car drivers of any country in the world.

The American-built Bendix radio is average — no better and no worse than others. The radio occupies the space where the glove compartment is usually located. Volvo's glove compartment has no door. There is a very roomy and useful shelf (or tray) located below the cowl and above the passenger's feet. There is a useful panel tray light. There is also a roomy side pocket in each front door panel.

Volvo has a feature that no U.S. car has and very few foreign cars have. They utilize an up-and-down roll curtain mounted in front of the radiator, to shut off or control the air going through the radiator. By pulling a cowl-positioned chain ring, the driver can shut off the radiator core from cold air, or adjust the curtain to certain locations during the warmup period. Such devices were common years ago in the United States, and even today are invaluable in cold countries. The driver of a Volvo in cold weather regions will find this radiator shutter control an excellent feature. He never has to use makeshift cardboard, as many do, to partially shut off cold air, thus keeping the engine warm enough to supply hot air for the heater to work properly. Californians may not know what I refer to, but the cold weather country drivers do.

There is a lot of leg room, which I especially like as I weigh 185 lbs. and am 5'11". The floorboard space for the left foot between the clutch pedal and the dimmer switch is fine for a big foot. The pen-

In my many years of car testing I have found some of the best spots to test any car on the Mojave Desert of California. There are thousands of rough, sandy, back roads that many times are nothing more than trails. There are also the scenic and interesting Joshua trees for miles on end in this arid and usually very hot desert, (which is also an excellent place to test the cooling qualities of any engine.) The day these photos were taken the desert heat was 112°.

One of the finest road racing courses on the Pacific Coast is the Willow Springs Raceway located west of Rosamond, Calif. on the Mojave Desert. Shown above is track owner Harold Mathewson (at left) and Clymer photographer Javier Lopez.

dulum clutch and brake pedals are well located, easy to operate. The foot throttle is also overhead-mounted and well located in relation to the brake. I especially like the long movement of the foot throttle, which is much less sensitive to foot movement than a pedal with a short movement.

The 3-stage electric windshield wipers do about the best job of fully covering the windshield of any that I have ever used. The hood is locked from the inside, which is a protection against underhood units being tampered with or stolen. The hood can be unlocked only by a pull lever located under the cowl and operated by the driver's left hand.

The front seats are adjustable, and I have never ridden in a more comfortable form-fitting bucket seat in any car, regardless of price.

The oil measuring stick showed no loss of oil during the test, and the 12-gallon gas tank seems ample in view of the economy of the car. I found the economy to be excellent (see my Economy Chart), and premium gas best to use in the Volvo. It has a fairly high compression engine, set up for maximum performance for its displacement, and part of its fine performance is due to the twin SU horizontal carburetors.

For a car of its size, price and horsepower, the Volvo has surprising acceleration, power and top speed. The speedometer showed 95-98 mph and, just like most U.S. speedometers, I found it 4 to 6% fast. The engine has a very pleasant hum as it is accelerating, which reminds me somewhat of the old Hudson "Super-Six" with the open throat Ventura carburetor.

Volvo has won many endurance contests, races and rallyes throughout the world. When anyone tests and abuses a car such as I did this one, it is easy to understand why some Volvo ads advise the owner to "Drive it like you hate it," and I did just that. I gave the Volvo as rugged a test as I have ever given any car and I was pleased with its operation, the many innovations, and I was fascinated by some of its exclusive features. Shortly after World War II, I saw my first Volvo in 1948 at Paris. Then in 1949 when one of the first Volvos came into the States I drove it a few miles in Los Angeles. Some time ago I also drove a P-444 model from the factory in Gothenburg, Sweden, across to Stockholm, about 300 miles. I was then impressed by the quality of workmanship and the performance of the car, though driving on the opposite side of the road in Sweden (as one does in England) did not give me the same opportunity to test and evaluate the car that I could for the same distance in this country.

The Swedes have long had the reputation of being

We turned many laps on the Willow Springs Raceway track in the Volvo 122-S at high speeds, and I found maneuverability and cornering characteristics to be exceptionally good.

excellent tool and die makers, and Sweden has the reputation for producing some of the finest steel in the world. The body seems to be very tight and assembled with good workmanship — no squeaks or rattles developed, nor did dust enter the body. When closely examining the workmanship, one will find that everything fits and there is no evidence of sloppy workmanship or poor assembly. And some Volvo features are exclusive and unique.

It has many unconventional features that are just not found in other cars. One surprising feature which I liked very much was the ease with which one could re-set the odometer (trip mileage indicator). Usually this is a rather tedious task involving twisting a lever only to find that one usually turns it too far. On the Volvo you simply pull a lever downward and, in a fraction of a second, the mileage is re-set at zero.

The manufacturers of Volvo are rated by Fortune Magazine as the 86th largest industrial firm in the world. They have many subsidiaries and own many other companies, and a total of 15 Volvo-owned and/or controlled industrial manufacturing enterprises.

Volvo is sound, substantial, and in business to stay. They have a well organized group of dealers throughout the world, and those in the United States are, generally speaking, experienced dealers with ability and finances to conduct business properly and as Volvo demands.

In summing up my evaluation of Volvo and basing my judgment on the tough testing that I have done in Volvos, I would pay it the highest tribute by saying that I consider it a small Mercedes-Benz.

CLYMER'S VOLVO 122-S ECONOMY TEST

Los Angeles to Lake Arrowhead; first half heavy traffic, last half winding mountain roads......................................23.9 mpg

Lake Arrowhead, via winding San Gorgonio Mountain roads, via Lancaster, to Mojave Desert............................26.8 mpg

Testing 200 miles over Mojave Desert roads........................32.4 mpg

Lancaster, via Angeles Crest snowy and slippery highway to Mt. Wilson, then to Los Angeles......................................23.3 mpg

At steady 30 mph...36.1 mpg
(Premium gas of various brands was used.)

SPECIFICATIONS

ENGINE. 4-cylinder, bore 75 mm (2.95″), stroke 110 mm (4.33″), displacement 1.95 litres (118.5 cu.ins.), output about 28 h.p. at 2000 r.p.m. At this engine speed, the speed of the car in direct gear is 60 km. p.h. (37 m.p.h.)

IGNITION. Battery ignition with Bosch distributor and automatic as well as manually adjusted ignition setting.

CARBURETTER. "Solex" with replaceable jet and choke adjusted from the instrument panel.

ENGINE LUBRICATION. A gear pump driven from the camshaft forces oil under pressure to connecting-rod and camshaft bearings. The oil is cleaned through a special strainer.

COOLING. Thermo-syphon (self-circulating) with fan. Honeycomb radiator. Capacity of cooling system about 15 litres (3 1/4 gallons).

SELF-STARTER. Bosch 6 volt starter motor operating on the flywheel and sufficiently powerful to start the engine in the coldest weather.

LIGHTING. Bosch charging and lighting systems, 6 volt, 2 headlights with Bilux dimming system, and parking lights. Rear light and internal instrument lamp.

BATTERY. "Bulldog", 6 volt, 85 amp/h.

HORN. Bosch with loud and soft tones. Horn button on steering wheel.

STEERING. Specially designed for balloon tyres and adjustment for subsequent backlash. Robust and elegant steering wheel of red beech with aluminium spokes. Left-hand steering.

CLUTCH. Dry-plate type with adjustment for subsequent wear.

GEARBOX. Ball engagement with 3 forward speeds and reverse. Controlled by a gear lever placed on the right-hand side of the driver (in the centre of the car).

SPEEDOMETER. Deuta with tripmeter, positive drive from the gearbox.

INSTRUMENT PANEL of brown polished red beech, fitted with the following instruments: Bosch switch box with starter button, charging control lamp, contact for full and dimmed headlights, fuse and lock, contact for parking lights, instrument lamp, oil pressure gauge, fuel gauge and speedometer.

WINDSCREEN. Nickel-plated brass frame with safety glass and automatic windscreen wiper.

TYRES. Balloon tyres "straight side cord" 29 × 4.75.

WHEELS. Robust and elegant wooden wheels of artillery type in natural colour, removable rims, spare rim with tyre fitted at the rear.

FRONT AXLE. I-section of drop-forged special steel. SKF taper roller bearings in the front wheels.

REAR AXLE. "Semi-floating" spirally cut crown wheel and pinion, Gleason system, journalled in SKF ball bearings. The rear axle housing is of pressed steel with a large inspection cover at the rear.

SPRINGS. Semi-elliptic, extra long and resilient. The rear springs are suspended under the rear axle.

BRAKES. The hand brake operates by means of a band on a drum on the propeller shaft. The foot brake operates on the rear wheels by means of brake shoes totally enclosed in drums.

LUBRICATION OF CHASSIS. Zerk system with grease gun.

FUEL TANK placed at the rear, capacity 45 litres (9 7/8 Imp.galls.). Petrol fed to the carburetter by means of a vacuum-tank.

EQUIPMENT. Jack, tyre pump, grease gun and set of tools.

TOURER BODY. Of sheet metal fitted on ash and red beech framework.

UPHOLSTERY. Real Swedish leather.

SALOON BODY of ash and beech externally covered with leatherette. Separate front seats which are adjustable relative to the steering wheel and pedals and both backrests can be lowered to the rear forming a bed for two persons. This makes the car very suitable as a camping vehicle. Standard fittings include a flower vase, ashtray, curtains on all windows and a luggage trunk.

SEAT CUSHIONS & BACK RESTS of Swedish horse hair.

DIMENSIONS

Wheelbase	mm.	2850	112.24″
Track width	,,	1300	51.2″
Overall length	,,	4050	159.5″
Overall width	,,	1550	61.1″
Length of frame	,,	3912	154″
Width of frame, front	,,	670	26.4″
,, ,, ,, rear	,,	1000	39.4″
Ground clearance	,,	220	8.7″
Weight of car with tourer body	kg.	950	(2090 lb.)
Weight of car with saloon body	kg.	1100	(2320 lb.)

Volvo 131 1,778 c.c.

AT A GLANCE

Full safety equipment.

Very smooth, quiet engine.

Excellent gear change and quiet transmission.

Comfortable seating, and firm suspension.

Very serviceable and complete interior finish.

Reasonable fuel consumption.

Efficient brakes but heavy pedal loads.

The instrument panel with ribbon speedometer is well cowled, and the gear lever is direct-acting

MANUFACTURER:

A.B. Volvo, Gothenberg, Sweden.

U.K. Concessionaires: 28, Albemarle Street, London, W.I.

PRICES:

Basic	...	...	...	£845	0s 0d
Purchase Tax	...	...	£177	12s 1d
Total (in G.B.)	...	...	£1,022	12s 1d

PERFORMANCE SUMMARY

Mean maximum speed	...	90·5 m.p.h.
Standing start ¼-mile	...	20·7 sec
0-60 m.p.h.	...	17·6 sec
30-70 m.p.h. in third gear	...	20·7 sec
Overall fuel consumption	...	26·5 m.p.g.
Miles per tankful	...	288

IMPORTED cars of about two litres capacity have tough competition these days from British manufacturers. The handicaps of import duty and import surcharge make it difficult for them to be competitive in terms of price and quality, quite apart from general design considerations. The Volvo range, still retaining a live rear axle located by radius arms and supporting the body on coil springs, has been remarkably successful, thanks to rugged construction, high quality finish and reliable high-speed performance from quite ordinary-looking pushrod four-cylinder engines. Now the addition of the 131, which with two-door bodywork and the single-carburettor 75 b.h.p. B18 engine costs £1,022 and is the cheapest Volvo offered to United Kingdom buyers, is likely to continue this success.

Autocar Road Test 2020

MAKE: **Volvo**

TYPE: **131 1,778 c.c.**

Speed range, ratios and time in seconds

m.p.h.	Top (4·1)	Third (5·6)	Second (8·2)	First (12·8)
10—30	—	7·2	4·9	4·0
20—40	9·7	6·7	5·0	7·0
30—50	10·1	7·2	5·8	—
40—60	11·5	9·2	—	—
50—70	14·7	13·5	—	—
60—80	22·9	—	—	—

TEST CONDITIONS

Weather ... Overcast with 0-5 m.p.h. wind
Temperature ... ... 4 deg C. (39 deg F.)
Barometer, 30·20in. Hg. Dry concrete and tarmac surfaces.

WEIGHT

Kerb weight (with oil, water and half-full fuel tank)
20·75cwt (2,324 lb—1,056 kg)
Front-rear distribution, per cent ... 53 F., 47 R.
Laden as tested ... ... 23·75cwt (2,660lb-1,204kg)

TURNING CIRCLES

Between kerbs ... ... L. 34ft 4in.; R. 34ft 1in.
Between walls ... ... L. 36ft 2in.; R. 36ft 0in.
Steering wheel turns lock to lock ... ... 3·25

PERFORMANCE DATA

Top gear m.p.h. per 1,000 r.p.m. ... ... 18·0
Mean piston speed at max. power ... 2,476ft/min
Engine revs at mean max. speed 5,030
B.h.p. per ton laden ... ... 63·4

FUEL CONSUMPTION

At Constant Speeds

30 m.p.h.	...	...	...	...	40·8 m.p.g.
40 "	...	...	...	...	37·0 "
50 "	...	...	...	...	33·9 "
60 "	...	...	...	...	30·8 "
70 "	...	...	...	...	27·2 "
80 "	...	...	...	...	23·0 "

Overall m.p.g. ... ... 26·5 (10·7 litres/100km)
Normal range m.p.g. 26-30 (10·9-9·4 litres/100km)
Test distance ... ... 1,661 miles
Estimated (DIN) m.p.g. 25·5 (11·1 litres/100km)
Grade ... ... Premium (95-97RM)

OIL CONSUMPTION

SAE 20 ... ... 3,300 m.p.g.

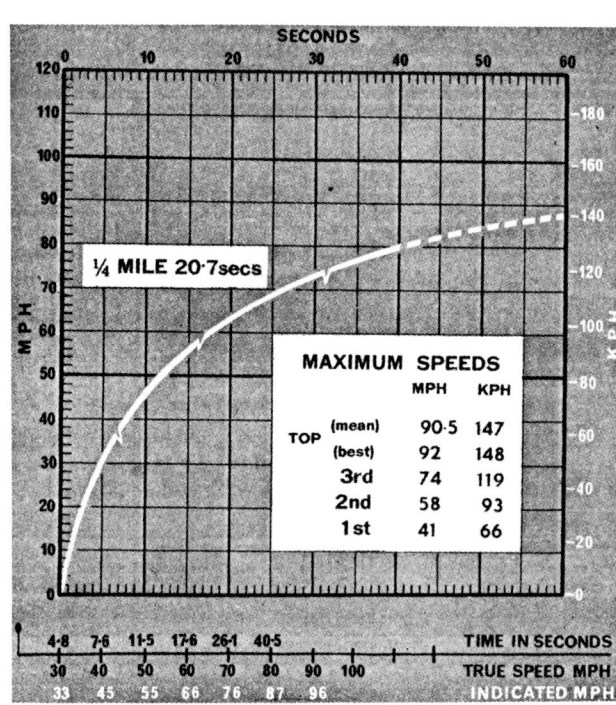

¼ MILE 20·7secs

MAXIMUM SPEEDS

		MPH	KPH
TOP	(mean)	90·5	147
	(best)	92	148
3rd		74	119
2nd		58	93
1st		41	66

BRAKES
(from 30 m.p.h. in neutral)

Pedal Load	Retardation	Equiv. distance
25lb	0·25g	120ft
50lb	0·45g	67ft
75lb	0·65g	46ft
100lb	0·80g	38ft
140lb	0·90g	33·4ft
Hand brake	0·37g	81ft

CLUTCH
Pedal load and travel—50lb and 4·75in.

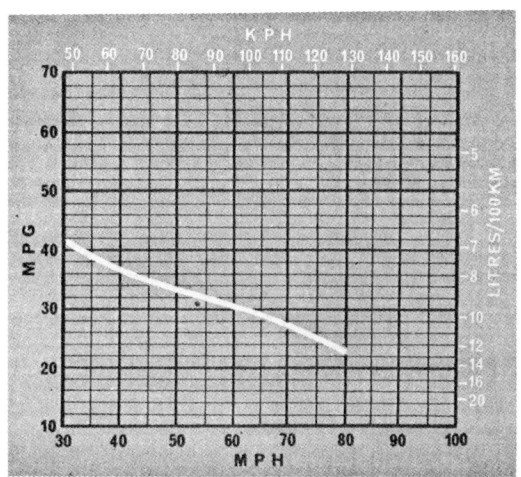

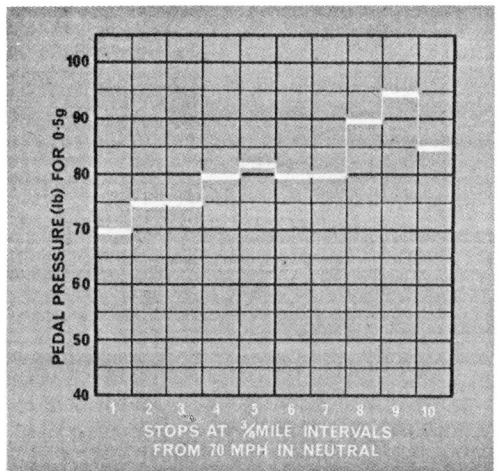

STOPS AT ¼MILE INTERVALS FROM 70 MPH IN NEUTRAL

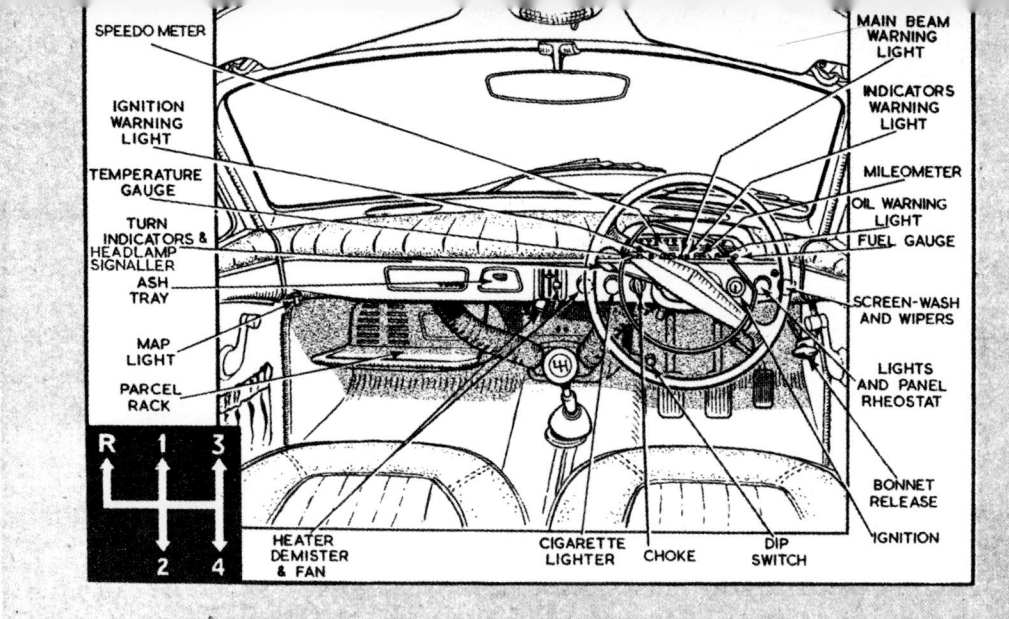

SPEEDO METER

IGNITION WARNING LIGHT

TEMPERATURE GAUGE

TURN INDICATORS & HEADLAMP SIGNALLER

ASH TRAY

MAP LIGHT

PARCEL RACK

MAIN BEAM WARNING LIGHT

INDICATORS WARNING LIGHT

MILEOMETER

OIL WARNING LIGHT

FUEL GAUGE

SCREEN-WASH AND WIPERS

LIGHTS AND PANEL RHEOSTAT

BONNET RELEASE

IGNITION

R 1 3
2 4

HEATER DEMISTER & FAN

CIGARETTE LIGHTER

CHOKE

DIP SWITCH

Although relatively inexpensive, it is by no means an austerity model. The specification includes Girling disc front brakes, a fresh-air heating system and the latest type of Volvo front seats with lumbar adjustment. Lap-and-diagonal seat belts are also standard.

While the 131 has the lower powered engine, its performance is very adequate—90 m.p.h. maximum and the standing quarter-mile in 20·7 seconds—and some prefer it to the two-carburettor model because of its refined performance and quietness. Indeed, the smoothness of the engine, and its willingness to pull away strongly from low speeds, leave an outstanding impression. With less need to get the last ounce of power out of the engine, an effective air silencer is fitted and the power roar as r.p.m. increases, peculiar to the twin-carburettor model, is absent, making long-distance motoring more restful.

Body

Walking round the cherry-red road test car, one realises that, despite the fact that the body is beginning to look high-waisted, the general appearance is quite sporting for an everyday saloon. Practical motorists will like the sturdy bumpers. The easy-to-clean body, with few nooks and no crannies, will also appeal to those who look after their own cars.

Two-door bodywork always has some advantages, other than the sporty style. There are two fewer doors to produce rattles and two fewer sources of draughts; small children can be carried safely in the rear seats without fear of the door being opened by inquisitive, tiny fingers.

Cold weather starting was immediate with brief use of the choke, and possible without it if the accelerator were pressed a couple of times to operate the accelerator pump. Within half-a-mile the engine was hot enough to pull smoothly. After a couple of miles warm air was coming

from the heater. The overall petrol consumption was 26.5 m.p.g., the rate being strictly related to speed and not showing abnormally high figures when driven hard. During the 1,661-mile test four pints of oil were used, equivalent to 3,300 miles per gallon.

A comforting thought is that this least-tuned and least-stressed of Volvo engines can be driven near to the maximum speed of the car for most of the time, without losing its edge. It was smoothest at a true 70 m.p.h. and reached that speed quickly. At higher speeds a slight out-of-synchronization hum, apparently from the transmisson, set in, but was only disturbing to the ears of the mechanically minded.

Transmission

In other respects the transmission was faultless. Gearchanging action was lighter than we recollect on previous Volvos and the synchromesh was equally unbeatable. The long, plated lever, which is spring-loaded over towards the third and fourth gear plane, is heavily reinforced at the base. This no doubt contributes to precise action, but the heavy lever sometimes jumped out of gear on rough pavé.

The ratios are fairly closely spaced, as the maximum speed of 40 m.p.h. in bottom indicates, although just low enough to allow starts on a 1-in-3 incline. On the road, third is particularly useful for passing, with its maximum of 76 m.p.h., although holding it above 70 m.p.h. did not offer better acceleration than was available in top. Like the accelerator mechanism, the clutch operating linkage is progressive, and helpful to tiros strange to the car.

Volvo have years of rallying behind them, including many successes by private owners driving more-or-less standard cars. Sure, predictable handling is therefore to be expected. The steering is moderately high geared, needing three-and-a-quarter turns between generous locks, and is

precise without feeding back road shocks. The tyres fitted to the test car were of the conventional, diagonal cord type, without shoulder treads, and tended to lose grip when cornering sharply on other than dry surfaces, although they were quite stable on fast bends.

The general toughness and one-piece nature of the car showed up well on rough pavé, the body remaining quite rigid and completely free from rattles. Its ability to hold a line when bouncing along at 40 m.p.h. was better than that of some independently-suspended cars we have tried. The performance on washboard corrugations was equally good, with a "smooth-out" speed range of between 45 and 55 m.p.h. On more normal going the ride was firm and well damped but comfortable, with complete freedom from pitching.

All Volvo cars now have Girling disc front brakes and drum rears. Heavy pedal loads are required to get the best results, 140lb pressure being needed for maximum stopping power; but in return the brakes are commendably free from fade. They withstood repeated 50 per cent g stops from 60 m.p.h., only 20 per cent increase in effort at the pedal being required after 10 applications. After immersion in a wading trough, the brakes briefly lost efficiency but gripped about 20 yards after being applied.

Noise

In regard to road noise, the Volvo is good without being exceptional, and one is only aware of the road surface when traversing newly-laid granite chippings or stone sets. The low level of wind noise makes cruising in the seventies restful, and partly-open front quarter-vents did not increase the noise appreciably.

The driving position is appropriately upright and commanding, but short-legged drivers might find the steering column too long, while long-legged folk could find the rearward seat adjustments insufficient. This

Left: Carefully profiled seats have a non-slip, ventilating pattern on their pleated panels. The front ones have adjustable backrests (over a rather limited range) with locks for the fold-forward mechanisms, and safety harness is standard. Right: in the back, beneath hinged side windows, are recessed armrests, and there is also a folding one in the middle

Volvo 131 ...

latter shortcoming is mitigated by the ability to rake the seat squab to give a more arms-stretched driving attitude.

The 15in. diameter, two-spoke steering wheel is smooth and thick-rimmed, with good finger notches on the underside, and imparts a subtle feeling of quality to the car. Purists might object to the heavy, out-of-proportion styling motif which forms the crossbar of the horn ring. Minor controls for choke, lights and two-speed wiper and washers have large flat knobs and are disposed in a row on the facia, on either side of the steering column. The pull-out light

switch and the combined wiper-washer switch are together on the right of the column, following American practice. It is easy to push off the light switch accidentally when turning off the wipers. The action of the washer is unusual, in that it operates when the wiper switch is pulled out past the full-speed position; the combination of high wiper speed and water gives very speedy screen cleaning.

Three neat levers working in vertical quadrants in the middle of the dash control the heater system; there is some lag in action. At night they are illuminated by a small light which can throw reflections in the screen in unfavourable conditions.

The improved type of driving seats

which, in addition to adjustment for tilt and squab rake, have provision for varying the degree of lumbar support, are a complete success, and must rank among the best we have tried Their virtue is in the combination of spring rate, squab and cushion shape and choice of covering material. The amount of up-and-down movement absorbed by the seat springs alone is almost uncanny. Attractive to look at, the trim is primarily functional. Tightly covered side panels make for firm edges, while the middle panels have loose transverse pleats of basket-textured plastic which is soft, comfortable, locates the occupants well and gives adequate ventilation. Trim material is black, two-way-stretch plastic.

Safety Features

A pleasing feature of the fore-and-aft seat adjustment is that it works freely without the necessity to jerk vigorously to move the seat. But the squab rake adjustment, by large plastic knobs to give about 15 deg movement, could not be operated with any weight on it. The folding seat backs, for access to the rear seats, have locks to prevent them folding-up the occupants in a crash. When passengers have to be released from the back, they present a slight social problem in that the levers operating the locks are on the outside of the seats, that on the passenger side being out of the driver's reach.

Volvo as a company have co-

In the quite roomy boot the spare wheel is upright and accessible. Reversing lamps are incorporated in the main tail lamp units, and the fuel filler cap has a lock

The 131 has a sporty appearance and is well finished. Those bumpers are as tough as they look, and flexible mud-flaps below all four wheel arches are an unusual consideration to following traffic

operated for some years with the Swedish safety authorities in designing the interior of their cars to give the best protection to passengers in the event of an accident. Thus the facia of the 131 is topped with a hard but crushable black plastic-covered roll of proper consistency, and the steering wheel motif is intentionally blunt to minimize chest injuries. Window winder and door handles are flat to the door and have the least possible protrusion, while the sun vizors and the parcels shelf are crushable. Safety belts are standard equipment, and the Volvo lap strap and diagonal pattern has been imitated widely. In use, the belts are fiddling to adjust, while the knack of attaching the main clip to the ring bolt attachment between the seats requires practice and some contortions.

Body dimensions are the same as those of the four-door car, and legroom in the front and rear compartments is identical. Rear seat knee room is adequate, partly because of curtailed front seat adjustment. Design of the seats has been given the same attention as those in the front; a centre armrest is provided, and recesses in the quarter panels give more elbow-room than in the four-door model. Access to the rear seats through the wide doors is easy, bearing in mind the previous remarks about the seat back locks. Attention to neatness of finish was good, with a notable absence of visible screwheads. A shelf under the dash on the passenger side is suitable for small parcels and maps—heavy articles vibrate on it—and there are elastic-topped door pockets. The window winders, working close to the trim pad, tend to bend or tear maps or brochures placed in the pockets and limit their capacity.

Attention to creature comforts has been well catered for; there are a cigar lighter and large-capacity ashtray on the facia, pile carpets on the rear floor and plastic pulls on the centre pillars.

Ventilation

Ventilation can be adjusted accurately by opening the forward-hinged rear windows slightly. Air is drawn through a slot at the base of the screen, and a two-speed fan is provided to help it on its way when the car is moving slowly. Temperatures from that of the outside air to fever heat are obtainable readily. Although there was not much flow of air through the windscreen demister slots with the fan turning over at its rather noisy top speed, the screen is cleared quickly of condensation.

The spare wheel is strapped vertically to one side of the luggage boot, the jack and tools being stowed between it and the body side. A large bulge over the axle and rear suspension reduces the floor area to about 3ft by 3.5ft, but there is still room in the irregular space which remains for the luggage of two or three people on holiday journeys. Torsion bars counterbalance the boot lid and

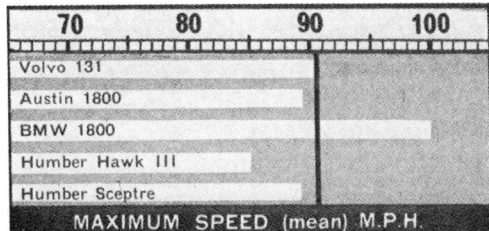

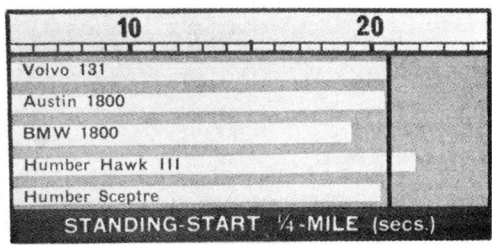

HOW THE VOLVO 131 COMPARES

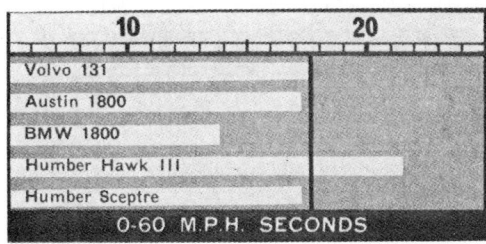

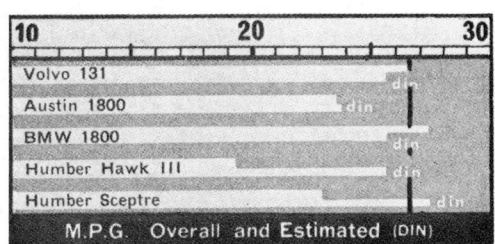

129

hold it open when loading. The floor and axle hump are covered with tough rubber material.

Under the bonnet there is lots of clear space round the engine and every ancillary is reached easily. Thanks to an efficient sealed panel surrounding the radiator block, the well-finished compartment keeps remarkably clean. In fact, the engine and side panels of the test car, after some 8,000 miles total during which it had obviously not been touched, could have been restored to showroom cleanliness with a duster.

Holding 9·75 gallons, the fuel tank gives a safe touring range of 250 miles before refuelling. The gauge has a white warning sector next to the empty mark, but neither a reserve cock nor a warning light remind the improvident. Petrol thieves might be tempted by the large filler neck giving directly into the tank, but this feature makes for quick refuelling.

The Bosch headlamps reminded one that a pair of large diameter units can be more efficient than two pairs of small ones. The beam was set rather low for high-speed night motoring, making the dipped range too short, but it was obvious that adjustment would have given a first-class driving light. Reversing lamps are built into the rear lamp clusters.

Honest is possibly the best single descriptive word for Volvo cars. Not flashy, and now rather dated in appearance, their natural quality comes from integrity of design and construction, coupled with attention to detail. After only a few miles at the wheel one feels that they would be nice cars to own and to cherish.

SPECIFICATION : VOLVO 131 FRONT ENGINE, REAR-WHEEL DRIVE

ENGINE
Cylinders	... 4, in-line
Cooling system	... Water; pump, fan and thermostat
Bore	... 84·1mm (3·31in.)
Stroke	... 80·0mm (3·15in.)
Displacement	... 1,778 c.c. (108·5 cu. in.)
Valve gear	... Overhead, pushrods and rockers.
Compression ratio	8·5-to-1
Carburettor	... Zenith
Fuel pump	... AC; mechanical
Oil filter	... Wix, full-flow
Max. power	... 75 b.h.p. (net) at 4,500 r.p.m.
Max torque	... 101 lb.ft. at 2,800 r.p.m.

TRANSMISSION
Clutch	... Borg and Beck single dry plate, 8·5in dia.
Gearbox	... 4-speed, all-synchromesh; central floor change
Gear ratios	... Top 1·0; Third 1·37; Second 2·0; First 3·12; Reverse 3·25.
Final drive	... Hypoid 4·1-to-1

CHASSIS AND BODY
Construction	... Integral with steel body

SUSPENSION
Front	... Independent, coil springs and wishbones; telescopic dampers; anti-roll bar
Rear	... Live axle, coil springs, radius arms and Panhard rod; telescopic dampers

STEERING
	... Cam and roller Wheel dia 15in.

BRAKES
Make and type	... Girling disc front, drum rear; no servo
Dimensions	... F, 10·88in. R, 9in. dia.; 2in. wide shoes
Swept area	... F, 226 sq. in.; R, 113 sq. in. Total: 339 sq. in. (287 sq. in.) per ton laden)

WHEELS
Type	... Pressed steel disc, 5 studs, 4in. wide rim
Tyres	... Goodyear G8 tubeless 6·00-15in.

EQUIPMENT
Battery	... 12-volt 60-amp. hr.
Generator	... Bosch 240-watt
Headlamps	... Bosch 40/45-watt
Reversing lamp	Standard
Electric fuses	... 4
Screen wipers	... Two-speed, self-parking
Screen washer	... Standard electric
Interior heater	... Standard, fresh-air type
Safety belts	... Standard
Interior trim	... Vinyl seats, plastic headlining
Floor covering	... Foam-backed rubber
Starting handle	... No provision
Jack (type)	... Screw-pillar
Jacking points	... 4, under body sills
Other bodies	... 4-door and estate car
Fuel tank	... 10 Imp. gallons (45·5 litres)
Cooling system	... 14 pints (including heater) (8 litres)
Engine sump	... 6·5 pints (3·75 litres) SAE 10W/20. Change oil every 3,000 miles; change filter element every 6,000 miles.
Gearbox	... 1·25 pints SAE 90. Change oil every 25,000 miles
Final drive	... 2·25 pints SAE 90. Change oil every 25,000 miles
Grease	... 8 points every 3,000 miles
Tyre pressures	... F, 20; R, 23 p.s.i. (normal driving); F, 20; R, 26 p.s.i. (fast driving); F, 20; R, 26 p.s.i. (full load)

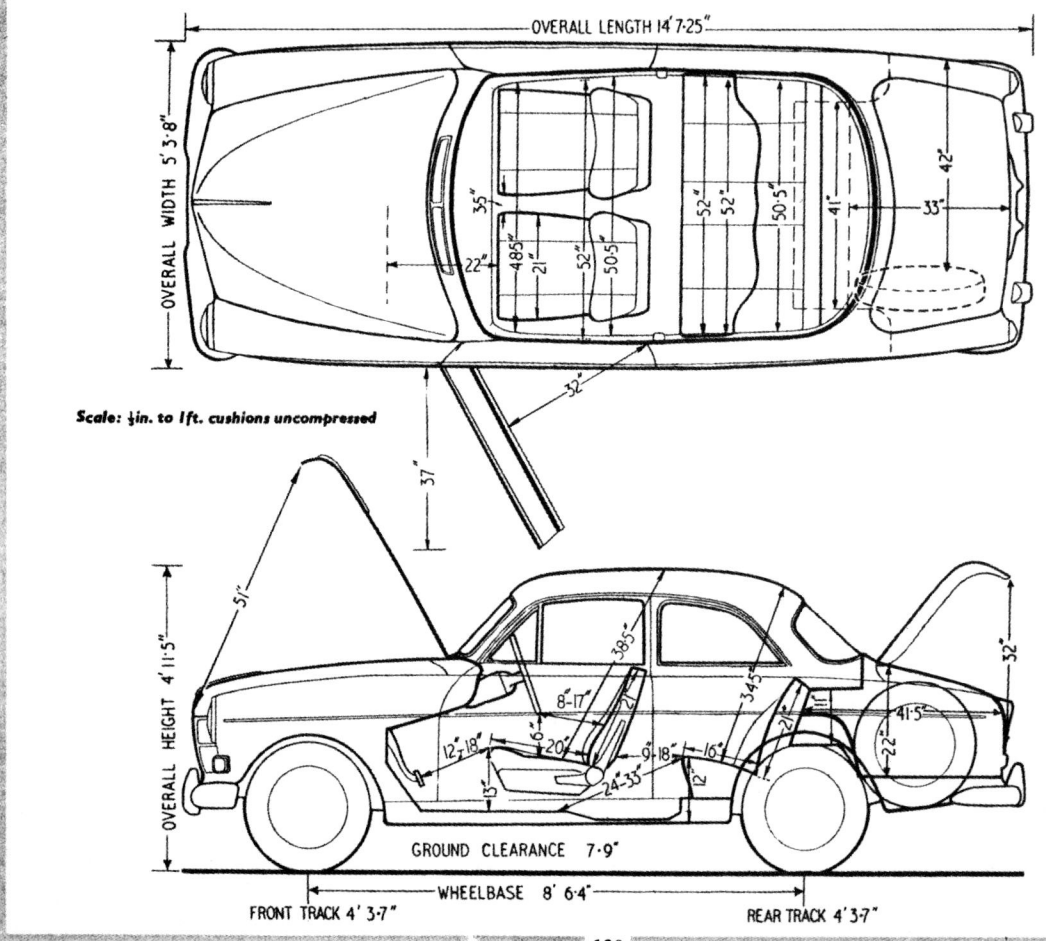

Scale: ½in. to 1ft. cushions uncompressed

A short table will give the essential differences at a glance:

	Saloon	Station Wagon
Length..	175 in.	176·5 in.
Height..	59·25 in.	60·25 in.
Weight	2,405 lb.	2,645 lb.
Lb./h.p.	31·2	34·4
Rear axle	4·1 to 1	4·56 to 1
Tyres ..	6·00 x 15	6·40 x 15

These are also clear pointers to the performance to be expected of the station wagon. With the durable and muscular B18D engine of 1·78 litres, the wagon with its lower gearing (offset slightly by the larger tyres) falls behind the saloon in acceleration through the gears, fuel economy and maximum speed, but it is a fairly close

VOLVO B18-122S STATION WAGON

LOOKING at the Volvo 122S saloon, it is not easy to visualize it as a station wagon. It does not look the type of car adaptable to utility purposes, but paradoxically, it becomes a most successful station wagon with the right treatment.

The Volvo saloon is not particularly spacious internally for a car of its size and weight. It is essentially a comfortable four-seater, taking three adults in the back at a pinch.

As regards seating accommodation, the station wagon version (also known as the 4 + 1, denoting number of doors) is substantially the same, except for a small reduction in rear headroom necessitated by the folding seat.

A CAR ROAD TEST

But behind the rear seat, an abundant load or child space becomes available. Measuring 40 cu. ft., this utility space is hard-pile carpet-trimmed throughout — including the sides. With the rear seat folded down, the utility space jumps to 65 cu. ft., with an overall length of just under 6 ft., which is impressive by medium station wagon standards.

In achieving this, the rear end is built up attractively with good design, its only failing being the long, fixed rear side windows, which look a bit unbalanced and afford no direct ventilation to the rear interior.

There are also some dimensional and mechanical changes in the station wagon. The engine and front suspension remain the same, but the station wagon has additional rubber cushions to serve as overload springs at rear — a wise measure.

The wagon is also somewhat bigger than the saloon overall, and has changed gearing by way of rear axle and tyre size. The gearbox is the same.

match in top gear and third gear flexibility.

The saloon, using 6·00 tyres, is geared for about 18·3 m.p.h. per 1,000 r.p.m. in top gear, against the wagon's 17·1 m.p.h., and the undergeared wagon also has a small gain in frontal area through its inch of greater height.

Again, a short comparative table will tell the story best (using figures from our August, 1962, Road Test of the 122S saloon):

	Saloon	Station Wagon
0—60 ..	14·7	17·5
0—80 ..	28·3	35·8
Speed ..	95·3	90·1
M.p.g. at 30 ..	42·3	39·2
M.p.g. at 60 ..	35·5	30·9
40—60 (3rd) ..	7·5	9·6
40—60 (top) ..	10·9	11·3

Although it has dropped down the scale of performance cars in the past four years because of the number of electrifying newcomers which have appeared, the Volvo saloon remains above-average in general performance and economy, and the station wagon inherits

★

Little girl finds a comfortable perch inside the split tailgate, which gives access to 65 cu. ft. of utility space.

INTERIOR NOISE LEVEL

S.I.L.
ROAD
WIND
MIN.

AVE. dB AT 60
80·2.

DECIBELS

MILES PER HOUR

ACCELERATION

Top
S¼
3rd
2nd
1st

MAXIMUM SPEED
90·1

M P H

TIME IN SECONDS

FUEL CONSUMPTION

MPG AT 60 MPH
30·9

M P G

MILES PER HOUR

ENGINE SPEED

MAXIMUM TORQUE
Top
3rd
2nd
1st

M P H

REVS. PER MINUTE

VOLVO B18-122S station wagon

SUMMARY OF SPECIFICATIONS:

ENGINE:
Cylinders: 4, in-line. Valves: Overhead. Cooling: Water.
Carburettors: Twin SU side draught.
Mounting: Front. Main bearings: 5.
Bore and Stroke: 84·14 x 80·0 mm. (3·313 x 3·15 in.).
Cubic capacity: 1·78 litres (109 cu. in.).
Compression ratio: 8·5 to 1. Fuel rating: 93-octane.
Maximum horse-power: 90 b.h.p., S.A.E. at 5,000 r.p.m. (77 net).
Maximum torque: 105 lb./ft. at 3,500 r.p.m.
Electrical system: 12-volt.

BRAKES:
Hydraulic, discs front, drums rear, servo-assisted.

SUSPENSION:
Front: Independent, coil springs and stabilizer. Rear: Rigid axle, torque rods and coil springs, with auxiliary rubber springs on station wagon.

TRANSMISSION:
4 Forward speeds, synchromesh on all. Column shift.

STEERING:
Type: Cam and roller. Turning circle 34·4 ft.

PRICES:
Coast: R2,700. Reef: R2,700. Rhod.: £1,340.

SUPPLIED FOR TEST BY: Farbers of Dock Road, Cape Town.

PERFORMANCE

INTERIOR NOISE LEVELS:
Idling: 49·5 dBA.

At speeds:	30 m.p.h.	45 m.p.h.	60 m.p.h.	75 m.p.h.	Full throttle
Minimum:	66·5	70·0	74·5	81·0	88·0
Open window:	68·0	73·5	80·0	86·0	94·0
Rough road:	76·0	80·5	86·0	See graph	

Ave. dB at 60 m.p.h.: 80·2.
Hooter at 20 ft.: 104·2. At 100 ft.: 94·4.
(Measured in decibels A. "Minimum" is with windows closed, car running at steady speed in top gear. Speech interference level inside a car is assessed at 85 decibels.)

ROAD PERFORMANCE DATA:

ACCELERATION THROUGH GEARS:

0—30: 5·1	0—60: 17·5	Quarter mile: 20·6
0—40: 8·2	0—70: 26·2	
0—50: 13·2	0—80: 35·8	

MAXIMUM SPEED: 90·1 m.p.h.

ACCELERATION IN 3RD GEAR:

20—40: 7·1	40—60: 9·6
30—50: 7·4	50—70: 14·4

ACCELERATION IN TOP GEAR:

20—40: 10·1	40—60: 11·3	60—80: 17·1
30—50: 10·1	50—70: 14·3	

(Measured in seconds, all figures average of two-way runs on a level road, with car carrying test crew of two and their equipment.)

FUEL CONSUMPTION:
At steady speeds in top gear:

30 m.p.h.: 39·2	75 m.p.h.: 24·0
45 m.p.h.: 35·2	Full throttle: 18·0
60 m.p.h.: 30·9	

(Measured in miles per Imperial gallon, at true speeds, averaging both ways on a level roadway.)
Fuel tank capacity: 10·0 gal.
Approximate cruising range: 210 miles at 80 m.p.h.

GENERAL PERFORMANCE

RATIOS AND GEARING:

	1st	2nd	3rd	Top
Gear ratios:	3·13	1·99	1·36	Direct
Speeds at peak r.p.m.:	27·4	43·0	63·0	85·5

Maximum gradients: 1 in 3·7; 1 in 5·0; 1 in 7·4; 1 in 10·6
Rear axle ratio: 4·56 to 1.
Road speed in top gear at 1,000 r.p.m.: 17·1 m.p.h.

BRAKING FROM 50 M.P.H.:
10 emergency stops: 3·1, 2·9, 3·1, 2·9, 3·1, 2·9, 3·1, 2·9, 3·1, 3·1. Average of emergency stops: 3·02 sec. Comment: Good.
Handbrake stop: 8·3 sec. Comment: Good.
(Measured in sec., with stops at half-minute intervals on a good bitumen surface).

SPEEDOMETER CALIBRATION:

Indicated:	20	30	40	50	60	70	80	90
True speed:	20	29	38	47	56·5	66	76	86

enough of this all-round ability to be a most satisfying drivers' car.

It has a strong feel, and the twin-carburettor engine is as willing as they come, going up to 5,500 r.p.m. if required.

It does not pay to over-rev., however, as we found in the 50–70 acceleration. The third gear time (using 5,500 r.p.m.) is fractionally slower than the top-gear time, in which the rev. range is from 3,000 to 4,100. So it pays in both time and economy to get into top at the 5,000 r.p.m. peak, and not to flog the engine.

A particularly strong feature in the Volvo station wagon is that it will cruise effortlessly at 80–85 m.p.h. returning 20 m.p.g. or just over. And few people travel that fast, even where the road permits it.

Like the saloon, the station wagon has rock-like stability at speed, with accurate steering. It also handles well, with a short turning circle and little tendency to tail-whip in cornering. In fact, in our hard-driven 90-degree corner for test purposes (see picture), the wagon tended to develop a mild four-wheel drift on conventional tyres without the slightest cause for anxiety.

With the modern round-shoulder tyres this would probably have been ironed out to pure neutral steer. It is a safe-handling vehicle by any standards.

In the safety department, there is even one aspect where it is clearly one up on the saloon — and that is braking.

The wagon has the same discs at front and drums at rear, but it is given the refinement of servo-assistance to reduce pedal pressure. This, coupled with the greater weight on the rear wheels, gave it an average stopping time of 3·02 sec. from 50 m.p.h. in our crash test, against 3·73 for the saloon. There was no tendency to fade, though the brakes were getting hot at the finish.

A peculiarity here was the gentle treatment needed with the servo brakes. Too much pedal pressure would lock the rear wheels, and too little would not give an optimum stop. We had plenty of chance to practice, but found ourselves getting one or the other alternately almost all the way through.

Sound level tests showed the Volvo station wagon to be very quiet-running, and little addicted to wind noise. Its ventilation system is inadequate for hot South African weather, so open windows are needed. It is fairly prone to road drumming, as the graph shows.

Because of its weight, large wheels and firm suspen-

The B18-122S engine is an efficient and long-wearing unit, with twin SU carburettors.

Volvo handling is safe and sure. Here a hard 90-degree corner is taken cleanly.

sion, it rides well on bad roads, and the dust-proofing proved unbeatable, even at the tailgate. This is the American type, breaking at the horizontal middle with the lower section dropping to extend the floor, and the upper hingeing upwards.

On the latest model, this upper section is counter-balanced to rise automatically, and can be locked in a variety of positions. (This can be done to the earlier station wagon as a bolt-on modification costing about R10, we are told).

The upper half can be push-button-opened from the inside, which is not a particularly good idea when young children are the passengers.

The folding rear seat is well engineered, with a positive lock in both up and down positions — a rare and important feature on station wagons for passenger safety.

The much-publicized new ripple seats proved every bit as comfortable as is claimed. They are form-fitting and pliant without being mushy. We were disappointed in the reclining backrests, which do not recline far enough to give a really comfortable full-reach driving position.

The front seats also do not slide far enough back for tall people, though this is a matter which could be workshop-adjusted to taste by adjusting the floor rails.

We noticed one peculiarity in the station wagon which is not present in the saloon — creaks in the body work on uneven surfaces, or when cornering. The rear side doors seemed to be the source, and it is probably caused by the extra weight high up at rear.

But these are small criticisms of what is basically a pleasant, sound and useful family car. The fully-trimmed utility space is a honey, and the car itself has a fair modicum of luxury in its furnishing comfort and useful extras, including a most effective heater, cigar lighter, and a general tasteful and well-finished air.

It is every bit what one would expect a Volvo to be.

MEASUREMENTS:
Rear capacity: 40 cu. ft., with rear seat folded, 65 cu. ft. Ground clearance: 8·0 in.
Licensing weight: 2,645 lb.
Annual licence cost: R24·00.
Seat widths: Front, 21 in. Rear, 48 in.
Leg space: Front, max. 37·5 in. Rear, min. 25 in.
Headroom: Front, 35·5 in. Rear, 35 in.

TEST CONDITIONS:
Test car's mileages: 2,720. Weight of test crew and equipment: 366 lb. Conditions: At sea level; barometer 30·07; fine and warm. ●

PRICE

List price.................$2875
Price as tested............$3015

ENGINE

No. cylinders & type....4 cyl, ohv
Bore x stroke, in......3.31 x 3.15
Displacement, cc...........1780
  Equivalent cu in............109
Compression ratio..........8.5:1
Bhp @ rpm............90 @ 5000
  Equivalent mph............88.1
Torque @ rpm lb-ft..105 @ 3500
  Equivalent mph............61.7
Carburetors, no. & make.....2-SU
No. barrels & dia........1-1.75
Type fuel required......premium

DRIVE TRAIN

Transmission type: Borg-Warner
  Type 35 automatic (torque con-
  verter with 3-speed planetary
  gearbox).
Gear ratios: 3rd (1.00).....4.10:1
  2nd (1.45)..............5.93:1
  1st (2.32)..............9.51:1
Converter stall ratio.......2.0:1
Differential type..........hypoid
  Ratio..................4.10:1

CHASSIS & SUSPENSION

Frame type.......unit with body
Brake type...........disc/drum
  Swept area, sq in.........339
Tire size................6.00-15
  Make & model....Goodyear G-8
Steering type.....cam & roller
  Turns, lock to lock........3.25
  Turning circle, ft...........34
Front suspension: independent,
  coil springs, tube shocks, sta-
  bilizer bar.
Rear suspension: live axle located
  by trailing arms, torque rods,
  and a Panhard rod; coil springs
  and tube shocks.

ACCOMMODATION

Normal capacity, persons........4
Occasional capacity.............5
Seat width, front/rear..2 x 19/52
Head room, front/rear.....42/36
Seat back adjustment, deg......8
Entrance height, in...........52
Step-over height...........13.5
Door width, front/rear....33/29
Driver comfort rating:
  For driver 69-in. tall........94
  For driver 72-in. tall........94
  For driver 75-in. tall........83
  (85–100, good; 70–85, fair; under
  70, poor)

GENERAL

Curb weight, lb............2570
Test weight................2760
Weight distribution (with driver),
  front/rear, %...........54/46
Wheelbase, in..........102.5
Track, front/rear..........51.7
Overall length, in.........175.0
  Width..................63.75
  Height.................59.25
Frontal area, sq ft.........20.9
Ground clearance, in........6.9
Overhang, front/rear...26.5/44.0
Departure angle (no load), deg..14
Usable trunk space, cu ft......9.2
Fuel tank capacity, gal........12

INSTRUMENTATION

Instruments: 120-mph speedometer, water temp., fuel, trip
  odometer.
Warning lights: ammeter, turn signal, oil pressure, high beam.

MISCELLANEOUS

Body styles available: 2-door and
  4-door sedans, station wagon.

OPTIONS & ACCESSORIES

Included in list price: 3-point front
  seat belts, heater, vinyl upholstery.
At extra cost: automatic transmission, radio, full range of
  accessories.

CALCULATED DATA

Lb/hp (test wt).............30.7
Mph/1000 rpm (3rd gear).....17.6
Engine revs/mi............3410
Piston travel, ft/mi........1785
Rpm @ 2500 ft/min........4760
  Equivalent mph..........84.0
Cu ft/ton mi...............77.2
R&T wear index............60.9

MAINTENANCE

Crankcase capacity, qt.........4
  Change interval, mi.......3000
Oil filter type...........full-flow
  Change interval, mi.......6000
Chassis lube interval, mi.....3000

ROAD TEST RESULTS

ACCELERATION

0–30 mph, sec...............5.1
0–40 mph....................7.7
0–50 mph...................11.1
0–60 mph...................15.8
0–70 mph...................22.3
0–80 mph...................31.7
50–70 mph (2nd & 3rd gears).11.6
Standing ¼-mi, sec.........20.6
  Speed at end, mph.........67

TOP SPEEDS

High gear (5100), mph.......90
  2nd (5100)................67
  1st (5000)................43

GRADE CLIMBING

(Tapley data)

High gear, max gradient, %...10.3
  2nd.....................18.3
  1st.....................26.9
Total drag at 60 mph, lb......111

SPEEDOMETER ERROR

30 mph indicated......actual 27.9
40 mph....................37.6
60 mph....................57.6
80 mph....................79.0

FUEL CONSUMPTION

Normal driving, mpg......20–23
Cruising range, mi.......240–275

ACCELERATION & COASTING

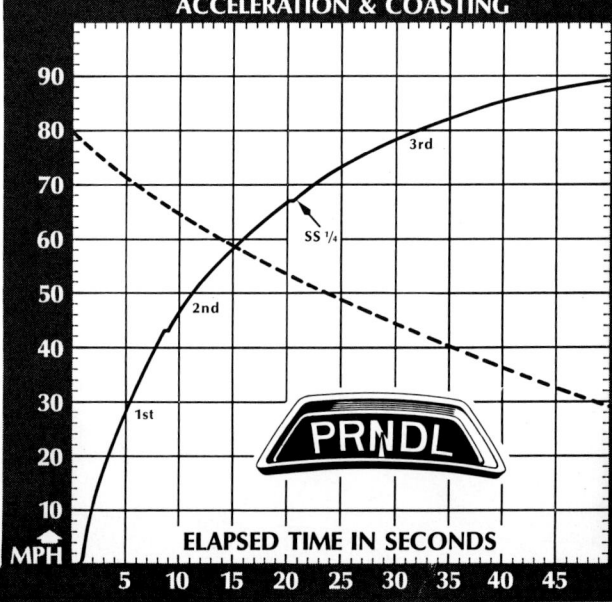

ELAPSED TIME IN SECONDS

VOLVO 122-S AUTOMATIC

The familiar Swedish sedan is now available with automatic transmission

 WHEN THE VOLVO 122 was introduced at the London Auto Show in 1957, it was a thoroughly new and interesting sedan that reflected the sound engineering practices and scrupulous attention to quality control for which the Swedish manufacturer was widely respected. These attributes, plus a pleasant appearance and better-than-average road manners, assured the new model a warm welcome. Since that time it has undergone a number of changes that have kept it mechanically up to date and has enjoyed continuing popularity among drivers to whom a car is more than a styling exercise. The latest change is to offer an automatic transmission as an option and this version of the 122-S is the subject of our test.

As it has been four years since we lasted tested a 122-S (the "S" stands for Sport, incidentally, and distinguishes it from the lower-output version sold in the home market), a brief examination of the basic machine is perhaps indicated. The 122-S is offered as a 2-door or 4-door sedan and as a station wagon. In overall size, with a wheelbase of 102.5 and a length of 175 in., the sedan is about the size we think American compacts should be. It is big enough to be practical in U.S. driving conditions, small enough to be easy to drive and yet not so tiny as to be accidentally stepped on. The body/chassis is a welded-up unit and consequently displays both the vices and virtues of this type of construction. On one hand it is strong, rattle-free and durable, but there is also the inevitable kettle-drum effect which results in considerable noise inside even though extra-thick padding is used on the floor.

The front suspension of the 122-S is conventionally inde-

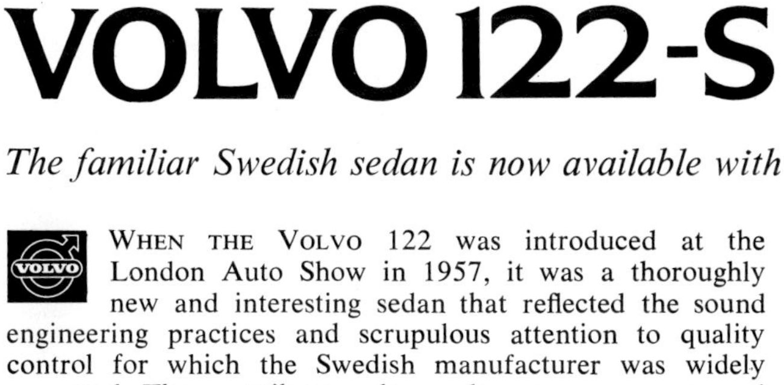

VOLVO 122-S AUTOMATIC

AT A GLANCE...

Price as tested	$3015
Engine	4 cyl. ohv, 1780 cc, 90 bhp
Curb weight, lb	2570
Top speed, mph	90
Acceleration, 0-60 mph, sec	15.8
50-70 mph (2nd and 3rd gears), sec	11.6
Average fuel consumption, mpg	22

VOLVO 122-S AUTOMATIC

pendent, with A-arms, coil springs, tube shocks and an anti-roll bar. There is a live axle at the rear, but a series of arms and links assures that the axle stays where it is supposed to be and it is consequently far more satisfactory than the average live-axle rear suspension arrangement. It behaves so well, even over rough roads, that it makes you wonder why anyone bothers with independent rear suspension on a front-engine sedan.

Since our last test the engine has been increased in displacement to 1780 cc (from 1586) by enlarging the bore, and there has been an increase in horsepower from 85 at 5500 rpm to 90 at 5000. In design the engine is a completely straightforward 4-cyl ohv with five main bearings and it is carbureted by a pair of 1.75-in. SUs. It is a beefy engine with reserves of ruggedness obviously built in. Other changes in the 122-S include the adoption of the now-popular disc/drum front/rear brake combination and these we found to be fully up to their job.

The driving position is good, the seats are high enough to afford a commanding view of what's going on and are adjustable enough to be comfortable for almost anyone. The steering is quick for a car of this size (3.25 turns lock-to-lock) and its accuracy contributes to the driver's feeling of rapport with the machine.

When the 122-S is driven hard there is considerable body lean and a pronounced understeer, but once the driver has become accustomed to these characteristics it is an easy car to handle at pretty near its limit.

Other features of the Volvo that we like include the over-the-shoulder-and-across-the-lap seat belts that are standard on all models, the impressive care with which everything is put together, and the heater which is one of the most effective in the business. We also heartily approve of the manufacturer's policy of making a genuinely useful range of accessories available. By this we mean that there is not only the usual assortment of sideview mirrors, floor mats, roof racks and convenience baskets, but also that one can obtain such items as a complete service manual ($15), a tourist kit that includes basic spares ($13.19) and even an emergency gas can that fits into the spare wheel ($7.50). Good practical stuff.

The automatic transmission that is now available in the 122-S is the Borg-Warner Type 35, a torque converter with 3-speed planetary gearbox. This is not the finest type of transmission ever built, in our opinion, but it is available to European manufacturers at a reasonable price ($180 more than the manual gearbox in the 122-S) and is adaptable to such widely different machines as the Sunbeam Alpine and the Jaguar 3.8-S sedan. From the enthusiastic driver's point of view, there's simply too big a gap between the three gears, the shifts are relatively slow and, when this transmission is used with a typically small-displacement, low-torque European engine, there is an annoying lurch and a noticeable loss of steerage way after each shift.

We covered a total of about 3000 mi in the 122-S automatic and were able to drive it in conditions that varied from downtown rush-hour creeping to hours of flatland cruising and hundreds of miles over an assortment of mountain roads. Only in heavy downtown traffic could we see any advantage to having the automatic, where it relieved the necessity of rowing through the gears. In highway cruising, where only high gear is used, the automatic was neither a plus nor a minus, but it demonstrated better than average efficiency as we consistently got 23 mpg in this kind of driving. On mountain roads we found the automatic a damned annoyance as it buzzed back and forth from gear to gear and we wished we had a manual box so we could stick it in third and leave it there mile after mile.

We realize that the manufacturer didn't add the automatic transmission to the option list expecting that the experienced enthusiast would become rapturous over it. The automatic is offered because there is an ever-growing segment of the auto driving public that has never learned to use a manual transmission and isn't going to learn. So the manufacturer sells cars that he would not have been able to sell otherwise. It's good business.

But don't let us give you the impression that we didn't like the 122-S automatic. It's just that we think the prospective shiftless buyer is missing part of the fun and pleasure that the 122-S can be.

We must loudly proclaim, however, that a Volvo 122-S with automatic transmission is far better than no Volvo 122-S at all.

Volvo 122-S Automatic

VOLVO'S FIRST OFFERING on the American market 10 years ago was the P-444 "fastback," which looks like a "Sanforized" version of post-war Fords. In late '59, the 120 series sedans were brought over to compete for more U.S. business, heeding the demand by U.S. buyers for an economical, middle-sized car. The Volvo offered what was demanded, but unfortunately for it, so did a clutch of domestic-built compacts which were making their debut that year. The compacts have since grown up, and with the absence of Detroit-ware from this category, Sweden's Volvo could very well be the answer to the customer who may someday wear a "Romney for President" button.

Our test car was a 2-door model of the 122-S series, but a 4-door sedan and station wagon are offered. By request, our car was equipped with Volvo's optional automatic.

This 3-speed automatic is a Borg-Warner unit, and consists of two main components: a hydraulic torque converter, and a hydraulically operated planetary gearbox with control system. The selector is conventional by most standards, having a P-R-N-D-L arrangement. Low or second gear can be locked in by means of the L position. A passing gear, engaged by pressing the accelerator to the floor, will remain in until about 55 mph but really doesn't do much good after 50 mph as the engine seems to peak out then and torque falls off. Falling into high gear, the engine will start to pull once more, and a top speed of 90 mph is possible.

You won't experience any neck-snapping starts with an automatic-transmissioned Volvo, but you will be treated to smooth starts and almost imperceptible gear changes. The slow starts, a virtue of the automatic when on ice or snow-covered roads, get a little quicker as the engine gets hotter.

The unit shares, with Jaguar and others, a Borg-Warner problem; *i.e.,* don't park uphill, shift to DRIVE and expect to go forward. You must first gently engage REVERSE, and vice versa when parked downhill.

Gas mileage isn't hindered greatly by the automatic transmission, as we averaged 24 mpg throughout our test.

Most of our fun in driving the Volvo can be attributed to first, Swedish know-how in designing for winter conditions, and second, what might be termed as the "world's most adjustable seats." Though the adjustments have to be made manually, each of the front seats can be moved up and back, tilted up or down, and higher or lower. The backrest can be angled to different degrees, and there is even a built-in lumbar (vertebrae) support. These "Relaxacisors" come standard, and optionally, there are headrests and a reclining seat. The covering on the seats is impregnated with a rather smelly substance that keeps you from slipping around.

We feel that the average American driver would be hard pressed to wear this car out, and neither would he sell it because he got tired of it. The car, after a few miles of living with it, turns you into a volunteer salesman. — *Steven Kelly*

FOR CAR SPECIFICATIONS AND PERFORMANCE CHARTS

Cornering at low speeds requires extra muscle due to stiff steering. Turn signals are visible to side as well as front.

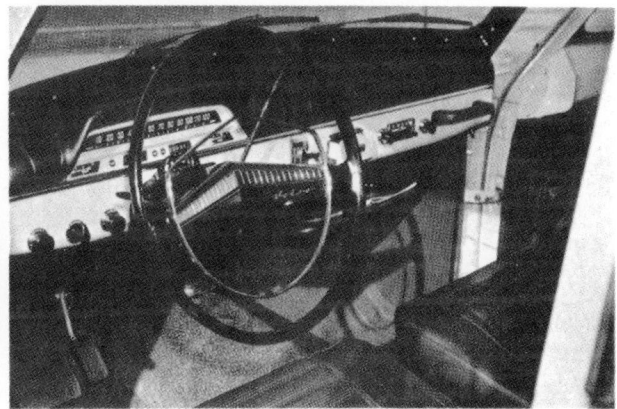
Hard-to-decipher, 9% fast speedometer is only flaw on well laid out dashboard. Wipers clean almost the entire windshield.

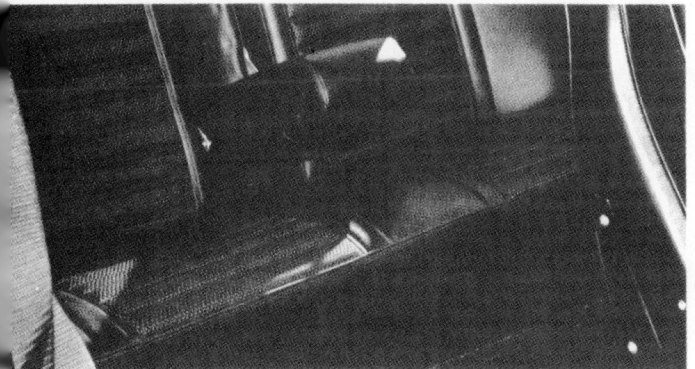
Front seatbacks have latch to prevent them from falling forward on hard stops. Rear seat arm rest is thoughtful touch.

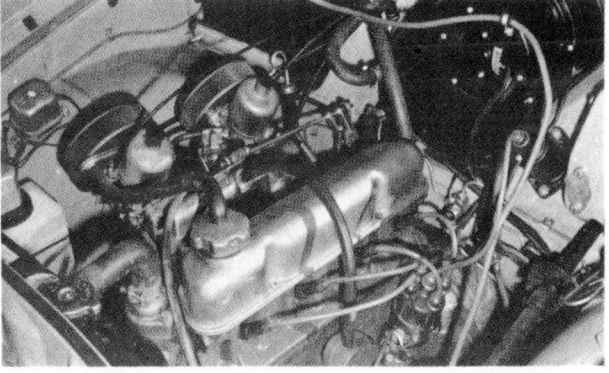
Wide, high-opening hood allows easy access to all parts of 95-hp 4-cyl. motor. Twin SU carbs emit a throaty roar when opened.

122S-B18 saloon (after 120,000 miles)

✱ And reporting on a Monza free-flow exhaust system, which was fitted at 6,500 miles, and remains fully serviceable after 113,500 miles—most of it over Namaqualand roads!

IT is well known by now that a sure recipe for mechanical longevity is high-mileage driving with regular servicing. So it is that reports are legion of commercial travellers, car hire firms and the like attaining tremendous mileages with their vehicles.

One such is the subject of this test: a Volvo B18-122S saloon in which a firm's representative has averaged about 40,000 miles a year for three years, without the engine ever being touched, and with remarkably little replacement necessary.

Perhaps even more news-worthy is the fact that for 113,500 miles of its life this car has had only one exhaust system — a Monza free-flow unit which was fitted in

A CAR ROAD TEST

its young days, and which is still going strong.

The car joined the Cape Town fleet of Glenton and Mitchell, tea and coffee blenders and packers (their products include Joko tea and FG coffee), on May 16, 1962, and was allocated to one of their representatives, Barry Lancellas.

His duties covered a rough route: Namaqualand. Mr. Lancellas told us that 100,000 miles of the car's life had been spent on those rough and dusty roads, which are hard on brake linings, tyres, suspension and filters.

Apart from routine replacements such as plugs, points and tyres, and several sets of dust-chewed brake linings, major replacements were as follows:

48,000 miles: front shock absorbers and springs.
72,000 miles: clutch plate.
86,000 miles: rear shock absorbers and springs.
98,000 miles: front shock absorbers and springs.

110,000 miles: steering tie rod, clutch plate, battery.
111,000 miles: front and rear universal joints.

The engine had never been opened and had never given him a moment's trouble, he said.

The car is now to be traded in — on another Volvo, of course — and has been valued at R1,000!

When we heard of this remarkable car's history we investigated further, and by courtesy of Glenton and Mitchell we were able to borrow it for a formal test.

The car itself, only three years old, looked like new. Apart from chips in the bonnet and fender paintwork from Namaqualand stones it was flawless. Internally, the seats were unbroken and comfortable, and mechanically the car was completely sound. We used peak revs and plenty of verve in performance tests without any protest.

All the instruments were working, the only flaw here being that the fuel gauge had developed a short circuit and the needle flickered badly with the tank full. The speedometer remained reasonably accurate, and even the cigar lighter was fully operational.

The car had lost some performance, naturally. After burning more than 4,000 gallons of high-octane in three

The Volvo "Monza" system after well over 100,000 road miles, photographed alongside a new one in the Powerparts showroom.

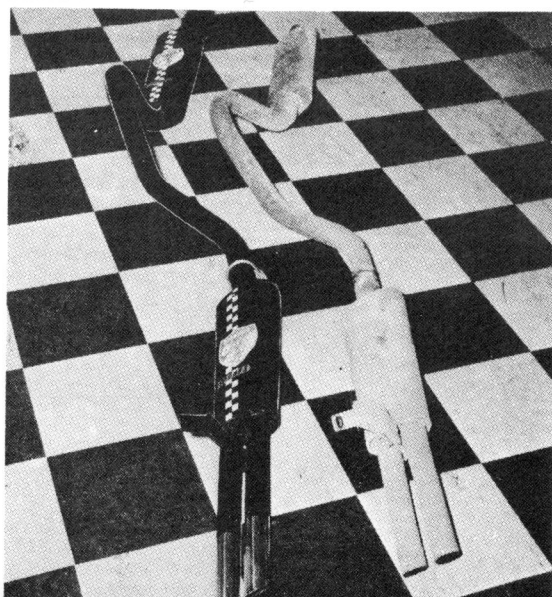

years it must have accumulated some carbon and lost some compression, but in spite of this it remained reasonably close to new-car performance right through — even to fuel economy (with the original carburettor jets).

Much of this excellent showing could be attributed to the long-lived Monza exhaust system, which was still functioning completely satisfactorily, and would have given the car enhanced performance and economy right through.

Internally and externally the car's noise levels were excellent, with sound remaining below speech interference level at full speed.

The exhaust system was scarred here and there from stones thrown up by the wheels, but it was still intact right through and retained a solid sound when tapped. The glass-wool packing showed no signs of failure, and on scraping the twin tailpipes, traces of the original chrome finish could still be seen.

As a matter of interest, this particular Monza exhaust was the first Volvo 122S one supplied by Powerparts, of Cape Town, when they started manufacturing these free-flow systems in 1962.

We join Mr. Lancellas and his colleagues in saluting a tough car and a durable South African accessory, both of which have weathered well over 100,000 miles of testing road conditions with striking success. ●

VOLVO 122S-B18 saloon

SPECIFICATIONS

MAKE AND MODEL: 1962 Volvo B18-122S saloon.

MILEAGE COVERED: 120,000 miles.

MODIFICATION: "Monza" free-flow exhaust system fitted at 6,500 miles.

OTHER SPECIFICATIONS: Standard.

PROVIDED TEST CAR: Glenton and Mitchell (Pty.) Ltd., Paarden Eiland, Cape.

PERFORMANCE

The accompanying tables and graphs compare the performance of the Volvo at 120,000 miles (solid lines) with the performance of the CAR Road Test of the Volvo saloon (August, 1962, issue, broken lines).

	August, 1962 Road Test	Volvo after 120,000 m.
ACCELERATION FROM REST (sec.):		
0—30:	4·0	4·9
0—40:	6·9	7·8
0—50:	10·5	11·6
0—60:	14·7	16·6
0—70:	20·5	24·5
0—80:	28·3	36·7
Standing ¼ mile	19·25	20·6
MAXIMUM SPEED:	95·3	90·5
ACCELERATION IN 2nd GEAR:		
20—40	6·9	7·9
30—50	6·8	7·9
40—60	7·5	9·2
ACCELERATION IN TOP GEAR:		
20—40	9·4	11·7
30—50	9·3	10·3
40—60	10·9	12·6
50—70	11·9	14·7
60—80	15·2	20·4
FUEL CONSUMPTION:		
30 m.p.h.	42·3	37·6
60 m.p.h.	35·5	32·9
Full throttle	15·8	13·9

MAXIMUM GRADIENTS IN GEARS:
1st: 1 in 4·0. 2nd: 1 in 5·1. 3rd: 1 in 7·6. Top: 1 in 11·8

MINIMUM INTERIOR NOISE LEVELS (dBA):
Idle: 51·5   60 m.p.h.:   74·0
30 m.p.h.: 66·0   75 m.p.h.:   76·5
45 m.p.h.: 71·0   Full throttle: 81·0

EXTERIOR NOISE LEVELS (at 40 ft.):
40 m.p.h. in 2nd gear: 85·8 dB. 90 m.p.h. in top: 88·3 dB.

SPEEDOMETER CORRECTIONS:

Indicated:	20	30	40	50	60	70	80	90	100
True Speed:	17	26	35	44	54	64	74	84	94

TEST CONDITIONS: At sea level; barometer 30·25; weight of test crew and equipment 382 lb.

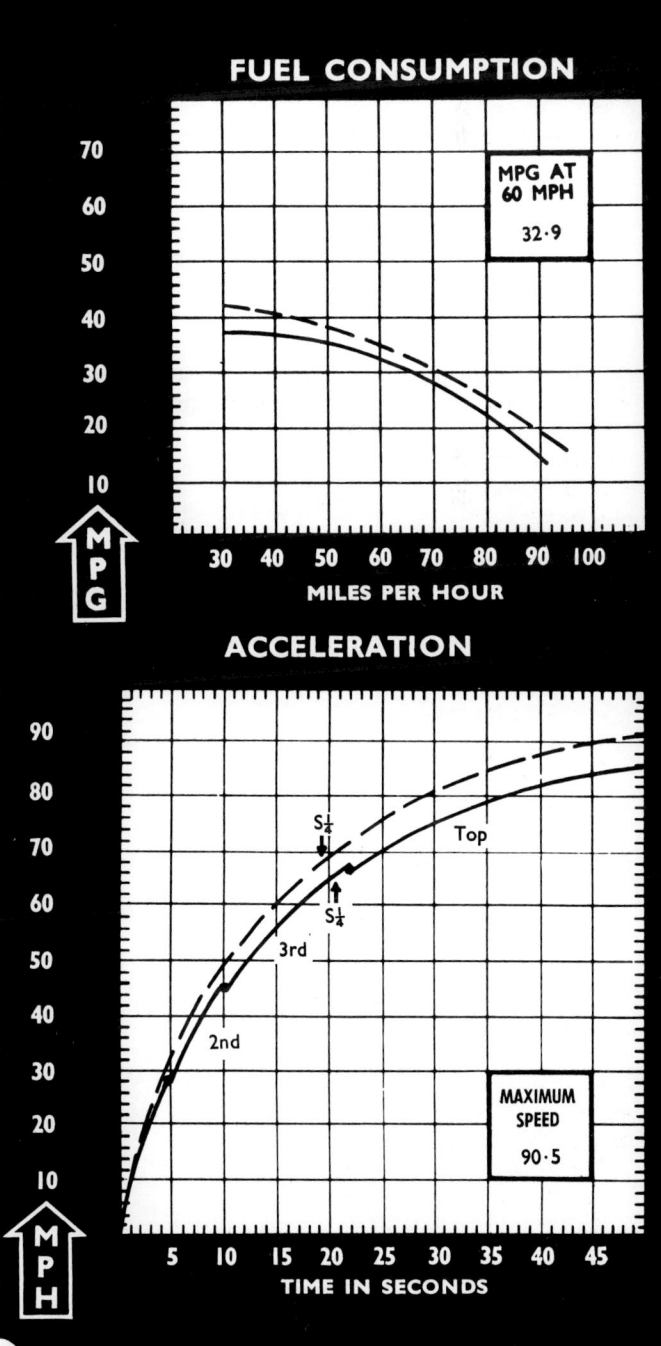

FUEL CONSUMPTION

MPG AT 60 MPH

32·9

ACCELERATION

MAXIMUM SPEED

90·5

STORY 122 S AUTOMATIC

BY JAMES W. GRAY

▶ Over the years Volvo has gained a reputation for building good, strong, reliable automobiles with definite sporting characteristics. Volvos have been hailed as family sports cars because of their sports car-like handling and performance. When the opportunity arose for us to do a driver's evaluation on the 122-S Automatic we expected the performance to be quite different from the stick shift Volvos we had driven in the past.

Fitting an automatic transmission to an under two litre sedan (and even some over two litre cars) usually results in a drastic reduction in performance, gas mileage and top speed. A goodly amount of power is used just to spin the fluid in the transmission. While the above applies to the 1.8 litre Volvo 122-S we were pleasantly surprised to find that there is ample power available and that performance losses were kept to a minimum.

The Borg-Warner automatic transmission mated to the Volvo type B 18 D engine is a good combination. Borg-Warner enjoys world acclaim

VOLVO 122S AUTOMATIC
(Four Door Sedan)

PERFORMANCE

Acceleration 0-60 m.p.h. (aprox.)	15.0 secs.
Maximum Speed (approx.)	95 m.p.h.
Gas Consumption (approx.)	25-30 m.p.g.

ENGINE & DRIVE TRAIN

Maximum Output (at 5000 r.p.m.)	90 b.h.p.
Maximum Torque (at 3500 r.p.m.)	105 lb. ft. SAE
Compression	8.5:1
Capacity	1,780cc/109 cu. in.
Number of Cylinders	4
Bore & Stroke	3.313″ x 3.15″
Carburetors	Twin SU Horizontal
Type of Valves	Overhead
Crankshaft (main bearings)	5
Cooling System	8 qts.
Engine Location	Front
Driving Wheels	Rear
Transmission	Automatic Borg-Warner type 35
Gear Ratios: 1st Speed	2.39:1
2nd Speed	1.45:1
3rd Speed	1:1
Reverse	2.09:1
9½″ hydrokinetic torque converter ratio	2:1 to 1:1

BODY

Chassis Construction	Unit
Length	175″
Width	63¾″
Height	59¼″
Wheelbase	102½″

SAFETY BELTS

Front — 2, standard equipment	3 point belt
Rear — factory installed anchorages	

GENERAL SPECIFICATIONS

Curb Weight (approx.)	2,390 lbs.
Gas Tank Capacity	12 gals.
Electrical System	12V
Tire Size	6.00 x 15
Steering Turns	3¼
Turning Circle Diameter	31′ 6″
Brakes — front	Disc
Brake Friction Area (rear wheels)	32.5 sq. in.

SUGGESTED PRICE

East Coast POE	$2,855

as the manufacturer of strong, reliable gearboxes, both standard shift and automatic. The transmission selector lever is mounted on the steering column and will immediately arrest the attention of standard shift Volvo enthusiasts, used to the long stick putting from the floor. The selector uses the well known P-R-N-D-L layout.

The B 18 D type engine is the same 90 hp, 1800 cc unit used in the regular 122-S and 544 Volvos. This five main bearing, four cylinder engine with its fine service and competition record is reputed to be next to unbreakable. Some of the sporting zip was lacking due to the automatic shifting of gears, we found. This can be overcome by shifting the transmission manually, but then that defeats the purpose of buying an automatic in the first place. Other basic characteristics remain. Steering is precise and the brakes (discs front and drum rear) are excellent.

In this day and age of full independent suspensions, the 122-S uses an independent front end with stabilizer and a rigid axle rear suspension, both utilizing coil springs. Rubber mounted torque arms locate the rear axle fore and aft. With this set up the Volvo proved to be one of the best handling sedans we've driven

in some time (better than some sports cars) and we can see why they haven't changed it.

No matter what comments they made about other aspects of our test car, everyone that, rode in, sat in, or looked into it, had something good to say about the new interior. The seating is superb. The front seats can be adjusted individually fore and aft and for seatback angle to fit drivers of all shapes and sizes. They look and feel as though they might have been designed for the space program. They support the body in all the right places especially the thighs which seem to be forgotten by most seat designers. A large fold-down armrest divides the rear seat and in the down position transforms this from a five to a four passenger sedan. The center portion of each seat is perforated to promote ventilation and prevent the body from sticking to the seat. Detroit could learn some lessons in seating here. In fact why not copy it lock, stock and barrel? The trunk is quite roomy and is easily loaded over the low lip. Accessibility of the vertically mounted spare will be appreciated when the trunk is fully loaded.

The controls on the instrument panel are well placed and easily reached. We would like to see the

ABOVE: No unnecessary do-dads here, only the functional essentials. Grille inserts are new. BELOW: The 122S retains its familiar "no nonsense" automobile look. Detailing has been cleaned up. The new wheels with brake ventilating slots and simple hub caps are very attractive.

long "ruler" type speedometer replaced with a circular unit such as the one on the 1800 S, and the idiot lights (generator and oil pressure) replaced with gauges like those now used for temperature and fuel. The crash padded dash, covered in a black glare-reducing material, and seat belts consisting of lap and chest straps with a common hook up, are only part of the safety features to come out of a co-op study between Volvo and Swedish safety authorities.

The exterior is familiar, having been around for several years. There are no unnecessary do-dads on the car, only the functional essentials. Detailing has been cleaned up with new grille inserts and "122-S" emblems on the front fenders. The new wheels incorporating large brake ventilating slots and simple hubcaps are very attractive. What with the wheel "craze", we're going through, it won't be surprising if these wheels show up on some other makes of cars.

The entire package reflects quality of materials and craftsmanship during assembly. The 122-S Automatic will more than prove suitable to the person choosing not to shift gears. ●

The new seats are well designed and most comfortable. The selector lever position on the 122S Automatic will arrest the attention of 544 and 122S stick shift enthusiasts immediately. The transmission, a reliable unit, is by Borg-Warner.

The test car, with 108 b.h.p.

VOLVO 122S-B18 station wagon MODIFIED

* Lawson's, BG and Speedparts join in giving the Volvo wagon plenty of extra steam for family cruising.

THE Volvo, with 90 b.h.p. in B18 form up to 1965, and getting 5 extra horses for 1966, has always been regarded as a rugged performer with plenty of solid capability.

Not all owners—particularly outside the Reef—realise that it can be given considerably more performance easily and at low cost, by applying P1800 specifications.

At Lawson's Corner, in Braamfontein, Johannesburg, where a stable of thoroughbreds is prepared each year for saloon car and endurance races (remember the Nine-Hour?) the P1800 conversion is offered in kit form, with the options of 100 or 108 b.h.p. SAE.

Many Transvaal owners have taken advantage of these kits, and one of them—Hannes Ganswyk, of

Alberton—broke a coastal holiday to give us the pleasure of testing his converted Volvo station wagon. Hannes ("I'm no racing driver") told us he found that the wagon, geared down for its greater weight, did not have long enough legs on those long family holiday trips, so he took it to Lawson's for the approved treatment.

P1800 CAMSHAFT

His is the full "108" conversion, which uses the P1800 camshaft, with compression ratio raised to 10.0 to 1 (from 8.5) and carburettors re-tuned for a warmer breathing mixture. At the same time, the vacuum advance is disconnected.

As an option, an oil cooler is fitted, and the total cost, with labour, came to a modest R120.

To go a stage further, he also fitted a Basil Green extractor exhaust manifold, teamed with a TruFlo exhaust silencer (by Speedparts) to give matching exhaust side improvement.

FLEXIBILITY UP

As the test figures show, these modifications really do the trick.

Most modifications are internal, but the wagon's engine shows the paper element cleaners and BG exhaust manifold. Major components of the "108" kit are high-lift camshaft and raised c.r.

They do no injury to the dignity of a quality car, yet they up the essential charateristics (flexibility, acceleration and maximum speed) by appreciable margins.

The modified wagon, with its lower gearing than the saloon, gains 8 m.p.h. in maximum and cruising speed, gets away from a standing start in times the 1966 saloon might envy—and with a small overall gain in fuel economy!

But most striking of all—and most useful to the family driver—is the remarkable improvement in top gear flexibility. There is no roughness low down, so that the wagon pulled away from 20 m.p.h. quite firmly, and the acceleration figures in top gear show a most unusual even pattern (20-40 in 11 sec., and 60-80 in 13.7 sec.) which speaks volumes for the torque improvement.

LITTLE EXTRA NOISE

Noise levels inside the car are up by a small margin, largely on the exhaust side at low and medium speed, and with the induction side coming in at high revs, but remain at reasonable levels right through. In 80 m.p.h. cruising, for instance, the level inside the car does not exceed the speech interference barrier.

The Volvo wagon, with Lawson's P1800 kit, pulls, climbs, cruises and tows better than before, by a good margin. This same kit, on the lighter and better-geared saloon, should show even more dramatic results.

In normal cruising, even at the coast, Hannes runs the car on standard premium fuel—93 octane —without distress.

RACING FUEL MIX

For this test, where full use was to be made of the throttle right alongside the Atlantic breakers— we judged it fairer to up the octanes a bit, so a mix of premium and racing fuel to give 97 octane was used. The car was courteously prepared and checked over for the test by Farber's, of Dock Road, Cape Town colleagues of Lawson's.

The car itself had done 15,000 miles at the time, the last 11,000 of them in modified form.

This is a good-value kit which achieves most satisfactory results, and represents quite permissible tuning of the very robust and willing Volvo motor.

As well he might be, the owner is proud of—and pleased with—his capable cruising wagon. ●

PERFORMANCE

The accompanying tables and graphs compare the performance of the modified Volvo station wagon (solid lines) with that of the standard station wagon (CAR Road Test, July, 1965, broken lines).

	Standard	Modified
ACCELERATION FROM REST:		
0–30	5·1	4·2
0–40	8·2	6·5
0–50	13·2	9·1
0–60	17·5	13·1
0–70	26·2	18·4
0–80	35·8	26·6
¼ Mile	20·6	19·1
ACCELERATION IN THIRD GEAR:		
20–40	7·1	7·0
30–50	7·4	6·8
40–60	9·6	7·8
50–70	14·4	9·0
ACCELERATION IN TOP GEAR:		
20–40	10·1	11·0
30–50	10·1	11·0
40–60	11·3	11·6
50–70	14·3	12·4
60–80	17·1	13·7

(Measured in sec., to true speeds, averaging runs both ways on a level road, test crew of two in car.)

MAXIMUM SPEED:

	Standard	Modified
True speed	90·1	98·4
Speedo reading	96·0	106·0

(Measured to true speed, averaging runs both ways on a level road.)

MIN. INTERIOR NOISE LEVEL:

	Standard	Modified
Idle	—	52·5
30 m.p.h.	66·5	71·0
45 m.p.h.	70·0	75·0
60 m.p.h.	74·5	79·0
75 m.p.h.	81·0	82·0
Max.	88·0	94·5

(Measured in decibels, "A" scale, at true speeds.)

FUEL CONSUMPTION:

	Standard	Modified
30 m.p.h.	39·2	39·5
45 m.p.h.	35·2	35·5
60 m.p.h.	30·9	31·6
75 m.p.h.	24·0	25·6
Max.	18·0	14·6

(Measured in miles per Imp. gallon, at true speeds, averaging runs both ways on a level road.)

MAXIMUM GRADIENTS IN GEARS:

	Standard	Modified
1st gear	1 in 3·7	1 in 3·5
2nd gear	1 in 5·0	1 in 4·9
3rd gear	1 in 7·4	1 in 7·7
Top gear	1 in 10·6	1 in 11·4

TEST CONDITIONS:

Altitude	At sea level
Weather	Hot, windy
Baormeter	30·20
Fuel used	97 Octane

SPECIFICATIONS

CAR TESTED:

Make	Volvo 122S-B18
Model	Station Wagon
Year	1965
Mileage	15,000

Owner: J. C. Ganswyk, Esq., of Alberton, Transvaal.

MODIFICATIONS

Kit	Volvo P1800
Cylinder head	Skimmed
Compression ratio	.10·0 to 1
Carburettors	Needles changed
Camshaft	"108" by Lawson's
Vacuum advance	Disconnected
Oil system	Cooler fitted
Cost, with labour	R120,

Modified by: Lawson's, Braamfontein Johannesburg.

EXTRA MODIFICATIONS

Exhaust manifold	Basil Green
Exhaust silencer	TruFlo
Aircleaners	Volvo paper element

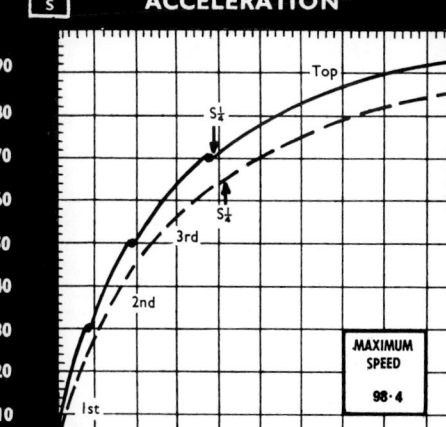

Volvo 121 Automatic

IT was way back in 1959 that we were first able to get our hands on the then new Volvo Amazon. This was before the cars were available in Britain, and we had to cross the North Sea to Holland, where what might have been a slightly better-than-standard car was provided for test. Its performance was outstanding, and even today, seven years later, the twin-carburettor 122S models are still among the faster medium-sized saloons.

However, the sporting image that Volvo built up may have frightened off some potential buyers—buyers who wanted the excellent Swedish engineering without the mythical bother of keeping the twin carburettors in tune. As a result, the two-door 131 was brought out, with a single carburettor version of the 1,778 c.c. engine. In this form the power drops from 98 b.h.p. at 5,400 r.p.m. to 75 b.h.p. at 4,500 r.p.m.; the drop in torque is far less marked—101 lb.ft. at 2,800 r.p.m. compared with 107 lb.ft. at 2,500 r.p.m. The engine is a pillar of convention—a cast-iron block, overhead valves operated by pushrods and a Zenith carburettor. The bore and stroke are slightly over square—84.1mm by 80.0mm.

The latest of the single carburettor models is the automatic, and the car we tested was the four-door version, called a 121. The Swedish industry, being fairly small, picks and chooses from the best that Europe can offer in the way of bought-out equipment,

and in this case Volvo have chosen the Borg-Warner Type 35 transmission. It is a 3-speed unit with torque converter, and it replaces the 4-speed all-synchromesh manual gearbox. On cars with automatic transmission the 4.1 to 1 rear axle ratio is used, as on the normal manual versions without overdrive. When overdrive is fitted, the ratio is lowered to 4.56 to 1.

With its very flat torque curve, the Volvo engine is well suited to take automatic transmission, and as the performance figures show, its speed is not affected as much as one might expect by the churning power losses. With the Type 35 transmission one has almost as much control over the gears as on a manual box, so that one never need be "taken for a ride" by the automatic brain. Maximum speeds in the two indirect gears are 37 and 58 m.p.h. with Drive selected, but they can be held by the Lock-up control to 44 and a very useful 69 m.p.h. The kickdown change from Top to Intermediate operates at below 53 m.p.h., and from Intermediate to Low at below 26 m.p.h. By using the Lock-up control, one can override the kickdown and hold on to Intermediate in traffic or through corners, and use full engine braking from quite high speeds.

With part-open throttle the changes are barely noticeable, the gearbox easing its way to Top smoothly. Full-bore acceleration is a bit more harsh, and the changes can be felt with an appreciable thump.

Volvo have made a very neat job of the conversion from manual to automatic. The selector is on the steering column, with the pointer and scale in a small hooded cowl, where they can be seen easily through the horn ring. The gating of the selector is positive, and there is little chance of inadvertently selecting the wrong gear. At night, the scale—as are the rest of the instruments—is lit by a ghostly green light. On the floor, where the manual gear lever used to go through the rubber mat, a neat "patch" and a "V" are the only remains of what might have been there.

As with all automatic transmission cars, the Volvo could be started only in Neutral or Park. Warm-up is quick, but it is best to keep the choke out just a little to maintain a fast idle until one is sure it will not stall in traffic.

Although the body design is now looking a little out of fashion, the Volvo gives one an impression of very thorough workmanship and immense strength. Individual front seats are fitted, with adjustable lumbar support pads built in; a breathing plastic with a coarse weave design is used on the wearing parts, and a small degree of rake adjustment is provided.

Volvo, of course, were the first manufacturer to fit front seat belts as original equipment. By today's standards, however, the adjustment has become very clumsy and it is difficult to get the belt tight enough. At the

Impressions and figures of the latest saloon with Borg-Warner transmission

Left: Somehow the Volvo looks as tough as it is. A label on the boot lid reads "Automatic" and there are mudflaps for front and back wheels. Reversing lamps are built into the tail lamp clusters and there are big, separate reflectors below

Above: There is no glove locker, but a shelf above the passenger's knees and elastic-topped pockets in each front door make do instead. The seats have very versatile adjustments

rear, the seats are individually shaped, with a drop-down arm-rest between them. In many ways, the Volvo is a very safe car; the facia top is protected by firm crash padding, and the parcel shelf over the front passenger's shins is collapsible. In other ways, it does not match up to current thinking, with projecting door handles, window winders and instrument knobs.

The years of rally experience which Volvo have behind them have produced a car with excellent handling. Goodyear G8 tyres were fitted on the 15in. wheels and in all conditions they gave predictable grip. The majority of Volvos we have driven have been fitted with more expensive radial-ply tyres and we know them to suit the car well. There is natural understeer, and at the limit of adhesion on a dry test track one back wheel lifted just off the road—but one has to be driving very fast for this to happen. The cam and roller steering is well suited to the car, with enough feed back to give the driver a good feel of what the front wheels are doing.

Damping is firm, and the Volvo never pitches or rolls unnecessarily. On sections of rough road one can put the suspension through its paces, and the big car—it weighs just over a ton—can be hurled across loose and treacherous surfaces with the secure feeling that it was born and bred to do just this sort of thing. In its home country, where so many roads are

PERFORMANCE DATA

Figures in brackets are for the Volvo 131 tested in AUTOCAR of 2 April 1965.

Acceleration Times (mean): *Speed range, gear ratios and time in seconds:*

m.p.h.	Top		(3rd)	Inter	(2nd)	Low	(1st)
10—30	(4·1-8·2)	(4·1)	(5·6)	(5·9-11·9)	(8·2)	(9·8-19·6)	(12·8)
	—	—	(7·2)	5·6	(4·9)	4·2	(4·0)
20—40	9·1	(9·7)	(6·7)	6·1	(5·0)	5·4	(7·0)
30—50	10·1	(10·1)	(7·2)	9·9	(5·8)	—	—
40—60	12·4	(11·5)	(9·2)	12·2	—	—	—
50—70	17·0	(14·7)	(13·5)	—	—	—	—
60—80	27·2	(22·9)	—	—	—	—	—

From rest through gears to:

30 m.p.h.	..	.. 5·2 sec (4·8 sec)
40 ,,	..	.. 8·2 ,, (7·6) ,,
50 ,,	..	..12·2 ,, (11·5) ,,
60 ,,	..	..18·5 ,, (17·6) ,,
70 ,,	..	..27·4 ,, (26·1) ,,
80 ,,	..	..44·9 ,, (40·5) ,,

Standing quarter-mile 21·0 sec

(20·7) sec.

Maximum speeds in gears:

		m.p.h.	k.p.h.
Top (mean)	..	87(91)	140(147)
(best)		88(92)	142(148)
(3rd)	..	—(74)	— (119)
Inter	..	69(—)	111(—)
(2nd)	..	—(58)	—(93)
Low	..	44(—)	71(—)
(1st)	..	—(41)	—(66)

Overall fuel consumption for 612 miles:
23·7 m.p.g.; 11·9 litres/100km
(26·5 m.p.g.; 10·7 litres/100km.)

Volvo 121 Automatic

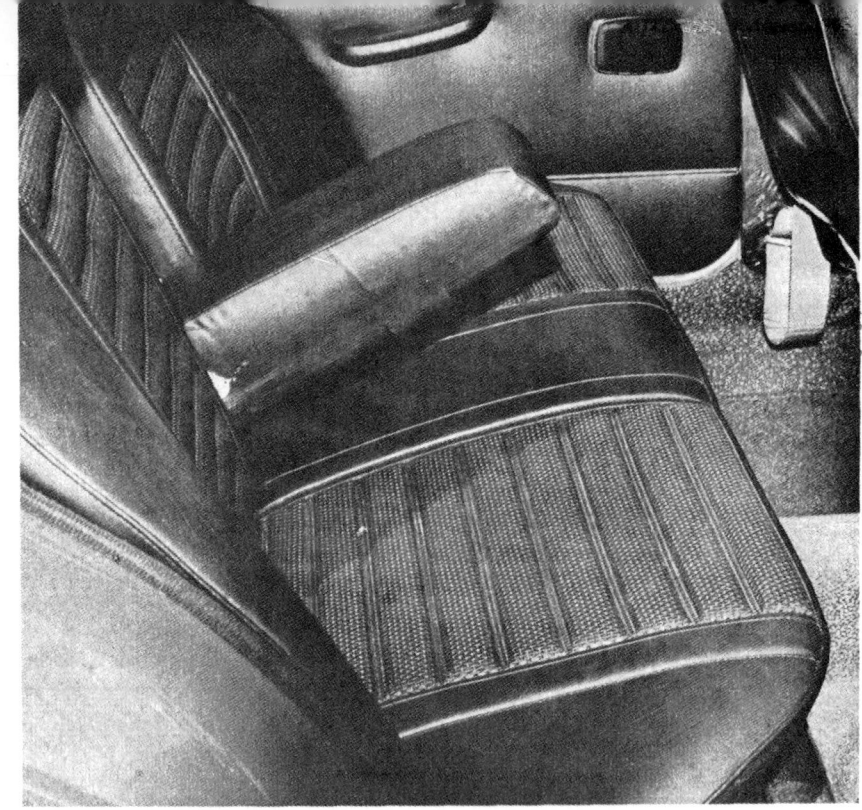

loose-surfaced, mud flaps have become a necessity, and they are fitted behind both front and rear wheels.

A mixed Girling disc front, drum rear braking system is used, without a servo. On the automatic car, the brake pedal is double width, so that either left or right foot can be used for braking. The handbrake, outboard of the driving seat, has its release button protected by a loop, so that it cannot be kicked off accidentally.

The Volvo 121 automatic comes with a very full equipment list—2-speed screen wiper, electric screen washers, a superb heating system with ducts to carry air to the rear, and the in-built engineering quality that all Swedish goods seem to have.

The basic price of the car is £985, and with purchase tax, this is raised to £1,212 8s 11d. The Volvo 121 Automatic is imported by Volvo Concessionaires Ltd., P.O. Box 7, Tower Ramparts, Ipswich, Suffolk. ∎

Back seats are well contoured and the centre armrest folds away flush. Ventilated upholstery front and rear breathes in hot weather and grips well. Below: The balanced lines have not aged too much over the years, and only the small windows indicate how long this Volvo has been in production

The high-waisted body style looks more dated from the inside than from the outside.

Quality and performance

' . . . a reputation for strength and durability . . .'

P EOPLE who are not closely familiar with the Volvo marque will probably be surprised to learn that the present series of saloons was first introduced—as the Volvo Amazon—almost exactly 10 years ago, in 1956. For a car which was not designed ahead of its time but merely to the best standards of the day, and in the fiercely competitive 1½-2 litre market, 10 years is a long time. But numerous rally successes plus quality have maintained sales at a high level and built up a reputation for strength and durability, while some pioneering work on the fitment and design of seatbelts (carried out while Ralph Nader was still at law school) has established a similar reputation for safety. And from the original Amazon (the name still used in Sweden) has sprung a range composed of various combinations of three different body-styles—four-door saloon, two-door coupé and estate car—and two different versions of the 1,780 c.c. engine—68 b.h.p. and 86 b.h.p.

By comparison with recent designs, however, the 132S Volvo that we tested (with the two-door body introduced early this year

and the 86 b.h.p. engine) is beginning to show its age a little. The high-waisted, high-scuttled body with its shallow windows and thick pillars gives a distinctly claustrophobic impression to anyone accustomed to the light and airy interiors of more modern cars, and the driving position, with the seat set too close to a high steering wheel, is of the sort that turns us all into little old men trying to peer through the spokes.

The handling of the Volvo, moreover, is indifferent when compared to one or two of its closest rivals: on corners initial understeer changes to roll oversteer and the live rear axle hops badly if the surface is not smooth. The ride, too, falls below the standards of its competitors, with pitching and the violent transmission of road shocks, while the engine emits a loud intake roar.

But, in compensation, the seats are extremely comfortable and adjustable in a wide variety of ways, the gearbox is pleasant, the car has a long stride and, at £1,101, the price is moderate for the quality and performance.

Performance and economy

When starting off in the mornings a little choke is needed, but more to eliminate hesitation and obtain clean pulling for the first mile

---

PRICE: **£910 plus £191 2s 11d purchase tax equals £1,101 2s 11d.**

Volvo 132S

Performance

Conditions

Weather: Cool and dry.
Temperature 53°–54°F. Barometer 29.6–29.7 in. Hg.
Surface: Dry concrete and tarmacadam.
Fuel 98 octane (R.M.).

Maximum speeds

	m.p.h.
Mean lap speed banked circuit	96.3
Best one-way $\frac{1}{4}$-mile	101.2
3rd gear	80.0
2nd gear	55.0
1st gear	36.0

"Maximile" speed: (Timed quarter mile after 1 mile accelerating from rest)

	m.p.h.
Mean:	93.8
Best	97.8

Acceleration times

m.p.h.	sec.
0-30	4.0
0-40	6.2
0-50	8.9
0-60	12.8
0-70	18.0
0-80	26.0
0-90	39.5
Standing quarter mile	19.4

m.p.h.	Top sec.	3rd sec.
10-30	—	7.3
20-40	10.4	7.2
30-50	10.4	7.2
40-60	10.5	7.2
50-70	11.5	8.8
60-80	14.5	13.3
70-90	22.8	—

Hill climbing

At steady speed		lb./ton
Top	1 in 9.1	(Tapley 245)
3rd	1 in 6.2	(Tapley 358)
2nd	1 in 4.2	(Tapley 515)

Brakes

Pedal pressure, deceleration and equivalent stopping distance from 30 m.p.h.

lb.	g	ft.
25	0.21	143
50	0.45	67
75	0.62	$48\frac{1}{2}$
100	0.77	39
125	0.94	32
Handbrake	0.36	$83\frac{1}{2}$

Fade test

20 stops at $\frac{1}{2}$g deceleration at 1 min. intervals from a speed midway between 30 m.p.h. and maximum speed (=63.2 m.p.h.)

	lb.
Pedal force at beginning	68
Pedal force at 10th stop	82
Pedal force at 20th stop	91

Fuel consumption

Touring (consumption midway between 30 m.p.h. and maximum less 5% allowance for acceleration) . . . . . . 28.8 m.p.g.

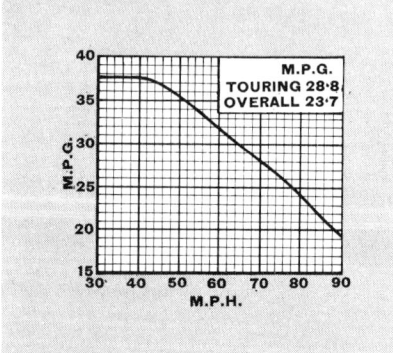

Overall	23.7 m.p.g.
	(=11.9 litres/100 km.)
Total test distance	1,140 miles
Tank capacity (maker's figure)	10 gal.

Steering

Turning circle between kerbs:	ft.
Left	$31\frac{1}{2}$
Right	$29\frac{3}{4}$
Turns of steering wheel from lock to lock	$3\frac{2}{3}$
Steering wheel deflection for 50 ft. diameter circle	1.1 turns

Clutch

Free pedal movement	$=\frac{1}{2}$ in.
Additional movement to disengage clutch completely	$=3\frac{1}{2}$ in.
Maximum pedal load	=43 lb.

Speedometer

Indicated	10	20	30	40	50	60	70	80	90
True	10	$17\frac{1}{2}$	27	36	45	55	64	$73\frac{1}{2}$	$82\frac{1}{2}$

Distance recorder . . . . . . 7% fast

Weight

Kerb weight (unladen with fuel for approximately 50 miles)	20.8 cwt.
Front/rear distribution	54/46
Weight laden as tested	24.6 cwt.

Parkability

Gap needed to clear a 6ft. wide obstruction parked in front

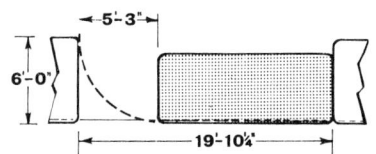

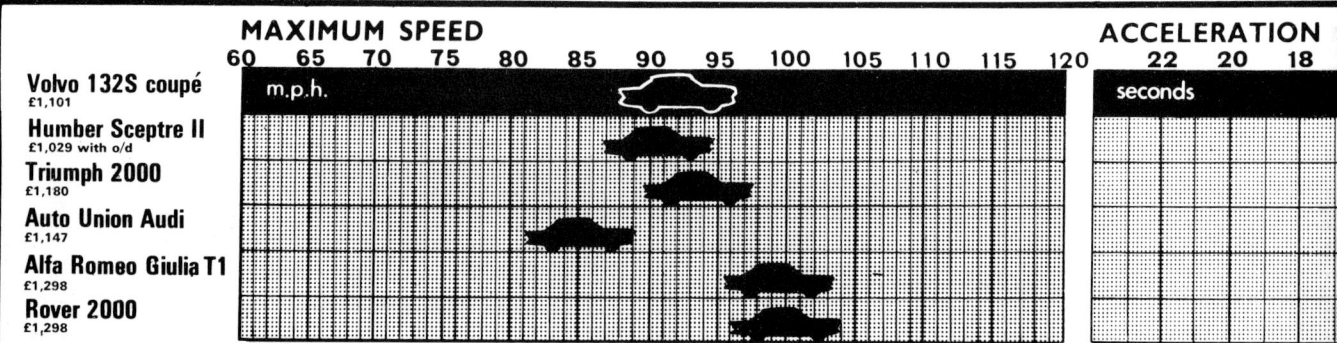

Front seats tilt forward for access to the rear and lock in position when upright.

Front seats have limited fore-and-aft adjustment.

or so than for actual starting. Once the engine is warmed up it shows itself to be lively and willing rather than refined—smoothness is only moderate for a five-bearing four—and there is considerable intake roar when the engine is revving hard. With a power increase from 80 (net) b.h.p. to 86 b.h.p. (introduced last August), and a little less weight, performance is very good and the 132S was considerably faster than the four-door 122S that we tested in 1962. Top speed was 96.3 m.p.h. compared with 94.8 m.p.h., while 60 m.p.h. was reached from a standstill in 12.8 sec., compared to the 14.2 sec. needed by the 122S. Performance of this kind took the effort from long journeys and allowed motorway cruising at 85 m.p.h., or over 90 m.p.h. on the optimistic speedometer, but after a time this was made tiring by the hard rasping noises from under the bonnet. One solution would be the fitment of an overdrive, but in the saloon range of cars this is currently only available in England on the 122S (it is standard on the P1800S sports car). Another— judging by their appearance— might be to redesign the aircleaners to make them more effective as intake silencers.

Despite hard driving fuel consumption remained a creditable 23.7 m.p.g. of premium grade, and most owners will probably get nearer to the touring consumption figure of 28.8 m.p.g.

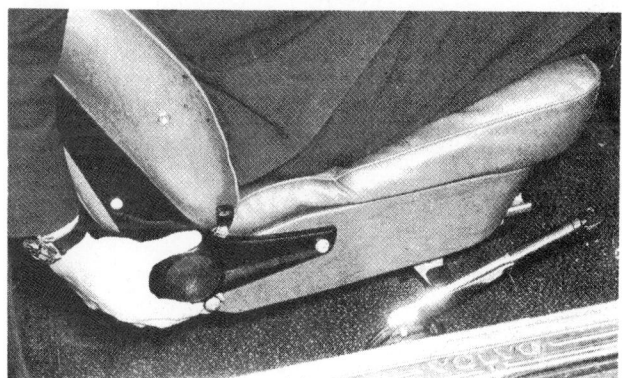

Front seats are extremely comfortable and can be adjusted in a variety of ways in addition to the normal fore-and-aft and rake. One of these—for 'lumbar support'—is accessible by screw-driver through the eyelet hole above the rake adjusting knob

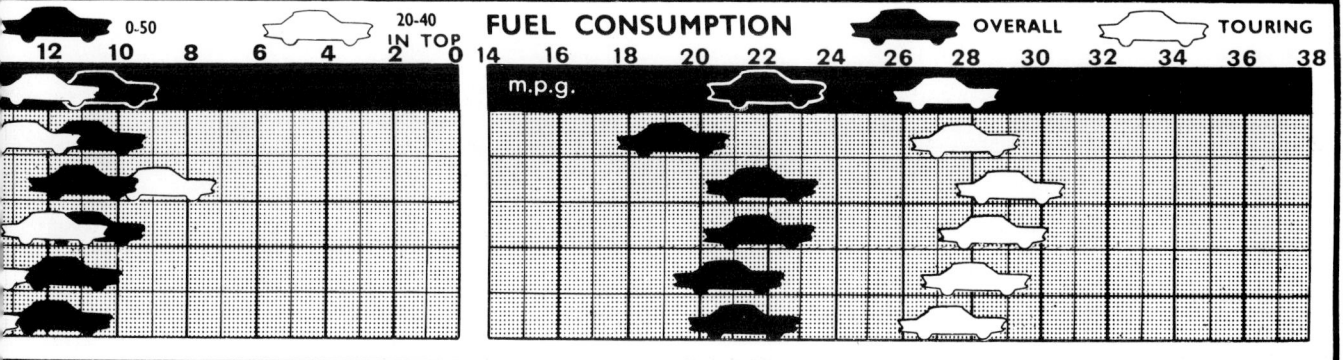

Rear head and legroom is adequate for six-footers with the front seat fully back.

Volvo 132S

Safety check list

1	**Steering assembly**	
	Steering box position	Ahead of axle line
	Steering column collapsible?	Designed to crumple and fitted with fabric coupling for jack-knifing
	Steering wheel boss padded?	No
2	**Instrument panel**	
	Projecting switches, etc.	Large round knobs for lights, wipers, fan and cigareet lighter. Flexible grab handle
	Sharp instrument cowls, etc.?	None
	Effective padding?	On top half of facia
3	**Ejection**	
	Anti-burst door latches?	No
	Child-proof door locks?	No, but only two doors
4	**Windscreen**	Laminated
5	**Door structures**	
	Interior door handles, etc.	Projecting
	Front quarter-light catches	Projecting
6	**Back of front seats**	Frame not padded: seats locked into position
7	**Windscreen pillar**	Unpadded metal frame
8	**Driving mirror**	
	Framed?	Yes
	Collapsible?	Yes
9	**Safety harness**	
	Type (on test car)	3-point Volvo
	Anchorage	Pillar and two floor mountings built in

Texture, colour and design of door trim has a rather dated appearance.

Transmission

One of the most pleasant features of the car is the gearbox, despite the unfashionably long gearlever. The change was light, with effective synchromesh, but occasionally a little notchy; the clutch, although rather heavy, was smooth and free of slip and easy to co-ordinate with the gearchange. Ratios were well-chosen and sporting. A maximum of 36 m.p.h. was possible in first—which is perhaps a little too "sporting" since a restart was only just achieved on a 1-in-3 slope—55 m.p.h. could be attained in second, and the astonishingly high true maximum of 80 m.p.h. ($86\frac{1}{2}$ indicated) in third. Overtaking lorries is *not* a problem in the Volvo.

Handling and brakes

The general handling characteristics of the Volvo can best be summed up by the designation "lurch": there is a good deal of roll on corners, and the transition from a large roll angle in one direction to a large roll angle in the other at every turn in the road soon begins to obtrude on the driver. This is particularly so because the initial understeer which the car exhibits changes to oversteer as the roll builds up, an effect which is especially noticeable in the wet. Despite location by double trailing arms and a Panhard rod, the live rear axle hops badly on moderately bumpy surfaces and wheelspin is easy to achieve in the wet. A heavy foot and a low gear, however, do not generally induce axle tramp.

A good deal of feedback can be felt through the steering which is light but without the positive feeling usually associated with kickback, although it is in no way slack or soggy. Pirelli Cinturato tyres were fitted to our test car, but Firestone or Goodyear tyres are now being fitted to this Volvo model.

Although an acceptable 0.94g maximum deceleration was achieved on braking, a pedal pressure of 125 lb. was needed to attain it. A more reasonable value would be 100 lb. pedal pressure for maximum deceleration, while the best modern systems require about 80 lb. pedal pressure for a 1g stop. In addition, fade was quite marked during our test, with pedal pressure increasing by over 20 lb. The watersplash, however, left the brakes unaffected, and the handbrake held the car on a 1-in-3 gradient.

Spare wheel and irregular shape makes boot stowage difficult.

Comfort and controls

Bumpy surfaces induced strong pitching, and road shocks were violently transmitted into the car, both as noise and as vibration, even on relatively smooth surfaces. These faults were partly redeemed by the comfort of the high-backed seats which, in addition to the conventional fore-and-aft and rake adjustments, could be altered in several other ways. First there is a screw adjustment for "lumbar support", which means varying the size of the bulge in the backrest, then the whole seat can be tilted from the front by another screw adjustment, or bodily raised or lowered by yet another.

Apart from the high scuttle and steering wheel already mentioned, and from the fact that the driver's seat will not move far enough back, the driving position is otherwise good, with the gearlever, pedals and handbrake all well located in relation to each other. As for minor controls, the flasher/indicator stalk and the horn ring are within fingertip reach, while the lights and wiper switches are on the facia. The wiper switch is pulled out one notch for low-speed working, a second notch for high-speed working and a third for wash-and-wipe, with the result that when the knob was pulled out from the low to the high speed position to clear the spray thrown up by a preceding vehicle, it was almost always pulled out too far, thus operating the washer and obscuring the windscreen still further instead of clearing it.

An inevitable consequence of the high-waisted body style is poor visibility of the extremities of the car and of the road close

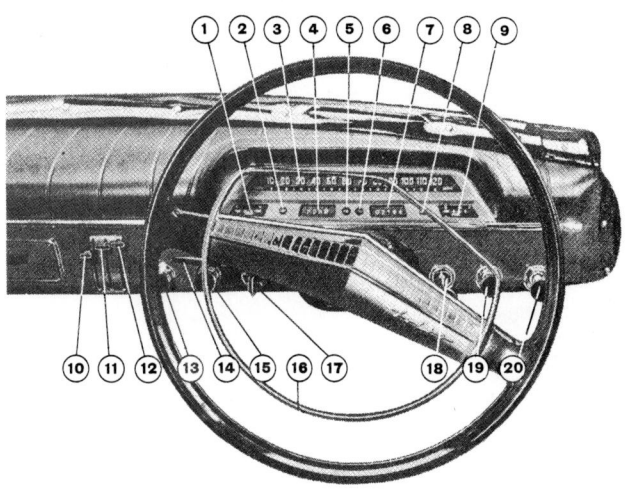

1, temperature gauge. 2, ignition warning light. 3, speedometer. 4, trip mileometer. 5, main beam warning light. 6, indicator warning light. 7, total mileometer. 8, oil pressure warning light. 9, fuel gauge. 10, front heater control. 11, windscreen and rear heater control. 12, heater temperature control. 13, heater fan control. 14, indicator/flasher stalk. 15, cigarette lighter. 16, horn ring. 17, choke. 18, ignition/starter lock. 19, lights switch. 20, wiper/washer switch.

Specification

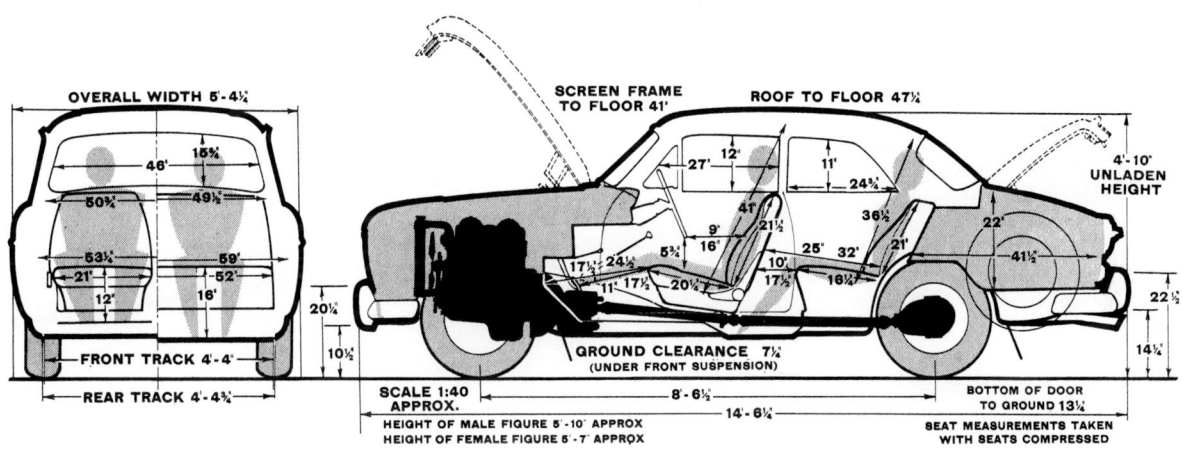

Engine

Cylinders	4
Bore and stroke	84.14 mm. x 80 mm.
Cubic capacity	1,780 c.c.
Valves	pushrod o.h.v.
Compression ratio	8.7:1
Carburetters	Two SU HS6
Fuel pump	AC mechanical
Oil filter	WIX/MANN full flow
Max. power (net)	86 b.h.p. at 5,000 r.p.m.
Max. torque (net)	107 lb. ft. at 3,500 r.p.m.

Transmission

Clutch	Borg and Beck 8½ in. s.d.p.
Top gear (s/m)	1:1
3rd gear (s/m)	1.36:1
2nd gear (s/m)	1.99:1
1st gear (s/m)	3.13:1
Reverse	3.25:1
Final drive	4.1:1
M.p.h. at 1,000 r.p.m. in:—	
Top gear	17.6
3rd gear	13.0
2nd gear	8.9
1st gear	5.6

Chassis

Construction	Unitary

Brakes

Type	Girling—disc front/drum rear, with pressure reducing valve for rear brakes.
Dimensions	10.8 in. dia. discs; 9 in. dia. drums.

Friction areas:

Front		28.7 sq. in. of lining operating on 238.8 sq. in. of disc
Rear		67.1 sq. in. of lining operating on 113 sq. in. of drum

Suspension and steering

Front	Independent by coil springs and unequal length wishbones
Rear	Live axle located by double trailing arms and a Panhard rod, and suspended on coil springs

Shock absorbers:

Front	} AC Delco telescopic
Rear	}
Steering gear	Gemmer cam and roller
Tyres	165-15 Pirelli Cinturato
Rim size	4J

Coachwork and equipment

Starting handle	No
Jack	Pillar screw type
Jacking points	Two each side under sills
Battery	12-volt negative earth, 60 amp. hour capacity
Number of electrical fuses	4
Indicators	Self-cancelling flashers
Screen wipers	Two-speed electric self-parking
Screen washers	Electric
Sun visors	2
Locks:	
With ignition key	Ignition only
With other key	Doors and boot
Interior heater	Fresh air

Extras	Radio
Upholstery	Vinyl
Floor covering	Rubber matting
Alternative body styles	Four-door saloon and estate car

Maintenance

Sump	7 pints S.A.E. 10W-30
Gearbox	1.3 pints S.A.E. 30
Rear axle	2.3 pints S.A.E. 90
Steering gear	Hypoid
Cooling system	14 pints (drain taps two)
Chassis lubrication	None
Minimum service interval	3,000 miles
Ignition timing	21-23° b.t.d.c. at 1,500 r.p.m.
Contact breaker gap	0.016-0.020 in.
Sparking plug gap	0.028-0.031 in.
Sparking plug type	Bosch W175T1
Tappet clearances (hot or cold)	Inlet 0.016-0.018 in., Exhaust 0.016-0.018 in.
Valve timing:	
Inlet opens	10° b.t.d.c.
Inlet closes	40° a.b.d.c.
Exhaust opens	38° b.b.d.c.
Exhaust closes	2° a.t.d.c.
Front wheel toe-in	0-0.16 in.
Camber angle	0 to +½°
Castor angle	0 to +1°
Kingpin inclination	8°
Tyre pressures:	
Front	20 p.s.i. light or heavy load)
Rear	24 p.s.i. (light load); 30 p.s.i. (heavy load)

Substantial-looking boot has little room inside.

Volvo 132S

to it, but long range visibility is fair except that the front and rear pillars are rather thick. The anti-dazzle darkening on the mirror is slightly overdone and makes long distance daylight rear vision a little trying when there is rain on the back window. Bright sunlight produced dangerous reflections from the shiny and garish centre piece to the horn ring. A tremendous blaze of light, both on main beam and when dipped, is thrown by the headlamps.

There was no occasion to use the heater during our test, but it produced a furnace-like blast of hot air when tried out, and if it can cope with a Swedish winter it will be able to manage anything British weather can hand out. It is controlled by three levers instead of the more usual two: one for temperature, one for the front seat passengers and one for the windscreen and rear seat passengers. When set to give ventilation during the warm weather of our test the system provided an adequate cooling breeze.

Wind and road noises are low and there is not much transmission noise except for a subdued whine from the rear axle. At moderate speeds engine noise is also low, but above about 70 m.p.h. in top or when accelerating hard, the intake roar becomes obtrusive.

Fittings and furniture

The dashboard is furnished with fuel and temperature gauges, trip and total mileometers, warning lights for main beam, indicators, oil pressure and charge, and a strip speedometer with a pointed end to its ribbon which introduced doubts into the minds of several of our test staff as to which part was supposed to be indicating the speed.

An ashtray on the facia and two more on the rear seat armrests provide for smokers, while odds and ends can be stowed in a facia parcel shelf, on another shelf at the rear, or in the small pockets in the front doors. Due to an irregular shape plus the presence of the spare wheel, boot capacity is poor—only 5.7 cu. ft. of our test boxes could be squeezed into it.

Most of our test staff found the safety belts (front belts supplied as standard) hard to adjust and use. This was partly because the ingenious pistol grip securing clamp needed some sort of "butt" to get hold of. Another standard safety feature of the car is the arch-shaped strengthening member in the roof between the door pillars, designed to give roll-over protection.

Servicing and accessibility

Servicing requirements, at 3,000 mile intervals, are simple and consist mainly of an engine oil change with no need for chassis lubrication. Accessibility is exceptionally good, with the brake and clutch reservoirs, carburetters, coil, distributor dipstick and radiator and oil filler caps all easy to get at. Jacking presented no difficulties once the undersealant had been cleaned from the jacking sockets under the sills.

M

Maintenance chart

A Engine. Every 3,000 miles—change oil. Every 6,000 miles—clean oil filler cap, replace oil filter, clean fuel filter, check valve clearance, fanbelt tension, sparking plugs, and compression. Check distributor, ignition timing and carburetter. Every 12,500 miles—change air cleaner element and sparking plugs.

B Transmission. Every 3,000 miles—check clutch fluid reservoir, check oil level in gearbox and back axle Every 6,000 miles—check clutch yoke travel. Every 12,500 miles—check propeller shaft. Every 25,000 miles—change gearbox and rear axle oil.

C Brakes. Every 3,000 miles—check brake reservoir. Every 6,000 miles—check pads, linings and adjustment.

D Steering and suspension. Every 6,000 miles—check front wheel alignment, ball joints and steering rods.

E Electrical. Every 6,000 miles—check battery charge and level. Check all bulbs and headlamp alignment.

1, brake fluid reservoir. 2, clutch fluid reservoir. 3, carburetters. 4, oil filler cap. 5, radiator filler cap. 6, coil. 7, distributor. 8, dipstick.

MAKE Volvo: MODEL 132S coupe: MAKERS AB Volvo Gothenburg, Sweden: CONCESSIONAIRES Volvo Concessionaires Ltd., 28 Albemarle St., London, W.1.

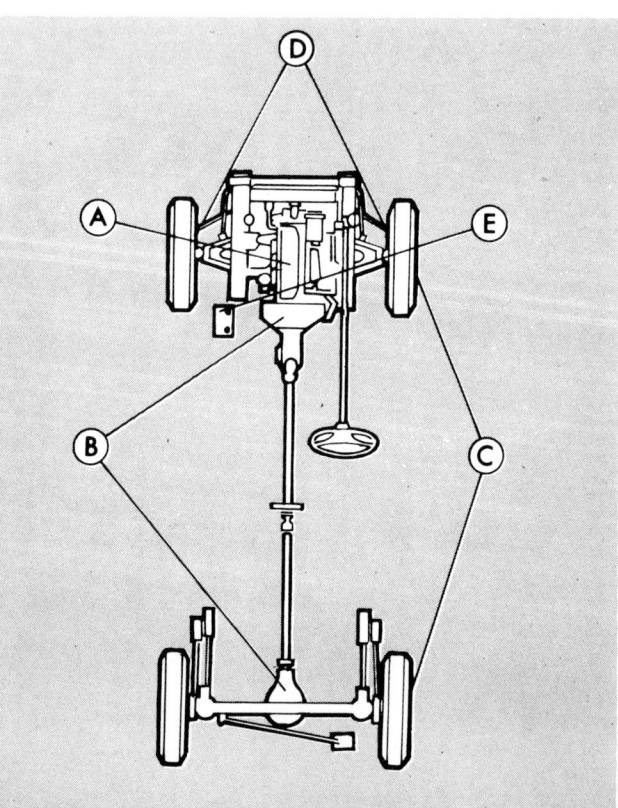

Volvo 221 estate car

OUR test of the Volvo 132S coupé described on the preceding pages was associated with a test of the 221 estate car which is fitted with the less powerful 68 b.h.p. engine (estate cars with the high power engine are not available in the U.K.). Driving this car after the coupé, a strong family resemblance is at once apparent. The facia and driving position are very similar, as is the handling. A servo makes the estate car brakes lighter but a little unprogressive in the middle of the pedal pressure range. More surprisingly the steering is also lighter, but this may be due to the slight rearward weight bias of the estate car compared with the coupé. In addition, the 221's engine is far less noisy.

The estate car has a large carrying capacity, even with the rear seat in its normal position, and is trimmed at the back with a tough, practical cloth—it is often so easy to tear carpets or trim with the corners of the sort of heavy object which needs to be carried in estate cars. There are also a number of thoughtful details. For example, the rear doors are arranged so that one folds up and the other folds down to increase the load space, and the door that lifts upwards is automatically opened and held open by a strut incorporating a spring of compressed gas. Similarly a rubber-covered step on the rear bumper facilitates access to the special Volvo roof rack that can be fitted. **M**

Performance

Weight

Kerb weight (unladen with fuel for approximately 50 miles)	23.1 cwt.
Front/rear distribution	48/52
Weight laden as tested	26.9 cwt.
Mean lap speed banked circuit	88.0 m.p.h.
Best one-way $\frac{1}{4}$-mile	90.9 m.p.h.
Maximile speed—Mean	84.9 m.p.h.
Best	88.3 m.p.h.

Acceleration times

m.p.h.		sec.
0-30		4.2
0-40		7.3
0-50		11.2
0-60		16.6
0-70		25.3
0-80		40.1
m.p.h.	Top sec.	3rd sec.
10-30	—	7.2
20-40	9.6	6.6
30-50	10.3	7.3
40-60	11.7	9.6
50-70	15.1	13.0
60-80	21.8	—

Fuel consumption

m.p.h.	m.p.g.
30	42.6
40	37.5
50	33.4
60	28.9
70	27.9

Touring (consumption midway between 30 m.p.h. and maximum less 5% allowance for acceleration) .... 28.0 m.p.g.
Overall .... 23.9 m.p.g.
(= 11.8 litres/100 km.)

Estate car body has clean lines. Upper door is automatically raised when lower door is opened and held by a strut with a spring of compressed gas.

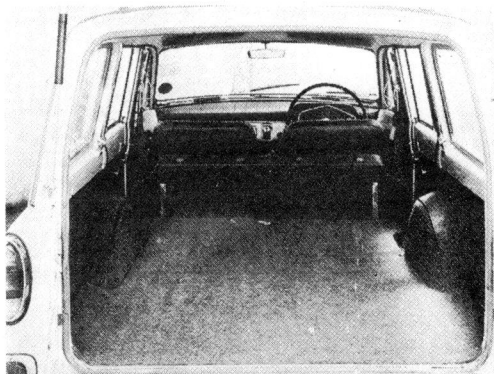

Rear luggage space is large with seat folded forward.

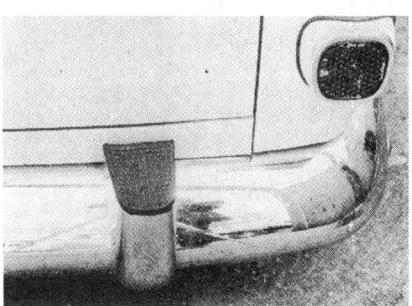

Thoughtful detail: step in bumper for access to roof to which special Volvo roof rack can be fitted.

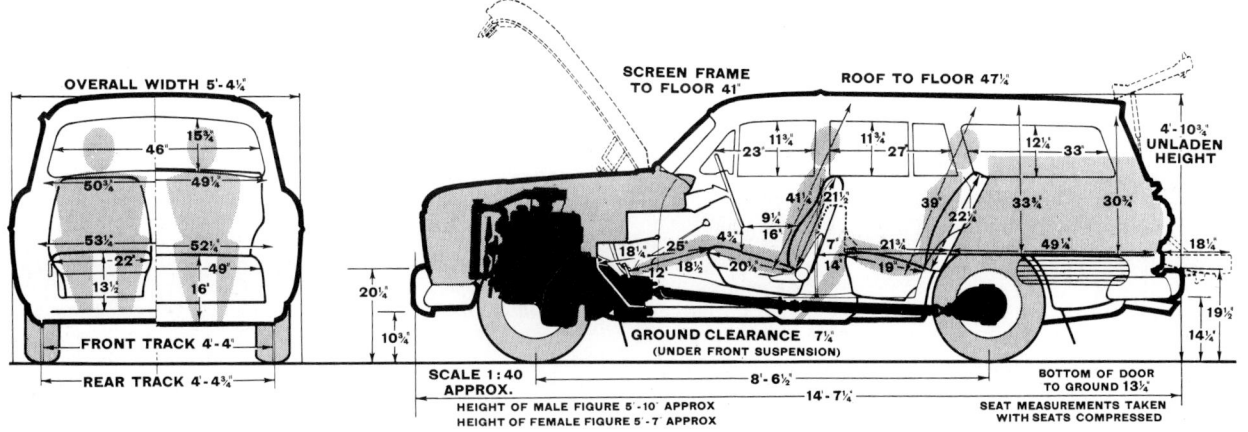

on the road with a

Volvo 132S

WHILE the American, and American-controlled car manufacturers push ahead with their policy of changing models as frequently as possible, there is still a strong feeling in Europe that designs should be conserved and improved over as long a period as possible ; economics, of course, play a large part. Volvo, the Swedish factory unusual in having one basic engine of 1780 cc to power the whole range, adhere strongly to the theme of improvement rather than innovation and while it may please some people to call their cars old-fashioned, they are at least pleasantly so. In the 132S two-door saloon we tested there is a combination of strength, performance and ruggedness ideally suited to a reasonably conservative market comprising people who want lively performance from a durable product.

Coming from a land of bad roads and hard winters (and by all accounts hard drivers), one can assume that the Volvo has strong suspension, a powerful heater, will remain free from corrosion, and prove generally trouble-free—this all seems to be true. It is also the biggest car to have performed consistently well in international rallies in recent years and this does not surprise us. Slow off the mark with high overall gearing, the 132S has a tremendous range in intermediate gears and covers the ground surprisingly quickly without feeling

the least bit strained. To sort out any confusion, the 132S is the 122 series, type 132, which explains the 122S badge on the side!

Performance and handling

Powering the 132S Volvo is a 95 bhp (gross) version of the four-cylinder engine, equipped in this case with two 1½-in SU carburettors. The unit has an iron block and head, overhead camshaft, and the crankshaft runs in five bearings ; it is not the most powerful version, that fitted to the P1800 Coupé developing 115 bhp gross, but a net ratio of 50 bhp/ton is still very adequate. The saloon has heavy, high-waisted lines and although not light, tipping 20.8 cwt at the kerb, it is quite efficient aerodynamically.

With just a little choke the engine starts easily and warms up quickly. Lightweight paper filters on the carburettors do little to silence the induction noise which takes on a rather harsh and well-tuned note under power, somewhat out of keeping with a medium-stressed five-bearing engine.

A 4.1 axle ratio gives 17.6 mph/1000 rpm in top gear, but an extremely high first ratio, which takes the car up to 35 mph, allows the gears to be closely spaced. It is almost essential to start in bottom, and the all-synchromesh gearbox is so nice that we found ourselves reaching for first before

any tight corner in town or in the country. Second takes the Volvo to 55 mph, third to 80 mph, and top speed is 95 mph though 100 is frequently seen in favourable conditions. As the engine will run happily to 6400 rpm, the car feels as though it will go on all day at top speed with a thousand revs in hand.

Although the performance figures through the gears are very good, the driver has to work the car hard to achieve them. The revs die for a few moments if a brisk start is attempted, unless the clutch is ridden, and such is the range in the gears that it is more natural to change up earlier and use the torque to its best advantage. It was noticeable on the test car, which may have had a hard life, that the clutch did not take kindly to hard upward gearchanges. The gearshift itself is smooth and positive, though with a long travel due to the lever being mounted well forward on the tunnel. The 132S is clearly not so much a sports saloon as a brisk touring car, with such weight and high gearing that inertia seems to keep it going rapidly without much effort.

The suspension may be very strong, suited to bad roads, but it is not ideal in this country. It is firm at town speeds and develops a long, floating motion on undulating surfaces when the car is driven fast suggesting soft springs and firm damping, an idea which is reinforced by its cornering characteristics. Marked understeer at low speed gives way to progressive roll, more noticeable at the back until it actually induces oversteer. However the whole process is very gradual and on wet roads the handling is entirely predictable and vice-free, with plenty of early warning when simple corrective action is required.

Worm and roller steering is fitted, again high-geared with 3¼ turns from lock to lock and not altogether free from kick-back on bumpy roads. But is it very quick and precise, contributing to the pleasure of driving the Volvo. For preference, a smaller wheel positioned a little further from the driver would be more pleasant for long-distance driving.

Big disc brakes at the front, 10.8 in diameter, and drums at the rear provide a superb, well-balanced braking system, well complemented on wet roads by the Goodyear G8 tyres. Pedal pressure is fairly high and it is rather surprising that a servo system is not optional, with women drivers in mind, but when accustomed to the system a driver could trust it implicitly from any speed, on any type of surface. The lights, too, are extremely good for high-speed travel, with a more penetrating dipped beam than most systems offer.

Comfort and equipment

You wouldn't really call the 132S a luxurious car, but comfort abounds as a product of careful planning. Carpeting is eschewed in favour of rubber matting throughout, so much more hard-wearing and practical in winter, and there is more paintwork around the fascia than one normally sees in an £1100 car. On the other hand the seats are amazingly comfortable, the heater is overpowering if given the chance, seatbelts are a standard fitting, and one can generally live in this car for hours, over long distances, without suffering.

Rather than use a mass of padding which contours itself, Volvo's seats have very pronounced (and adjustable) curvature in the lumbar region and extra padding to support the thighs, just above the knee. Without wishing to oust the anatomical experts of Harley Street, we suggest that the seats are ideal for anyone suffering from back trouble, especially since they have reclining backrests and a good range of height and reach adjustment to keep most customers happy.

The 132 is a two-door saloon, but access to the back is quite reasonable and once installed, there is plenty of legroom and headroom for a couple of adults, who also have the comfort of a central armrest. The rear windows are flexibly hinged at the leading edge, and propping them open is the easiest and quickest way of ventilating the car at high speed. Generally the wind noise is very low, though an anoying whistle was traced to the heater intake and disappeared if the inlet vent was opened slightly.

Individual controls on the fascia distribute air to the front, rear and the screen, with volume and temperature adjustment. The heater churns out warmed air very shortly after a cold start, and the blower is quite unnecessary when the car is moving.

We have enthused before about the seatbelts, which clip direct to a ring on the tunnel in a one-handed action. The handbrake is situated on the floor at the right

of the driving seat, well within reach but ideally situated to rap an unwary ankle as the driver gets out of the car. The pedals are unusually large and well spaced, a feature which is appreciated by drivers more used to cramped surroundings. Still the accelerator is conveniently placed for heel-and-toe gearchanges.

Instruments are contained in a binnacle directly ahead of the driver—a ribbon speedometer (rather optimistic) is flanked by a temperature gauge and a fuel gauge, with the usual warning light for oil pressure, dynamo charge, indicators, etc. Two-speed wipers are controlled by a pull-out knob, which on the third notch operates the washers as well. Although the knobs are labelled, it would be easier for a driver new to the car if the lighting switch was on the outer end of the row, instead of second to the wipers. A cigarette lighter is provided, and a trip mileometer is also installed. Indicators, and the headlamp flasher, are worked by a lever from the left of the steering column.

Boot space is pretty adequate for normal purposes though at holiday time the upright spare wheel might be a nuisance. The rear-hinged bonnet has an automatic stay, and engine accessibility is first class.

There are no grease points to worry about on the Volvo, and apart from changing engine oil and checking round at 3000 mile intervals the main service period is 6000 miles. This is the sort of car which builds up terrific customer-loyalty, irrespective of new models which come and go from rival factories, basing a reputation primarily upon reliability and good workmanship.

M.L.C.

SPECIFICATION

ENGINE
Four cylinders in line; bore 84.14 mm, stroke 80 mm, cubic capacity 1780 cc. Compression ratio 8.7:1. Maximum bhp (net) 86 at 5000 rpm, maximum torque (net) 107 lb ft at 3500 rpm. Overhead valves, pushrod operated. Two SU HS6 carburettors. AC mechanical fuel pump, tank capacity 10 gallons. Water cooling with pump, fan and thermostat, capacity 14 pints. Sump capacity 7 pints. 12V 60 amp/hr battery.

TRANSMISSION
Four-speed, synchromesh on all ratios, floor change. Gearbox ratios: 1st—3.25;   2nd—3.13;   3rd—1.99; 4th—1. Final drive via hypoid live axle, ratio 4.1; gearing 17.6 mph/1000 rpm in top.

CHASSIS
Integral steel two-door saloon body. Suspension, front—independent by coil springs and wishbones; rear—live axle located by double trailing arms, Panhard rod, coil springs. Brakes: front, 10.8 dia discs; rear, 9 in drums. Steering by Gemmer cam and roller. Pressed steel disc wheels, 4J-15, tyres Goodyear G8 165-15.

DIMENSIONS

	ft	ins
Wheelbase	8	6.5
Track, front	4	4
Track, rear	4	4.75
Overall length	14	6.25
Overall width	5	4.25
Overall height	4	10
Ground clearance		7.25
Turning circle	31	2
Kerb weight	20.8 cwt.	

PERFORMANCE

mph	secs
0 - 30	4.0
0 - 40	6.6
0 - 50	9.8
0 - 60	14.0
0 - 70	19.8

SPEEDS IN GEARS

	mph
1st	35
2nd	56
3rd	81
4th	95

Overall fuel consumption: 23.6 mpg

Price as tested: £1120.

ENGINE is a 95 bhp version of Volvo's famous four-cylinder engine, which seemed very adequate for the job.

HOT SEAT. The seats are, in fact, extremely comfortable and many British manufacturers could take note of this!

THERE IS plenty of room in the back for two adults, and access to the rear is quite reasonable.

You CAN
judge a book
by its cover,
after all . . .

VOLVO'S BANK – VAULT B 18

Rugged, masculine, efficient . . . these are the adjectives one finds one using about the Volvo. And the thing is that they do not do the car justice.

MORE than any other car, the Volvo has a great reputation among people who know nothing more about cars than that a Holden is made in Australia. For a car which sells in small numbers in this country, it has built an astonishing legend for itself. Perhaps it is because it looks so strong and rugged, or because the resale value is very high, or because the name keeps cropping up in enthusiasts' conversation; whatever it is, the name Volvo stands high in the lists.

This is all the more remarkable because the Volvo has the air of being uncompromisingly old-fashioned. The body style, more than nine years old now with its thick pillars and high-waisted look, makes no concessions to styling fashion, mainly because the whims of body design have little place in the efficient Swedish motor industry. Big wheels, smallish windows of flat glass, and high wing lines are more of the late 50s than the mid-60s. Yet one cannot deny the car's distinctive eye appeal; it has a masculinity, a ruggedness about it; an air of engineered efficiency.

In this case you can judge a book by its cover. The Volvo is one of the best-built cars in the world — certainly the best built for its price — or anywhere within $1000 above it. The Volvo system of "buying out" components from the best suppliers hand-picked from around the world and the tradition of the company ensure that their cars last and last and last.

By the same token you have to drive a Volvo for some distance to appreciate it fully. Not for this car the-once-around-the-block. Like the Peugeot 404, its real virtues — the orthopaedically accurate seating, the delightful gearbox, the precise roadholding — only develop when the driver has settled in to his job.

The current series of Volvo includes changes which were made to the range late last year. These include the new seats, a move to eliminate greasing points on the chassis, minor alterations to the exterior brightwork, a slight increase in engine power, and the addition of a pressure-limiting valve in the disc/drum braking system. All of them are minor changes — but all are designed to extend the sheer expertness of the car's behavior.

The sedan sold in Australia is the B18 122S series

Mud flaps come as standard equipment, and are originally fitted because a high percentage of Sweden's roads are loose gravel surfaces. We have some here, too.

WHEELS FULL ROAD TEST

in two door, four-door and station wagon form, using the twin carburettor B18D engine. Our test car, from Truck Sales and Service, the NSW distributors, was with us for about 1500 miles; the photographic car came from British and Continental Cars, a Sydney Volvo dealership.

We had remarkably few complaints about either car. The most serious was that the radial tyres fitted to the two-door Truck Sales car did not — as happens occasionally with all RP tyres — complement the car's suspension and handling characteristics very well. The braking and adhesion on wet roads was quite alarmingly affected, and gave our drivers several quite nasty moments. On dry roads all systems were go.

The most fascinating thing about the Volvo is how hard it goes. On the surface of it the 1780 cc engine is a simple four-cylinder unit with five main crankshaft bearings, two 1¾ in. SU carburettors, a cast-iron cylinder head and an average compression ratio of 8.7 to 1. It produces 95 bhp (SAE) at 5400 rpm, and 108 lb/ft of torque at 3500 rpm, but there is little to explain its very good performance. The engine spins very sweetly and well, and is as smooth as any six we have driven. There are no vibration periods anywhere in the range, and the only real indication that it is working hard is a carburettor intake roar, which is more or less standard equipment on the Volvo engine. All we can suggest is that the engine must be put together like a Swiss watch, with very close attention to head preparation, matching of ports for gas flow, camshaft timing, and balancing. But it is by no means overstressed, as warmer versions in the rally cars and the Marcos 1800 GT will comfortably deliver another 40 bhp without fuss. The performance figures are even more remarkable considering the ostensibly high drag factor of the body. What can be deceptive, however, is that the car is nowhere near as heavy as it looks, at around 2400 lb kerb.

It runs happily all day at 80 mph, although the short (4.1 to 1) final drive does tend to exaggerate the intake roar at these high crankshaft speeds, but is incredibly flexible in top gear, for this same reason. It will pull away smoothly on any throttle opening from 10 mph in top — a habit which should endear it to the average Australian driver.

The engine will spin to 6500 rpm, but some experimentation on our test strip showed that there was no real gain in taking it beyond 5700. The gearbox ratios are fairly close, and inclined to be high, with first running to well over 30 mph and third to more than 70.

This remarkable engine is housed in a very strong integrally-constructed steel body, with closed box sections at peak stress points. The body is thoroughly rust and corrosion proofed — particularly the underbody, as Swedish roads carry heavy doses of salt during the winter — and the exterior paintwork is as hard as a rock and retains its finish for years.

Volvo has used coil springs all round, with upper and lower control arms up front, torque rods and track rod at rear, the lot heavily insulated with rubber bushings. We were able to provoke wheel tramp in heavy acceleration, with slight wheel lifting in tight hairpins taken hard in a lower gear, but all in all the system works extremely well. There is little noise and roughness transmitted through either the suspension or the Gemmer cam and roller steering.

TECHNICAL DETAILS
OF THE
VOLVO 122S 2-Dr SEDAN

SPECIFICATIONS

ENGINE:

Cylinders	four in line
Bore and stroke	80.14 mm by 80 mm
Cubic capacity	1780 cc
Compression ratio	8.7 to 1
Valves	overhead, pushrod
Carburettor	two 1¾ in. SU
Power at rpm	95 bhp (SAE) at 4500 rpm
Maximum torque	101 lb/ft at 2800 rpm
Piston speed at max. bhp	2362 ft/min

TRANSMISSION:

Gearing	18.0 mph per 1000 rpm
Type	manual, all synchromesh
Gear lever location	central, floor
Ratios, overall—	
First	12.83
Second	8.16
Third	5.58
Top	4.01
Final drive	4.01 to 1

SUSPENSION:

Front	wishbones, coil springs, anti-roll bar
Rear	live rear axle, coils, radius arms, Panhard rod
Dampers	telescopic

Operating theatre: All the controls and instruments are properly grouped, and the big wheel at the right angle, if a little too high. Seat belts are standard.

The turning circle is only just over 31 ft, which is ridiculously small for a car with a 102.5 in. wheelbase. There are fairly high castor angles on the steering, giving strong return action, but the mechanism is very light and completely accurate.

The car has consistent slight understeer, with very faint dodging-about, possibly caused by slightly high rear damper rates. The ride is certainly better in the front than the rear. There is some feedback into the steering on sharp bumps, but only when using a lot of lock. Roll movement is small, although it is more apparent to the driver than from outside. But the Volvo can be hammered around corners with great confidence and can be got into extravagant attitudes without any worries. It is a very safe, predictable and essentially viceless handler.

The boot has a big lid, counterbalanced on torsion bars, which rises to a good height. The loading lip is quite low and the spare wheel strapped into place upright in the left wing. The boot floor is covered with hard-wearing materials.

The same sort of painstaking quality control is most noticeable inside the car. Everything fits perfectly; the interior paintwork and trim materials are absolutely first class, combining obvious durability with high quality — a rare characteristic these days.

The new seats are undoubtedly the best change to the current series. Very similar to those used in the P1800S coupe, they take the form of tall buckets in the front and a curved bench at rear with folding centre armrests. The trim is a type of perforated pvc that both ventilates the seat and has good grip qualities. The squab and cushion of each bucket is beautifully formed, curved in all the right places, giving complete support to the small of the back and under the knees. The squab tends to narrow towards the top and the shoulders could use more support, but this is a minor criticism. Also, we would have liked a little more sideways location in hard cornering, but Volvo buyers are expected to use the massive standard equipment three-point seat belts which clip onto a strong hook on the transmission tunnel.

There is little wind noise at speed, even through the big quarter vents, and despite the smaller-than-fashionable glass area the car is well ventilated. A very powerful (for Swedish winters, we would guess) heater/demister system comes as standard. The wiper/washer system is just as efficient, and the headlights are right up to the car's performance.

Other equipment includes reversing lights, mud flaps, cigarette lighter and grab handle for the passenger, who should seldom need it.

The bucket seats have three positions of rake for the squabs, and a little spanner work can raise or lower the seat itself. Even so, we found the steering wheel a little too big in diameter and perhaps a fraction high, even for our 6 ft staff members, who also complained of being a bit too close to the wheel for really comfortable work.

The driver finds the pull-up handbrake on his right, with a neat loop over the release button to prevent snagging when climbing in and out, and three strong pendant pedals underfoot. A horizontal ribbon speedometer, very steady and accurate, has the gauges underneath it and is topped by a big shroud which cuts out all windscreen reflection. The controls are mounted on the lower edge of the facia on each side of the wheel, with the heater controls in the centre. The clutch has a long but progressive action. The sun visors are crushable, the facia padding is very thick and colored to stop windscreen reflection and there are no projections to cause injury in a collision. The safety features that have always been so obvious in the P1800 have been carried into the sedan.

We were not very happy with the brakes on the two-door test car, as the pedal travel was quite long and we were unable to record really good figures, but this must have been due to the tyres, as the four-door was much better.

We have kept the gearbox until last, because such a marvellous unit needs more words than usual. It is controlled by a pleasantly-thick, chromed lever coming from high up on the transmission hump and longer than we are used to these days. It directs an all-synchromesh gearbox with short, positive movements yet it is very light and all the gears come in instantly and smoothly. It is spring-laden to the top-third segment, and a stranger to the car can tend to go from first to top gear; but the expert turns this to his advantage and the light pressure needed to get the right gate turns every movement into a charming diagonal slice with the lever. The whole thing is completely charming, and you find yourself changing gears far more than is necessary.

Reading back, this sounds like a panegyric of praise for the Volvo. If it is, we make no apologies. We have not suddenly fallen in love with the make; ours has been a long and constant affair that will continue to stand the test of time.   #

STEERING:

Type	cam and roller
Ratio	15.5 to 1
Turns, 1 to 1	3¼
Circle	30 ft

BRAKES:

Type	discs front, drums rear
Swept or rubbed area	345 sq in.

DIMENSIONS:

Wheelbase	8 ft	6 in.
Track, front	4 ft	3 in.
Track, rear	4 ft	3 in.
Length	14 ft	7 in.
Width	5 ft	4 in.
Height	4 ft	11 in.
Fuel tank capacity	10 gallons	

TYRES:

Size	5.90 x 15
Make on test car	Olympic GT radial

WEIGHT:

Kerb (with fuel and water)	21 cwt

GROUND CLEARANCE:

Unladen	6¼ in.

PERFORMANCE

TOP SPEED:

Fastest run	95.8 mph
Average of all runs	95.2 mph

MAXIMUM SPEED IN GEARS:

First	40 mph
Second	60 mph
Third	88 mph
Top	95 mph

ACCELERATION:

Standing quarter mile—

Fastest run	18.8 secs
Average of all runs	19.2 secs
0 to 30 mph	4.0 secs
0 to 40 mph	6.5 secs
0 to 50 mph	9.8 secs
0 to 60 mph	13.8 secs
0 to 70 mph	19.0 secs
0 to 80 mph	24.4 secs
20 to 40 mph	9.1 secs
30 to 50 mph	9.0 secs
40 to 60 mph	8.5 secs

BRAKING:

From 30 mph	31 feet
From 60 mph	130 feet
Handbrake from 20 mph	34 feet

GO-TO-WHOA:

0–60–0 mph	16.9 secs

SPEEDO ERROR:

Indicated	Actual
30 mph	27.3 mph
40 mph	36.1 mph
50 mph	46.1 mph
60 mph	55.9 mph
70 mph	65.2 mph
80 mph	76.2 mph
90 mph	NA mph

FUEL CONSUMPTION:

Overall for test	27.6 mpg
Normal cruising	29.33 mpg
Fuel used on test	Premium grade

TEST CONDITIONS:

Surface	Hot mix bitumen
Weather	Fine, slight breeze

PRICE:

Including tax	$3158

Carpet covers the floor in the wagon's out-back. It is neatly held in place by nylon "hook" strip and completely conceals spare tire well with extra space for tools and other oddments.

RIGHT: Wagon's rear seat folds forward forming solid cargo stop and conceals handy cubby space either when up for a seat or down for a load.

ABOVE: Thick, comfortable back on front seat adjusts for cushiness. Back seat is also roomy and comfortable.

RIGHT: Dash in both two-door and wagon (shown here with air conditioner) was relatively free of sharp knobs.

ABOVE: Spare tire lives under the floor in the rear of the wagon with lots of extra room for tools, and gear that live in any enthusiast's car.

RIGHT: Engine compartment looks full of five-main bearing B-18, but everything you need to reach is accessible.

VOLVO 122-S | COMPARISON

By ANGUS & CHRISTINE LAIDLAW — SOLID SEDAN VS. SPACIOUS WAGON

▶ When you finally get around to making up your mind about a new car, you pick the maker as well as the individual automobile and the dealer selling it. So before you make your final decision about color and other details, you most probably pick the make. Then comes the fight over sedan, wagon or sports model. Usually it comes down to sedan vs. wagon. And the balance between these in the Volvo 122-S line makes this decision particularly difficult.

Money-talk favors the two-door. It looks smaller, a plus if you now drive a small car. But the space in the wagon is only 1½ inches longer

The comparison becomes even harder when you take a close look at the specs. At 176.5 inches overall the wagon is only 1½ inches longer than the sedan, the width is the same, and the sedan is an inch lower. For comparison, these cars are less than a foot longer than the 1600 VW and 16 inches longer than

the Beetle making them only minimally harder to park in small spaces. So your first surprise is that the Volvo is not the large car it looks, feels, and drives as if it were. And it boasts many of the endearing small car traits like the small turning circle. It will easily U-turn inside a Beetle by about 5 feet making it in 31.6 feet instead of 36.

Your surprise at the Volvo's small-car virtues is a product of the large-car feel it gives you. The seats

Straight side view makes 122-S sedan look like a much bigger car than it really is. Smooth lines are pleasant and easy-to- live-with without being garish like many U.S. cars. Pictures cannot show the solid feel this car gives its driver.

Wagon side view shows that it is basically a four-door body with a squared off back. Again the styling effect is one of being neat but not gaudy making this a car to keep in the family for a long time.

VOLVO COMPARISON

Stern-view of two-door shows low ledge over which suitcases must be lifted into the trunk and high, solid bumper.

High, square trunk volume makes loading easy and space most useful. Tire at side can be reached without unloading.

Wide-open wagon is easier to load because four doors mean one can pull from the front while the other pushes from the rear when inserting heavy bulky cargo.

Folding front seat in two-door locks seatback solidly in position for added safety.

are large and sinfully comfortable with built-in recliners giving just about any angle from flat to bolt-upright. The backs can even be adjusted for cushiness or support to suit your taste and physique. And in both the sedan and the wagon, even the back seat is roomy and extremely comfortable. You can carry people you like back there instead of reserving it for small children and pet enemies. Front and rear passengers find the ride quiet without the jouncy bumps of the small car or that sea-sick-making upward surge which feels so much like putting a small boat's bow into a big wave that is typical of softly sprung larger cars.

Driving feel and performance are difficult subjects to describe for any car because what one driver likes another one hates and a lot of your taste depends upon the qualities of the car you are accustomed to driving. For us, used to faster, lighter steering, the Volvo steering seemed slow. But both the wagon and the

sedan went where you pointed them. And the sudden necessity to make a sharp swerve from one lane to another at 60 mph caused neither fuss nor feathers. The harder you pushed the car through a turn, the better it did. Both the sedan and the wagon seemed to say, "Now this is the kind of driving I was built for. Why have you been pussyfooting so?" And thinking back on Swedish drivers we have known, perhaps the car knows whereof it speaks. On a hard turn it will lean, but the wheels stay on the ground and there are no handling tricks, twitchiness, or unsignaled slides. And on the straights you just don't feel the side winds as

you do in most small cars—particularly VWs.

Since the engine is larger than many economy imports, and the compression high enough to require high test fuel, you expect better performance. And you get it. Maybe you won't outdrag a Chrysler Hemi, but you'll find more than enough power for jumping out into fast-moving expressway traffic or passing on a two-lane road. Of course it does not die on hills the way many under powered small cars seem to.

If it goes very nicely, the Volvo stops even more nicely. Power assisted front disks and rear drums

Continued on next page

162

stop the wagon quickly and straight. And brake feel is among the nicest either of us remember. There is none of the grabbiness of excessive power assist common on Detroit drum brakes which can bring you to a screeching halt when you really meant to slow down for a driveway, or making driving on a slick surface alarming rather than amusing and interesting. These power brakes give good pedal feel, good stopping distances, and straight-line controlled stops. The sedan uses the same disk and drum system without the power assist to achieve equally good braking.

The brakes are great, but the shift is merely good. There is a smooth if rather long throw for the four forward speeds, but reverse can be tricky. It takes a hard throw across the gate to over-ride the detents designed to keep you out of reverse when down-shifting from third to second. Once or twice it was easier to unfasten the shoulder harness to lean forward far enough to get it in. The automatic is a typical three-speed Borg-Warner unit which works soothly enough, but we preferred the manual, perhaps out of cussedness and prejudice against all automatics.

The over-the-shoulder safety harness may be a bit more awkward than the lap belt, but it will do a far better job of protecting you in an accident or sudden stop than the lap strap. And with the Volvo belts, we found we quickly got used to the diagonal strap.

As you would expect from a manufacturer long interested in safety, The Volvo 122-S is very solidly built with box sections commonly used. Galvanizing and rust-proofing keeps any moisture which might get inside them from causing rust and

eventual weakening. The smooth and padded dash is relatively free of projecting knobs. Instruments and controls make sense and are easy to read or operate. One thing we did not like was the point on the end of the ribbon speedometer indicator. Even understanding that the tip of the point indicated the speed, it was annoying and we liked the square-end ribbon as used in the MG 1100 sedans much better. But this is a small quibble.

Finish, inside and out, was beautiful. The paint job was one of the best either of us has ever seen, and was matched by fitting and workmanship inside the car. Anything that is well made is a pleasure to use on that count alone, and both of these cars lived up to this rule. And any car enthusiast will go for the many thoughtful design details built with such precision. Front seat backs in the two-door latch in position so they won't fold forward in a sudden stop. The rear seat back in the wagon latches very solidly to retain the load in the rear, something most other wagons do not do nearly so well. In the engine compartment only one small wire leads from the coil to the distributor and the input wire from the ignition switch is out of sight behind the dash in an armored cable. This makes it much harder to jump the ignition to steal the car or run it without the key. It can be done, but not without leaving evidence of tampering.

For the demon-home-mechanic, Volvos are relatively easy to work on with the very straight-forward water cooled in-line four OHV engine. Five main bearings make it smooth-running and should keep it relatively free of major rebuilding. Hardened crank journals, and other heavy-duty design features help.

And for tuneup work, the plugs and distributor are handy on the left side of the engine with the carbs (twin S.U.s) on the right leaving room to remove the valve cover without having to disconnect a lot of stuff in the way. Even the battery had connected cell caps to make checking cell level a one-shot operation without juggling six elusive screw tops.

We split again (as we did on the VW Fastback and Squareback) on the question of whether the sedan or the wagon was the better car. One of us went for the wagon with its greater load-carrying capacity and general all-round usefulness; the other picked the sedan because it felt smaller even if it isn't really much shorter. The sedan has the out-of-sight lockable trunk to protect luggage and valuables. Both cars had usable space under the rear seats and the wagon had room for tools and small things under the rear floor with the spare tire.

It all adds up to very good value for money at prices a bit higher than the cheapest imports. Both the sedan and the wagon offer the unenthusiastic car owner trouble-free, comfortable and quiet motoring while the enthusiast gets all this plus a machine he can appreciate for its safety, design, and performance. And if you want more power and are willing to pay for it, Volvo will supply completely engineered hop-up kits which can be bolted on by the dealer's mechanics or any well equipped amateur mechanic. So you can either appreciate the car as it comes, or improve its performance without killing the warantee. And we are still trying to make up our minds between the advantages of the wagon compared with the sedan. Either could be fun. ●

dominant features of the Volvo image.

Performance is still there, in third place, followed by comfort and solid construction.

Roadholding is down to a one-in-four rating, with handling, and minor praise goes to cruising ability, the useful gearbox and functional appearance (once again, no-one calls the Volvo beautiful!)

A/T SUGGESTED!

A few owners suggest a Volvo-based dream car:

"Automatic transmission would make this an ideal car," a Bellville (Cape) owner suggests.

From Mufulira, Zambia, comes this idea: "The ultimate would be a

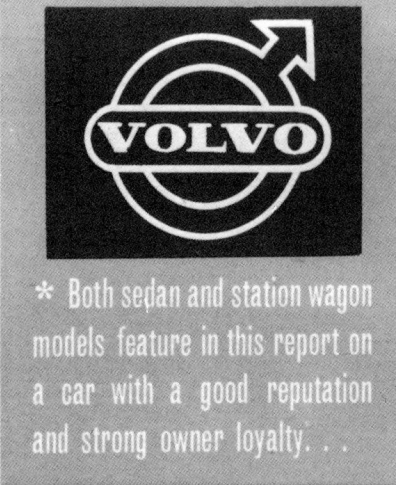

Both sedan and station wagon models feature in this report on a car with a good reputation and strong owner loyalty...

A CAR OWNERS' REPORT

A THICK pile of completed questionnaires awaited us as we started compiling this second Owners' Report by CAR readers on the Volvo.

And the compliments! — "My Volvo seems to be getting better as the miles pile up!", said a Pietermaritzburg man. "There may be faster, but there ain't any better," a Johannesburg business man reported. "Best car on the road at its price," from an Alberton owner. . . .

This looked like a whitewashing operation, but as we delved deeper a balanced picture emerged of a characterful car and a responsible strata of ownership.

TOTAL: 100 YEARS!

Volvo owners are interested enough in their cars to send in a high proportion of completed reports, and this summary is based on nearly 100 forms, totalling one-and-three-quarter million miles and almost 100 years of ownership experience!

122S-B18 models

All are 122S-B18 models, with a few B18-D's and a good sprinkling of station wagons.

● Also illuminating was the number of modified cars included—almost invariably fitted with the P-1800 conversion which has been incorporated in the 115-b.h.p. 1967 models as standard equipment.

FUEL AVERAGES

This gives us three basic categories for fuel consumption averages, and we have done them separately (see "Report Details").

Average monthly mileage has risen slightly since our previous Volvo Owner Report in May, 1963, and fuel consumption averages have changed now that the lighter Sport models are not included.

The station wagons, though heavier and lower-geared, are not much heavier on fuel on the road than the sedans, perhaps because they are not usually driven quite as hard. But the modified cars show a substantial drop, particularly in about-town use. This, of course, is because owners of the 108-b.h.p. models revel in using the extra power to the full!

ECONOMY TOPS

There is also a shift of emphasis in the features which owners praise most highly. Four years ago it was performance and roadholding: today the laurels go to economy and reliability, which have become

Volvo six-cylinder model with independent suspension all round and sealed grease points, etc!"

INSTRUMENTATION

Critical assessment by owners comes up with features which we have commented on in Road Tests: instrumentation needs up-dating, say one in three, and ventilation is insufficient for a hot country, say one in five.

Comparative smallness of rear space and trunk are mentioned, and noise levels, usually involving windows—and hence ventilation.

Body seal has deteriorated, in comparison with the previous report. In 1963, 92 per cent of owners reported good body seal (both dust and weather), while in this report on models ranging from 1963 to 1966, approval of dust-proofing is given by only 53 per cent of owners, and weather-proofing by 69 per cent.

GOOD SERVICE

The number of owners reporting no trouble at all has remained fairly constant, and the troubles reported are confined to pinion oil seals, clutch (thrust bearing and master cylinder), shock absorbers and instruments, all in small volume.

No major engine failures are reported, and other minor failures are in insignificant volume.

More than half of owners use Volvo service facilities, and the average rating is "good".

SOME PRAISE . . .

"My fourth Volvo since 1958."

"The only station wagon which appealed to my wife". (Owner of the wagon model).

"Best car for dirty farm roads." (A Rhodesian).

"Absolutely fantastic!" (Several owners).

"Best out of 14 cars I have owned."

"Got 22,000 miles out of first set of tyres (Pirelli) and have already done 21,000 on second set (Goodyear G8) . . ."

". . . Can do over 100,000 miles with little expense."

"Engine as smooth as a six — really pulls uphill . . ."

"Well worth its extra cost . . ."

"Wonderful to drive . . . Can take a hammering."

"Quality throughout."

"A pleasure to drive over long distances." (Several owners).

"Amazing stamina for such a small motor."

"Wonderful car in every respect."

"Feels, looks, and is, solid and dependable."

"Volvo's have given me excellent service."

"I recommend Volvo to anyone who needs a real car."

"Outstandingly safe and trouble-free."

"Goes like an Indian pony!"

SOME CRITICISM:

"Suspension needs improvement — wider track and lowering . . ."

"Should have more agents in country towns."

"Would like an overdrive, as there is plenty of power in top gear."

"Expensive for a four-cylinder car." (Several owners).

"Ventilation is insufficient in summer."

"Needs rev.-counter, oil gauge and ammeter."

"Interior is regrettably small . . ."

"Spares are expensive."

"Paintwork on my car was pathetic . . ."

"Hooter should be louder."

"Luggage capacity should be bigger." (Several owners).

"Disappointing and over-priced — I have had endless minor trouble."

"No space for fitting extra gauges and accessories."

WHY A VOLVO?

"It has a reputation for reliability." (Many owners).

"A tough car, good for tough conditions."

"Quality of engineering."

"Excellent reports from Volvo owners." (Several).

"Good tyre mileage."

"Best trade-in on my old car from the Volvo dealer." (Several owners).

"High resale value." (Several owners).

"Happy with my previous Volvo." (Several owners).

"Reputation for reliability in rallies."

★ A great many dealerships get "excellent" ratings: Kimberley, Johannesburg, East London, Salisbury, Durban, Bindura, Bloemfontein, Springs, Lusaka, Bellville, Cape Town and Blantyre, with many others getting consistent "good" ratings, and only a handful rating criticism.

One owner in five does his own servicing, and most of the others go elsewhere for geographical reasons.

OWNER LOYALTY

Volvo is one of those *marques* which inspires strong owner loyalty, very likely because of the reported reliability of both car and dealerships.

So we find that one Volvo owner in four considered no other car when buying—certainly an impressive percentage, and including many who had owned Volvo's before. Multiple ownership is also noticed—several owners reporting that they owned two Volvo's.

Other cars considered by owners range Alfa Romeo Giulia, Mercedes-Benz models, Valiant, Ford "Z" models, Citroën and Peugeot 404 to Vauxhall Velox/Cresta and Fiat 2300. At least 13 other makes get mentioned once or more.

VOLVO AGAIN: 81 PER CENT!

What is quite remarkable is the number of owners who specify Volvo again in giving their probable next choice.

Nearly 60 per cent list the 122S sedan, while a significant 11 per cent want the station wagon next time for its greater versatility.

One in 10, with an eye to the future, want either the P-1800S version as a special import or are prepared to wait for the 144S model which has already been announced overseas, and will be coming here as well.

The only real rival is the Jaguar "S" type, which gets 11 per cent of votes, while only 6 per cent say they are undecided at this stage.

"DEFINITELY. . . ."

Actually, we could not help smiling at some of the answers to Question 16: "Probable choice next time?"

Several had put "Volvo" with the word underlined, or followed by "definitely".

But a Transvaal owner, reporting on 40,000 miles in his first Volvo, take the biscuit for this one: "Volvo 122S, of course!" ●

MAKE AND MODEL:

Make	Volvo
Model	B18-122S, up to 1966

MILEAGES:

Total	1,730,193
Average monthly	1,460

FUEL CONSUMPTION:

In town (sedans)	27·8
(with conversion)	24·0
(station wagon)	27·6
Open road (sedans)	32·4
(with conversion)	30·4
(station wagons)	31·9

MOST-LIKED FEATURES:

Economy	56%
Reliability	48%
Performance	42%
Comfort	39%
Solid build	33%
Roadholding	26%
Handling	24%
Cruising ability	15%
Gearbox	14%
Appearance	12%

DISLIKED FEATURES:

Instrumentation	36%
Ventilation	18%
Rear seat space	12%
Small trunk	11%
Noise levels	10%
Body seal	10%

TROUBLE REPORT:

No trouble encountered	57%
Pinion oil seals	12%
Clutch	12%
Shock absorbers	11%
Instruments	10%

BODY SEAL:

Dust-proof	53%
Weather-proof	69%

DEALER SERVICE:

Use Volvo service	54%
Average quality	Good
No — own service	21%
No — geographical	14%
No — poor service	6%
No — expensive	5%

OTHER CARS CONSIDERED:

No other	26%
Alfa Romeo	18%
Valiant	17%
Mercedes-Benz	17%
Ford Zephyr	15%
Citroën	12%
Peugeot	12%
Fiat 2300	10%
Vauxhall Velox/Cresta	10%

PROBABLE NEXT CHOICE:

Volvo 122S Sedan	59%
Volvo 122S station wagon	11%
Volvo P-1800S	6%
Volvo 144S	5%
Jaguar 3·8S	11%
Undecided	6%

VOLVO
122S

* South Africa gets a special "super-Volvo" for 1967 — high-performance machinery, but no body alterations...

A CAR ROAD TEST

IF sports sedans qualify by power-weight ratio — which is as appropriate a means of classification as any — then South Africa's Volvo for 1967 is right back in the field with a special engine package that all but lends it wings.

We call it "South Africa's Volvo", because as far as we can ascertain it is unique. It is the only 122S model

B18-B sedan

which, as standard equipment, is fitted with the 115-b.h.p. P-1800 sports engine in production, to give it 1 b.h.p. net to every 25 lb. of body weight.

S.A. EXEMPTED

The ruling from the parent company in Sweden, apparently, is that the P-1800 engine may not be fitted in the 122S sedan and station wagon in any country where the P-1800 sports coupe is also sold, for obvious reasons.

As the sports model is not available here except as a special import, South Africa qualified for automatic exemption.

To distinguish it from previous models — the B18 (90 b.h.p.) and 1966 B18-D (95 b.h.p.) — this 1967 super-Volvo is listed as the B18-B, and is available in both sedan and station wagon form.

TIMING IRONY

This is the Volvo to end all arguments and break the "ton", but it is

ironic that it should be announced just after South Africa's general speed limit came into being, so owners will not be able to see for themselves.

They can take our word for it that this model will exceed 100 m.p.h. true speed both ways on a level road, though we have to be content with scientific calculation methods and could not satisfy our burning curiosity on the road, with the fifth-wheel giving the verdict.

ENGINE DETAILS

The 115-b.h.p. engine is basically that always used in the Volvo we have known, remaining at 1·78 litres, but compression ratio goes up from 8·7 to 1 to 10·0 to 1, and both induction and exhaust systems are changed.

The twin SU carburettors are re-tuned, and have reverted to the attractive and efficient Volvo pan-cake-type aircleaners, though these have the addition of plastic-foam covers to reduce induction noise and

SPECIFICATIONS

ENGINE:

Cylinders	Four in line
Carburettors	Twin SU
Bore	3·313 in. (84·1 mm.)
Stroke	3·15 in. (80·0 mm.)
Cubic capacity	109·0 cu. in. (1,780 c.c.)
Compression ratio	10·0 to 1
Valve gear	o.h.v., pushrod
Main bearings	Five
Aircleaners	Twin paper element
Fuel rating	Premium
Cooling	Water
Electrics	12-Volt AC

ENGINE OUTPUT:

Max. b.h.p. S.A.E.	115·0
Max. b.h.p. net	98·0
Peak r.p.m.	6,000
Max. torque/r.p.m.	112/4,000

TRANSMISSION:

Forward speeds	Four
Synchromesh	All
Gearshift	Floor
Low gear	3·13 to 1
2nd gear	1·99 to 1
3rd gear	1·36 to 1
Top gear	Direct
Reverse gear	3·25 to 1
Final drive	4·1 to 1
Drive wheels	Rear
Tyre size	165 x 15, sports nylon

BRAKES:

Front	Discs
Rear	Drums
Total lining area	N/S
Boosting	Vacuum Servo
Handbrake position	Inside driver door

STEERING:

Type	Cam and roller
Lock to lock	3·5 turns
Turning circle	31·2 ft.

MEASUREMENTS:

Length overall	175·0 in.
Width overall	63·75 in.
Height overall	59·25 in.
Wheelbase	102·5 in.
Front track	51·75 in.
Rear track	51·75 in.
Ground clearance	8·5 in.
Licensing weight	2,450 lb.

SUSPENSION:

Front	Independent
Type	Coils, stabiliser
Rear	Live axle
Type	Coils

CAPACITIES:

Seating	4/5
Fuel tank	10·0 gal.
Luggage trunk	17·5 cu. ft.

SERVICE DATA:

Sump capacity	5·75 pints
Change interval	1,500 miles
Oil filter capacity	0·7 pints
Change interval	6,000 miles
Gearbox capacity	1·25 pints
Change interval	6,000 miles
Diff. capacity	2·25 pints
Change interval	6,000 miles
Air filter change	7,000 miles
Greasing points	Nil

(These basic service recommendations are given for guidance only, and may vary according to operating conditions. Inquiries should be addressed to authorised dealerships.)

TYRE PRESSURES:

Crossply: Front	26 to 30 lb.
Rear	28 to 34 lb.
Radial ply: Front	28 to 32 lb.
Rear	30 to 36 lb.

Warranty:
Six months.

BASIC PRICES:

Coast	R2,650
Reef	R2,650

PROVIDED TEST CAR:
W. G. Thompson Motors, Buitengracht, Cape Town.

protect the paper elements from damage by flying grit.

Camshaft is the P-1800 high-lift, which does not cause any roughness and we were pleasantly surprised to find how tolerant this high-compression unit is of our 93-octane fuel, even at sea level. On a well-tuned Test car, no side effects could be detected.

A free-flow exhaust system with a pleasantly crisp note is used to round off the new breathing, without creating any decibel problems.

PERFORMANCE FACTORS

All Volvo models now come with high-performance tyres, and we are told that there is a choice of either radial-ply or sports-nylon makes.

The Test car was equipped with Goodyear G8-S sports-nylons in 165 x 15 size, and in this form the gearing came out at 17·85 per 1,000 r.p.m., compared with 18·07 on the radial-shod B18-D sedan tested in March last year, and 18·2 for the previous model on standard tyres (Car Road Test, August, 1962).

In addition to more power, there is a higher revs. peak, which on paper makes the new model substantially over-geared. A comparative data table is the best way of presenting the changes, which include small variations in weight:

	B18	B18-D	B18-B
Weight	2,405	2,425	2,450
C.R.	8·5	8·7	10·0
B.h.p. S.A.E.	90	95	115
Peak	5,000	5,400	6,000
Power/wt.	31·2	29·8	25·0

M.p.h./1,000	18·2	18·07	17·85
M.p.h./peak	91·0	97·6	107·2

Without even considering torque (which is much-improved but has moved up to a higher peak) it is obvious that performance capabilities are considerably changed.

PERFORMANCE

The change is obvious on the road.

The Volvo B18-B has a new vitality on the move which can be felt quite clearly, and which puts it near the top of the two-litre field in spite of its very heavy coachwork.

It just fails to break 13 sec. in 0–60 time, and covers the quarter-mile from rest in a smooth 18·7 sec.

The improvement is right through — in flexibility in the gears, cruising ability and gradient ability.

(continued overleaf)

To compare once again:

	B18	B18-D	B18-B
0-60	14·7	14·7	13·1
Max. speed	95·3	97·0	101·5
¼ Mile	19·3	19·8	18·7
30-50 (3rd)	6·8	7·8	7·2
50-70 (top)	11·9	12·9	12·3

There is plenty of range in the gears (to 35 in low, 55 in second and 80 in 3rd) and acceleration is accompanied by a business-like sound from the air intakes and exhaust which is well-modulated and exciting, and no source of annoyance.

FUEL ECONOMY

An improvement in fuel economy is logical with the higher compression ratio and free-flow system, but what is surprising is the amount that the new model gains:

	B18	B18-D	B18-B
M.p.g. at 30	42·3	40·7	46·9
M.p.g. at 45	39·6	37·9	39·8
M.p.g. at 60	35·5	32·5	33·8

This is an important feature of the new model, giving expectation of up to 30 m.p.g. in sustained 70-m.p.h. cruising.

BOOSTED BRAKING

The Volvo's have always had good brakes (discs at front) with no fade problems, but needing fairly hefty pedal pressure for a good stop.

This was particularly noticeable on the 1966 B18-D model, which introduced the very good feature of a pressure-limiting valve in the braking system to inhibit rear-wheel locking.

The 1967 model has the welcome addition of vacuum-servo boosting for the brakes, so that the car stops easily with quite light pedal pressure and sensitive control.

Actual stopping times recorded this year are not as good, principally because the front discs are taking more of the load with boost, and can be locked in hard stopping — requiring a full release of the pedal to release them. But stability is good, and the general stopping ability very good for a heavy car.

HANDLING AND RIDE

With its solid weight and firm, progressive suspension, the Volvo handles beautifully, though steering through the big and high-set steering wheel is heavy when stationary or at low speed.

But on the open road handling is precise and accurate, and there are no vices in the wheel geometry.

The new model has an extension thump at the front suspension that we do not recall in earlier models,

DATA AT 70

Min. Noise level	82·0 dBA
0-70 Through gears	18·4 sec.
M.p.h. at 70	28·2
Braking from 70	4·6 sec.
Reserve power at 70	0·067 g.
Max. gradient at 70 (top)	1 in 15·0
Speedo error at 70	7% over
Speedo at true 70	75
R.p.m. at 70 (top)	3,910

possibly caused by shock absorbers reaching full drop.

VENTILATION AND NOISE

There is a good, boosted heating system, but the cool ventilation system is below par by today's standards, even when the two-speed fan is used. But as the car is not too wind noise-prone, open windows can be used at speed without much trouble.

Mechanical noise levels are not much affected by the engine change, except at higher speeds, and road noise with the sports-nylon tyres is quite fair, though not as good as with radials. Generally the car is on the quiet side right through.

COMFORT AND STYLE

The Volvo is a quality car both mechanically and in furnishing, and the atmosphere inside is one of quiet good taste. The ripple seats are well tailored and provide a high degree of comfort for four passengers, with space for an extra one at rear at a pinch. Reclining seats, a centre arm-rest at rear which retracts, and other good features are in evidence.

In common with some other cars in this class, the Volvo has some old-fashioned features, such as a high-set steering wheel and small glass area. These are not serious criticisms, apart from the fact that there is no view of the rear wings in reversing because of the rear window height.

After the light and airy feel of cars in the modern idiom the first reaction on closing the Volvo door is a feeling of being cooped-in, but this has advantages in minimizing the sun's heat and providing the security of more solid metal round the occupants, instead of fragile glass!

SUMMARY

Reports are that this new Volvo is proving very popular, and we are not surprised.

It is a conservative car, but one with many solid virtues and all the marks of good breeding. With the vivid extra performance of the 115-b.h.p. engine it has become an exciting car to drive, and able to hold its head high in any company!   ●

This is a sports car engine, with foam plastic covers to silence and protect the pancake-type paper element air-cleaners.

The neatly finished trunk has 17·5 cu. ft. of space and a medium sill, with spare wheel vertically mounted.

Hard cornering is no problem to the firmly sprung Volvo. Picture is of a 90-degree turn at 30 m.p.h.

Instrumentation is unchanged and has not kept pace with the mechanical improvements. Upper half of the hooter ring is obtrusive on the high-placed wheel.

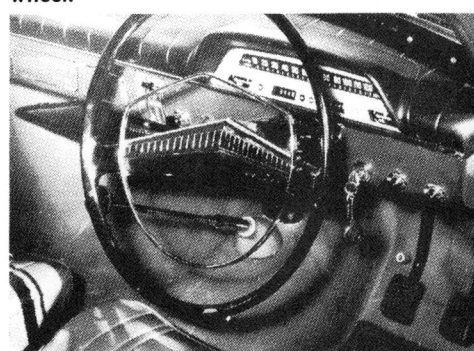

MAKE AND MODEL:
Make . . . . Volvo
Model . . . . . 122S-B18 B sedan

PERFORMANCE FACTORS:
Power/weight (lb./b.h.p.) . . . 25·0
Frontal area (sq. ft.) . . . . 26·1
Drag at 60 m.p.h. (lb.) . . .117·5
M.p.h./1,000 r.p.m. (top) . . . .17·85
(Calculated on licensing weight, gross frontal area, gearing and net b.h.p.)

INTERIOR NOISE LEVELS:

	Min.	Wind	Road
Idling . .	49·5	—	—
30 m.p.h. .	66·5	68·0	78·0
45 m.p.h. .	72·5	74·5	82·0
60 m.p.h. .	77·5	80·0	86·0
Full throttle			See graph
Average dBA at 60 . .			81·2

(Measured in decibels, "A" weighting, averaging runs both ways on a level road; "Minimum" with car closed; "Wind" with one window fully open: "Road" on a coarse gravel surface.)

ACCELERATION FROM REST:
0–30 . . . . 4·2
0–40 . . . . 6·6
0–50 . . . . 9·2
0–60 . . . . 13·1
0·70 . . . . 18·4
¼ Mile . . . . 18·7

OVERTAKING ACCELERATION:

	3rd	Top
20–40	.7·6	10·6
30–50	7·2	10·4
40–60	7·9	11·4
50–70	8·8	12·3

(Measured in seconds, to true speeds, averaging runs both ways on a level road, car carrying test crew of two and standard test equipment.)

MAXIMUM SPEED:
True speed . . . . . 101·5*
Speedo reading . . . . . 108·00
Calibration:
Indicated:
   20 . 30 . 40 . 50 . 60 . . 70
True speed:
   19 . 28·5 . 38 . 47·5 . 57 . 66
(*Calculated to the nearest 0·5 m.p.h.)

FUEL CONSUMPTION:
30 m.p.h. . . . . 46·9
45 m.p.h. . . . . 39·8
60 m.p.h. . . . . 33·8
Full throttle . . . . . See Graph
(Measured in miles per Imp. gallon, averaging runs both ways on a level road.)

BRAKING TEST:
From 50 m.p.h.:
First stop . . . . 2·9
Tenth stop . . . . 3·0
Average . . . . 3·0
Handbrake stop . . . . . 7·8
(Measured in sec., with stops from true speeds at 30-sec. intervals on a good bitumenised surface.)

GRADIENTS IN GEARS:
Low gear . . . . . . 1 in 3·4
2nd Gear . . . . . . 1 in 4·8
3rd Gear . . . . . . 1 in 6·9
Top gear . . . . . . 1 in 10·6
(Tabulated from Tapley g. readings, car carrying test crew of two and standard test equipment.)

GEARED SPEEDS:
Low gear . . . . 34·2
2nd Gear . . . . 54·0
3rd Gear . . . . 78·6
Top gear . . . . 107·2
(Calculated to true speeds, at engine peak r.p.m. — 6,000).

INTERIOR NOISE LEVEL

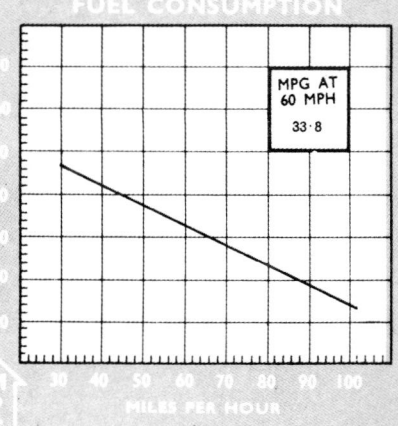

AVE. dB AT 60
81·2

ACCELERATION

MAXIMUM SPEED
101·5

FUEL CONSUMPTION

MPG AT 60 MPH
33·8

ENGINE SPEED

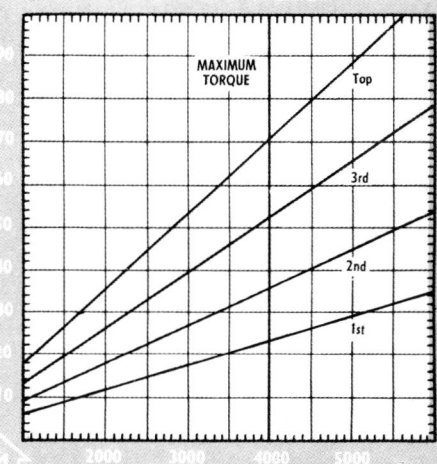

MAXIMUM TORQUE

SILENCE LEVELS:
Mechanical . . . . . . Good
Idling . . . . Good
Transmission . . . . Very Good
Wind . . . . Fair
Road . . . . Fair
Coachwork . . . . . Good
Average . . . . . . Good

ENGINE:
Starting . . . . . Very Good
Response. . . . . . Very Good
Smoothness . . . . . Good
Accessibility. . . . . Good

STEERING:
Accuracy . . . . . Good
Stability at speed . . Very Good
Stability in wind . . Very Good
Steering effort . . . Fair
Road shock . . . . Good
Road feel . . . . . Good
Centring action . . . Good
Turning circle . . . Good

BRAKING:
Pedal pressure . . . Very Good
Response. . . . . . Very Good
Fade resistance . . . Very Good
Directional stability. . Very Good
Handbrake position . . Good
Handbrake action . . Very Good

TRANSMISSION:
Clutch action . . . . Good
Pedal pressure . . . Good
Gearbox ratios . . . Very Good
Final drive ratio . . . Good
Gearshift position . . Good
Gearshift action . . Very Good
Synchromesh . . . Excellent

SUSPENSION:
Firmness rating . . . Very Good
Progressive action . . Very Good
Roadholding . . . . Good
Roll control . . . . Good
Tracking control . . Very Good
Pitching control . . . Good
Load ability . . . . Very Good

DRIVER CONTROLS:
Hand control location . Very Good
Pedal location . . . Very Good
Wiper action . . . Very Good
Washer action . . . Excellent
Instrumentation . . . Poor

INTERIOR COMFORT:
Seat design . . . . Very Good
Headroom front . . . Very Good
Legroom front . . . Good
Headroom rear . . . Good
Legroom rear . . . Good
Door access . . . Very Good
Lighting . . . . . Very Good
Accessories fitted . . Good
Accessories potential . . . Poor

DRIVING COMFORT:
Steering wheel position . . Fair
Steering wheel reach . . Good
Visibility . . . . . Poor
Directional feel . . . Good
Ventilation . . . . Poor
Heating . . . . . Good

COACHWORK:
Appearance . . . . Fair
Finish . . . . . . Good
Body seal . . . Very Good
Space utilisation . . . Poor
Trunk capacity . . . Good
Trunk access . . . Good

TEST CONDITIONS:
Altitude . . . . . At sea level
Weather . . . . . Fine and warm
Barometric reading . . 30·08
Fuel used 93-octane
Test car's mileage . . 6,350

How to get there quickly and comfortably: Go by Boeing 727 or go buy a Volvo 123 GT.

Seven-League Boots

We can't blame you if you haven't heard of it. They probably don't want the word to get around in case this stunning new Swedish screamer becomes a habit that's hard to kick.

THESE days, you've got to be on your toes. Volvo concessionaires Swedish Motor Importers snuck the 123 GT Volvo on to the market without even a roll of muffled drums. We found out about it only in casual conversation with Tony Applebaum, who is — even for a Volvo salesman — a singularly Volvophilic rep for Sydney dealers British and Continental. We should,

of course, have expected SMI to be quick off the mark in importing the first of these new hot ones; company chief Max Winkless and service operative Bill Nolan are this year bombing the local rally scene with works Volvos, ably helped by brilliant newcomer John Keran.

It is reasonable to assume that the 123 GT has been born directly from Volvo's extensive rallying experience, even though the company no longer competes in the European championship with a works team. Essentially, it is the 122S series with the engine from the P1800S sports coupe beloved of fans of The Saint television series. The body is the two-door version, and the four-speed gearbox with overdrive on top is fitted. To this is added all the other gear one would expect on a rapid sports sedan like this.

This is one of the toughest, most resilient engines in the world, and will stand an incredible amount of abuse. In fact, it is very probably one of the best production engines ever built. It was originally a 1.4 litre developing 40 bhp, which went up to 80 bhp being enlarged to 1.6 litres and 85 bhp. The P1800 version puts out 115 bhp (SAE) at 6000 rpm on 10 to 1 compression; it will run on straight local super, but prefers

170

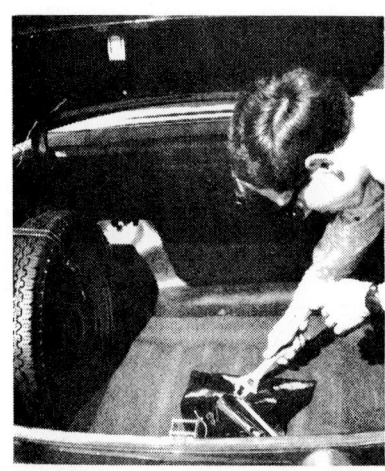

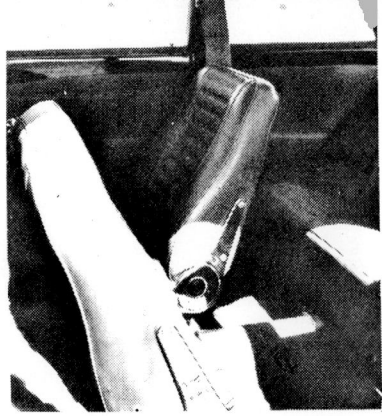

Left: Tool kit includes shifter and screwdriver, has its own plastic holder. Boot will stow a surprising amount of luggage, with spare mounted upright for easy loading.

a one-in-three methyl-benzine or benzol mix.

But even at that, its highest production power output, the engine is still not over-developed. Volvo has competition options for the engine that are used to put together special racing versions in Europe and America. These take the power output to around 125 bhp and enable the engine to crank to 7000 rpm. The block can be bored to raise the displacement to 1860 ccs, a higher overlap cam added, and special valve springs and twin-choke Weber carburettors used on a modified head.

The significant thing about this is that even the factory stormers use standard P1800 bearing shells and clutch, and the standard combustion chambers are not touched, as these leave the factory fully machined. About the only thing done to the engine apart from fitting "hot bits" is to lighten the flywheel and do a complete balance job.

So in the engine the basis is there to make the 123GT a first-class fast touring car. And this is what it is. We have never been out-and-out Volvo enthusiasts, feeling that both the 122S and the 144S are admirable cars but a little heavy and stodgy to be real open road fliers. The new one changes all that. Max Winkless asked us to take it to Surfers and return because the mileage would be welcome on an engine which always

takes some time to free up. At the end of more than 2000 miles of exceptionally fast motoring we felt that this was one of the fastest and safest point-to-point cars anybody could possibly buy.

All the normal 122S Volvo characteristics are there — the best seats in the world — without exception — ball-joint front suspension with roll bar, live rear axle beautifully located by support arms, torque rods and a track rod, and slung on coils, Girling disc front brakes, drum rear with pressure-limiting valve in the system. That, and the legendary Volvo construction and rust-proofing.

As well, the 123GT comes with laminated windscreen, higher rated dampers, radial ply tyres (Cinturatos as standard) matched quartz-iodine fog and driving lights complete with very pose-worthy white Volvo covers, two exterior rear vision mirrors, louder horn, alloy-spoked steering wheel, reclining front seats, and a small tachometer. These are all the things most fast drivers regard as essential to their work, but seldom get in one car. What's more, they would be useless if they weren't on a car as beautifully integrated and well-balanced as this one.

With the overdrive comes the 4.56 to 1 final drive ratio which is normally used with the wagon against the 4.1 on the non-overdrive car. A 4.88 ratio is available as a factory competition option. The overdrive has a ratio of 0.756, and at this is a little too tall for the car's frontal area; by the same token it is very good on the open road because the step is wide enough to make it a true fifth gear.

The B18 engine has never been particularly quiet, and we even had to comment on this in the luxury 144S road test. The two HS6 SU carburettors have only small pancake air filters, and the induction roar is very pronounced. But somehow this seems appropriate for such a sporty car and is matched by a very healthy exhaust note. Even after 4000-odd miles the engine on

the test car was still fairly stiff, and would not spin easily to 6000 rpm, which is the start of the red line. On the other hand, it felt quite relaxed and fairly quiet cruising at 5000 rpm in overdrive top. The noise only became oppressive in really hard acceleration, but the car goes so quickly on the open road that most of the accelerating seems to be done in top.

The tall overdrive gearing meant that a lot of ordinary main road hills called for a drop to fourth, but the overdrive switch works instantly, helped always by a quick stab at the clutch; after a while it became a very pleasant practised technique to snap back to fourth for the slightest check — such as a long, sweeping corner to be taken over 80 — and this contributed to some staggeringly high averages as fourth could be picked up at almost 90 mph without fear of over-revving. In any case, third gear is good for well over 70 mph; the result of all this is that we could comfortably average 70 mph between towns on the NSW coast road and never bother a soul.

The 123GT uses the same rim width as the normal sedan, but braced tread tyres and the re-rated shockers have lifted the handling levels — already excellent — to an astonishingly high degree. The car understeers less than the 122S, almost back to a neutral balance, and the rear end "stickability" produced prodigious cornering speeds. The steering is accurate and sensitive, with a little too much feedback over rough surfaces and some lost motion at top dead centre, but contributed a lot to this sort of cornering.

SPEC. PANEL OVERLEAF

We have nothing but absolute, unstinted praise for the brakes. A servo unit is fitted, and pedal pressures are light and progressive after overcoming an initial dead feeling in the pedal on first application. But the car would slam down to a stop time after time without any hooking or sign of fade — just a hot smell of Fried Ferodo. The combination of superb brakes and the versatility of the fourth-overdrive gearing means that a good driver can commit himself completely to what would otherwise be very dodgy situations. The handbrake, on the right side of the driver's seat, works well.

Another of the beauties of this exceptional car is the way it rides over any surface at all. The Banga-low-Lismore short cut off the Pacific Highway is patched and lumpy for most of its length, but the Volvo kept all feet on the ground at all times, even when accelerating hard in second gear out of tightish mountain hairpins. The car just whistles smoothly over the most appalling roads, and is outstanding on loose gravel and dirt. Roll movement is small and we could not produce wheel hop or axle tramp no matter how hard we tried.

The lighting was right up to the car's considerable performance, particularly with the QI light cut in. It is rigidly mounted, and on the test car was set too low. It was then that we found it needed a spanner to re-set. Oh well, you can't win 'em all. Dipping is by floor switch, but the headlight flashers pull in high beam in daylight. The horn, although louder than the standard unit, is not good enough; we suggest 123GT buyers should fit a set of air horns, because they will find themselves running over the top of people.

The test car was in dark green, almost a British racing green, and the paintwork (as on all Volvos) was magnificent. It looked about 20 coats deep, pure and lustrous, and we couldn't find a blemish in the finish anywhere. Inside the trim is deep tan with rubber floor matting.

You sit well down in all Volvos, and the glass area in the old-fashioned high-waisted body is not generous. The wing mirrors are helpful — more so than most of their kind — because the body does have blind spots despite a generous wraparound in the rear window. You are surrounded by shoulder-high padding and the renowned Volvo crash-proofing in design and as usual the three-point belts latch to a central mounting with two red quick-release levers. The belts are fussy to adjust and hard to lock onto the mounting, but they can be applied one-handed, which is a blessing.

The tri-alloy-spoked wheel is large and set high, as in the other cars, but the excellent range of adjustment in the seat meant that the only complaint was in the height of the wheel. There could be a fraction more rearward adjustment on the seat slides for six-footers. The seats are, of course, beyond criticism; orthopaedically-designed and nicely shaped, they range right up behind the shoulders. Both recline right back, but a built-in headrest would be better for a passenger trying to sleep.

Instrumentation is the same as in the 122S, except for a swivelling tachometer mounting high on the dash next to a special odds-and-ends tray (which is needed, as there is no glovebox, only a parcels shelf).

You get a strip speedometer, gauges for fuel and water temperature, and warning lights for everything else, including a red overdrive light which is far too bright, at night reflecting up into the driver's window. The tachometer is too small to be really useful, and is too dimly-lit at night.

There is the usual excellent Volvo heater/demister system, map pockets in both doors, and armrests which are far too hard and which gave most of our drivers sore elbows after a while. The long, sturdy gearbox lever is precise, despite longish throws and a slower-than-usual movement; you have to go against quite strong spring-loading to get back to the first-second layer, and we occasionally picked up top when going from first to second.

Access to the rear seat is not easy, as one has to step around the seat squab, which flips forward, and dodge the seat belt webbing which clips back onto the central pillar. Once in, things are very comfortable, with forward-hinged rear windows and a central armrest, although the seat cushion is a trifle short under the thighs. But the rear seat ride is just as good as in front, which is rare.

Windscreen wipers cover a good area, but the washers can only be brought in by pulling the wiper knob fully out, which means that the first stroke of the blades is on a dry screen. Not good.

We were surprised to find that we were averaging only around 21-22 mpg on the interstate trip, where the car returned about 24 mpg in normal use. The only way we can explain this, despite the overdrive, is that we used the car to its fullest extent, and kinder driving would certainly raise the cruising figure to over 26 mpg, which is excellent. We added one pint of oil in 2000 miles, and the only other trouble was a loose baffle that developed in the mufflers. We also found a lot of clutch slip during our acceleration runs, but this is common in Volvos. Recovery is always instantaneous, and slip never intrudes in normal driving.

During our time with the 123GT we drove two different examples. The first was Swedish Motors' own car, the unit that impressed us all the way to Surfers Paradise but it was still stiff by the time we came to run performance figures and would better no more than 98-100 mph. So to prove the Volvo factory's claim of 107.5 mph we set out in another 123GT from Brayson Motors of Rockdale, NSW, with over 8000 miles up, 4000 more than the SMI car. To our satisfaction we recorded a flat 108 mph which proves something about running in. The SMI car had been soft shoed all the way from new and nothing over four grand whereas the Brayson Motors car had been winning traffic light grand prix since leaving the showroom.

Maybe we're getting blase, but it takes a lot to make us really enthuse about a car these days. The 123GT Volvo caused more of a stir among the road test staff than almost any other car this year. It is by far the best Volvo yet, and this must be a real compliment, for the whole Volvo range is first-class. Yet the 123GT has real character and a sort of enduring spirit that will probably ruin an owner for anything else. We will have to go out and drive Holdens for a while to restore our sense of values. #

TECHNICAL DETAILS

MAKE	Volvo	COLOR	Dark green
MODEL	123GT	MILEAGE, START	1744
BODY TYPE	2-door sedan	FINISH	3922
PRICE	$4160	WEIGHT	22.6 cwt
OPTIONS:			Nil

FUEL CONSUMPTION:
Overall .... 23.4 mpg
Cruising .... 22.1 mpg

TEST CONDITIONS:
Weather: fine, cool; surface, hot-mix bitumen; load, two persons; fuel, premium grade.

PERFORMANCE

Piston speed at max bhp .... 3150 ft/min
Top gear mph per 1000 rpm — Overdrive 21.3; 4th 16
Engine rpm at max speed — Overdrive 5000, 4th 6750
Engine rpm cruising .... O'u 3525; 4th 4700
Lbs (laden) per gross bhp (power to weight) 22 lb/bhp

MAXIMUM SPEEDS:
Fastest run .... 108 mph
Average of all runs .... 104 mph
Speedometer indication fastest run .... 112 mph
In gears:
1st 35 mph; 2nd 52 mph; 3rd 74 mph; 4th 108 mph
Overdrive: 95 mph at 4500 rpm.

ACCELERATION:
(through gears)
0-30 mph	3.4 secs
0-40 mph	5.4 secs
0-50 mph	7.7 secs
0-60 mph	11.2 secs
0-70 mph	15.2 secs
0-80 mph	22.0 secs
0-90 mph	30.2 secs

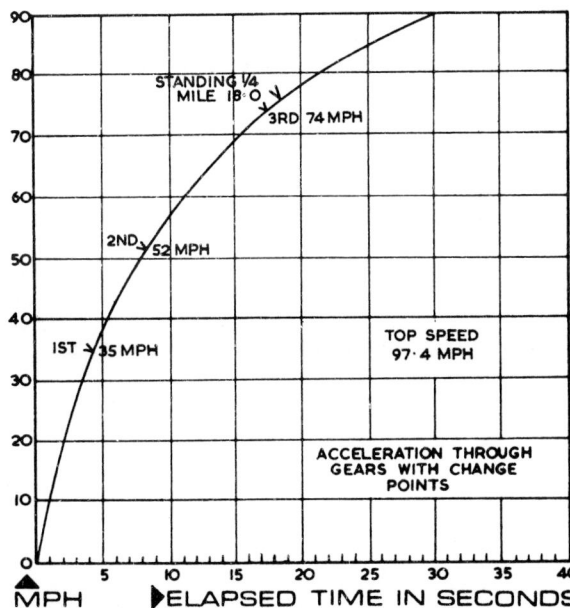

ACCELERATION THROUGH GEARS WITH CHANGE POINTS

	3rd gear	4th gear
20-40 mph	5.9 secs	9.4 secs
30-50 mph	5.9 secs	8.0 secs
40-60 mph	6.6 secs	8.5 secs
50-70 mph	7.0 secs	9.5 secs

STANDING QUARTER MILE:
Fastest run .... 17.8 secs
Average of all runs .... 18.0 secs

SPEEDOMETER ERROR:
Indicated mph:	30	40	50	60	70	80
Actual mph:	29.0	38.9	47.9	58.6	68.0	78

SPECIFICATIONS

ENGINE:
Cylinders .... ohv, four in line
Bore and stroke .... 84.14 mm by 80 mm
Cubic capacity .... 1778 cc
Compression ratio .... 10 to 1
Carburettors .... dual downdraft
Power at rpm .... 115 bhp at 6000 rpm
Torque at rpm .... 112 lb/ft at 4000 rpm

TRANSMISSION:
Type .... four speed all syncromesh with overdrive
Clutch .... diaphragm spring clutch
Gear level location .... floor, s/column o/d lever

OVERALL RATIO:
1st	14.3	4th	4.56
2nd	9.00	Overdrive	3.42
3rd	6.70	Final drive	4.56 to 1

CHASSIS AND RUNNING GEAR:
Construction .... integral construction

SUSPENSION:
Front .... wishbones, coils, a/roll bar
Rear .... live axle, coils torque rods
Shock absorbers .... telescopic

STEERING:
Type .... cam and roller
Turns 1 to 1 .... 3¼
Turning circle .... 31 ft
Steering wheel diameter .... 16 ins.

BRAKES:
Type .... servo-assisted disc front, drum rear
Dimensions .... 10.8 ins. disc, 9 in. drums

DIMENSIONS:
Wheelbase .... 8 ft 6½ in.
Track, front .... 4 ft 3 in.
Track, rear .... 4 ft 3 in.
Length .... 14 ft 7 in.
Height .... 4 ft 10 in.
Width .... 5 ft 4 in.
Fuel tank capacity .... 10 gallons

TYRES:
Size .... 165SR by 15
Pressures .... 26 front, 28 rear
Make on test car .... Pirelli Cinturato
Ground clearance: Registered .... 6 in.

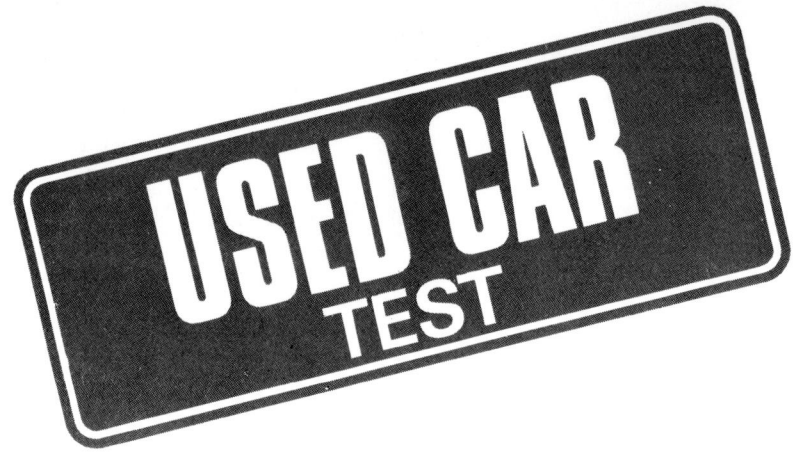

277 : **1964 Volvo 122S B18**

PRICES

Car for sale at Aldenham at	£625
Typical trade advertised price for same age and model in average condition	£630
Total cost of car when new including tax	£1,099
Depreciation over 3½ years	£474
Annual depreciation as proportion of cost new	12½ per cent

DATA

Date first registered	29 July 1964
Number of owners	1
Tax expires	31 March 1968
MoT expires	2 February 1969
Fuel consumption	23-26 mpg
Oil consumption	400 mpp
Mileometer reading	43,928

TYRES

Size: 165—15. Pirelli Cinturato on all five. Approx. cost per replacement cover £8 6s 6d. Depth of original tread 9mm; remaining tread depth: 8mm (front); 5mm (rear); 4mm on spare.

TOOLS

Jack, wheel nut spanner and hand tools complete; handbook in door pocket.

CAR FOR SALE AT :

Roundbush Garage Ltd., Aldenham, nr. Watford, Hertfordshire. Tel.: Radlett 6320.

PERFORMANCE CHECK

(Figures in brackets are those of the original Road Test, published 4 May 1962)

0 to 30 mph	4.5 sec (4.0)		
0 to 40 mph	7.8 sec (6.4)		
0 to 50 mph	10.6 sec (9.8)	In top gear :	
0 to 60 mph	15.3 sec (14.4)	10 to 30 mph	9.5 sec (—)
0 to 70 mph	22.2 sec (21.4)	20 to 40 mph	9.8 sec (8.8)
0 to 80 mph	36.8 sec (30.8)	30 to 50 mph	9.2 sec (8.9)
0 to 90 mph	56.2 sec (47.4)	40 to 60 mph	9.7 sec (9.4)
Standing ¼ mile	19.9 sec (19.8)	50 to 70 mph	12.5 sec (11.4)
Standing Km	31.7 sec (—)	60 to 80 mph	15.8 sec (14.9)
		70 to 90 mph	28.1 sec (26.0)

VARIETY in the choice of cars for test in this series is ensured by not repeating any actual model, yet we have tried a previous Volvo 122S. That test was published exactly three years ago, and dealt with the predecessor to the 122S B18, which was powered by the 1,582 c.c. engine. By a coincidence, the mileage and age of the car at the time of testing were much the same as here—43,000 miles-odd and just under four years—and it is interesting that the price asked in 1965 was only £50 less than the £625 which will today buy the 122S with B18, 1,780 c.c. engine and all the improvements described opposite. It has proved equally impressive in condition and performance.

Starting is reliable, though usually two or three revs on the starter and a lot of rich mixture are necessary when cold. The engine is always reluctant to pull at low revs, and although it remains very smooth when turning slowly on part throttle, it is almost as though it needs two or three seconds to "think" what has happened when the throttle is opened. This delayed response may be due partly to carburation leanness, and is noticed most when cold. The warm-up can be assisted by use of the radiator blind, a standard fitting, and it certainly seems that without it the running temperature is never really high enough.

At normal change-up speeds accelerator response is immediate, and the performance is impressive when the engine is made to work hard. A deep snarl of exhaust and induction roar adds to the feeling of zest, and the acceleration continues strongly right up to 90 mph, reached in 56.2 sec, but as this is even a little longer than the time for the 1,582 Volvo it is likely that careful tuning of ignition and carburettors will make it even better. From its mechanical quietness, smoothness, low oil consumption and the lack of fumes from the breather, it seems that there is a lot of service still to come from the engine, confirming a reputation for very long life.

Mechanical condition elsewhere is up to scratch, with the exception that some rattles and bonking noises from underneath on bad surfaces suggest that attention is needed, particularly to the dampers. The car retains its impression of solidity but there is too much reaction over bumps and quite a lot of wheel patter. The radial ply tyres may accentuate this, and are also responsible for making the steering very heavy at low speeds; but when the car is under way the control becomes lighter and the tyres hold on extremely well when cornering hard on greasy surfaces. At speed, the cam and roller steering lacks the precision one would like, but the Volvo is directionally stable even in side winds.

Adoption of Girling front disc brakes was one of the improvements for the model, and they are much more progressive than the snatchy drum brakes of the earlier car; there is no servo, and heavy pedal loads are needed, but good efficiency is available. The pull-up handbrake beside the driver's seat has a safety guard around the button release, and can be trusted to hold securely on steep gradients.

Clutch take-up is very smooth, with fairly light action; it can be engaged abruptly under full power for timed standing start getaways, and absorbs the power without any trace of clutch spin. Overdrive was available but was not added to this car. The four-speed gearbox retains a delightfully light, smooth and positive change, and the synchromesh is fully effective on all gears. The ratios are fairly well spaced, and third can be held to nearly 70 mph. Top gear cruising is very easy and relaxed at 70 mph, and still unstressed up to a true 80 mph.

With its rather low driving position, high scuttle and thick pillars, the Volvo seems a bigger car than it really is, but it is manageable. With its combination of really heavy duty construction, very good performance and fair economy, it is certainly "one to consider" for the man looking for a used car of quality at about the £600 mark.

Externally there is little to identify the later 1,780 c.c. B18 Volvo from the 122S with 1,582 c.c. engine

The bodywork and interior condition of this Volvo are extremely good. The durable kind of finish used does not deteriorate much with hard wear and tear. This has, in fact, been a doctor's car. Right: It is a very simple and accessible engine layout

Condition Summary

Bodywork

It is fairly easy to see that the car has at some time been involved in some small frontal accident, but the indication that only the centre section of the bumper had to be replaced (this can be guessed at when seen from underneath), and that only partial respraying was carried out, make it safe to presume that the damage was not substantial. The condition of the all-white paintwork is very good throughout, and there is little in the way of blemishes or rusted edges. One expects a Swedish car to be extensively rust-proofed underneath and it is a surprise to find that the body steel exposed to the road has very little protection. There is some rust, but it is not extensive, and the exhaust system is sound.

All the chromium and brightwork are in very good condition. Maroon pvc seats are rather hard, and initially very cold to sit on, but they are comfortable and well-shaped; the front back-rests are adjustable for rake through a small range. The rear seat appears to have been used very seldom, and apart from one or two marks on the perforated pvc roof lining the whole interior condition is clean and shows little sign of wear. Moulded rubber mats cover the floor.

Equipment

The only thing we can find which is not working perfectly is the indicator switch which occasionally does not self cancel after a left turn. Everything else is in sound working order, including the ribbon speedometer, which gives a steady reading throughout the range; but it does read rather optimistically, and indicates 86 at 80 mph. It is noticed that occasionally the ignition telltale does not completely extinguish, but charging is satisfactory.

Accessories

There is no radio, but the initial equipment of the Volvo was generous and included single point fixing lap-and-diagonal safety belts, an efficient heater and vigorous windscreen washers. Two-speed heater fan and wiper controls are fitted and, as mentioned, there is also a radiator blind.

About the 122S

The Volvo 120-series has been with us for over 11 years, and the basic engine range even longer. The very first model in the line was called the Amazon, had the familiar styling which is still unchanged, a four-door shell and the 1,583 c.c. four-cylinder engine developed from that of the PV.444. Power output was then 60 bhp at 4,500 rpm, c.r. only 7.5 to 1, and there was a three-speed gearbox.

By the time the car was marketed in Britain, a four-speed gearbox had been adopted, and all export cars had twin SU carburettors and an 8.2 c.r. helping to produce 85 bhp at 5,500 rpm. It was now called the 122S. Front seat belts became standard in 1959—this was the first European car to have them.

There were big changes for 1962. The 122S with B18 engine became an additional model. This was the 1,780 c.c. unit first shown in the P1800 coupé, developing 90 bhp (gross), and having a five-bearing crankshaft. Compression ratio was 8.5 to 1, and the cylinder head and manifolds were new. Girling front disc brakes were standard, and a Laycock overdrive optional. A 12-volt electrical system was adopted.

A single carburettor, 75 bhp, version of the car called the 121 had already been released, and this engine was fitted in the new 221, four-door estate car, announced in February 1962. To confuse everything further, a two-door saloon had been fed into the options, with either engine tune available. The 1,583 c.c. engine was discontinued during 1962. Two years later, the options and fittings were shuffled once again, with disc brakes being given to the 121, a servo to the estate car, interior re-styling to others, and 122S power up to 100 bhp (gross).

During 1966, the 140 series cars were introduced and to rationalize fittings with them, the 120 series had their engine power increased to 85 bhp (single carb) and 115 bhp (twin Stromberg carbs) respectively.

Once the 140 series had become well established, the 120 series was reduced, with all four-door saloons being discontinued during 1967. The two-door models are still produced, and a sporting version of the 132S (as the 122S was called in two-door form) was introduced—the 123GT. The latest engines are fully interchangeable with up-to-date 142 and 144 engines which are in full production. The two-carb 120 series car now being made is called the 133.

We drive the . . .

VOLVO 123 GT

SPORTS CAR WORLD · ROAD TEST

The Volvo 123GT is not the cheapest high-perform-ance car in Australia; and it's not the fastest. It may be one of the most economical, it's almost certainly the most durable and it's very likely the best all-rounder . . .

THE man said he thought it was really just a '38 Ford made in Sweden, anyway, and I felt like feeding him a face full of signet rings. But I don't wear any signet rings.

One still meets people like that: no substitute for cubic inches; God is a square-edged slushbox; bow before the V8 and all that garbage. Me, I just like a man's motor car. It doesn't have to have "lots of cubes" (although that can be nice) and space-age styling isn't a prime requisite. It does have to be functional, efficient, appealing and responsive. The driver of a Yank tank who gets a sample of the familiar inane driveway attendant jealousy may be able to render the offender ineffectual by raising the window glass at

the push of a button, but winding on six and a half turns of lock to get out of the man's garage will inevitably bring on a flushed face that destroys the whole image of indifference. With a quality car (read Volvo) you can at least do it with composure and dignity.

In fact you can do everything in a Volvo with composure and dignity: you can recover from a diabolical wet-road slide with no more effort than the simple ceremony of crossing wrists or stop for an unexpected obstacle while nonchalantly gazing out your own window — with confidence.

The average man might just revel in the sheer pleasure of driving a quality car that does exactly what he directs: but the average man would not drive a Volvo 123GT. For him there are more mundane (and less expensive) versions. The average man, too, might find a crude reference to shape in his 120 series car hard to answer, but a 123GT owner would likely treat the affront with total disdain, and its speaker to a lesson in road manners. After all Volvo does have living proof of its modern styling ability in the 144 series — the 123GT is a car to be coveted in the tradition of the modern classics.

The 123GT isn't spectacular or sensational as a high performance car — it's simply undramatically efficient. When you toss it at a corner hard enough to tie the boy racer behind in knots, it doesn't scream tyres and hang out on lots of lock. It just leans over on its chassis a bit to take up some of that G load off the suspension and goes round without protest. This is all part of the business of getting a good ride-handling balance. If you're driving on a wet road very fast and you're pretty observant you may note the front wheels in very fast corners sometimes point the way your head is facing — if it's looking out the side windows. But you'll note it without alarm. It's not worth going into further details on this car's handling — it's sufficient to say it handles. So does its tamer brother, the 122S, but in the GT (what a horrid label), the adhesion limits appear to have been upped to coincide with a power increase by slightly firmer damper settings and (possibly) slightly firmer spring settings. This makes for increased stickability, probably more body roll, and some extra noise (only slight) on rough surfaces.

Of course that roll-oversteer sensation at the

limit is delayed a little longer, but it's still just as predictable and controllable when it comes. For a good driver one of the most pleasant sensations of speed can be obtained on high speed sweeping corners, when the rear end tends to crab a little at the limit of adhesion. It's very gentle, doesn't need any steering correction and doesn't introduce rear-end steer, but it does become accentuated on slightly uneven surfaces and should be treated as a warning of approaching on-the-limits cornering.

Driving in a straight line isn't nearly so complicated: it's just a matter of poking it with the steering dead straight. There's no wheel tramp, axle dance or spinning wheels, unless you're very careless with the clutch, but there is strong straight acceleration especially after 3500 rpm where the cam and howling exhaust come in together with a blast.

That doesn't mean the car's a fussy competition machine either. You can mumble around town in any gear with a minimum of clutch slip and drive-line clatter, although we did notice at low rpm gearchanges did introduce the occasional dull transmission jolt when not handled with absolute smoothness. It felt more like a loosening universal joint than a driveline engineering fault and we haven't noticed it in other Volvos.

It's significant though, that when you leave the city limits behind in this car everything starts to smooth out. This is a trait common to only the best Continental high performance machinery and the sort of guy who buys this car will revel in the way the car can be sent down the highway at maximum speed mile on mile with a smoothly connected series of cockpit movements. The Nuvolari-sized steering wheel, occasionally too large in tight spots around town becomes a delight to swing on, and the low-speed feeling of front end weight on the steering disappears too. The brakes, a little heavy to operate around town and quite savage when they're used hard, are perfect on the open road. The top end performance is first class and body roll movement decreases with speed. Briefly, the car is completely sensitive.

The sort of man who buys extra performance

inevitably demands extra sophistication as well — that's just buying basics — and the Volvo gives it to him. AB Volvo, Gothenburg, Sweden, knows all about fitting out a high performance car for owner one-upmanship purposes.

It starts externally with twin QI spotlamps at the front, supplied as standard equipment with foam backed plastic protector covers. One is a long-range spot lamp, the other has low-angled light diffusion for good lighting of the verges of the road when cornering or driving in fog. There is more: chromed hub caps on wheels shod as standard with radials (Pirellis on ours) and fine profile windscreen wiper blades for high speed stickability. Of course there are GT badges all round. Quite out of place I feel, are the twin wing mirrors fitted to the front guards. Even when well adjusted (which is hard because they have a lock-tight base with built-in spring kit to deflect on impact — presumably with a pedestrian) the mirrors offer at most a token guide to action at the rear of the car, due mainly to the convex glass fitted and awkward location. A single racing-type mirror fitted on the driver's door would be at least twice as effective.

But the interior is near perfect. Volvo has added a deep-dished alloy-spoked plastic rim steering wheel that is both a delight to use and look at. It carries the 123GT emblem on the horn boss. This new performance-style wheel is the same huge diameter as fitted to the normal 120 series sedans and you might be excused for wondering why this giant was not reduced in size until you remember that at parking speeds the front end concentration is heavy enough to make the larger rim size necessary. In any case it soon becomes very pleasant to use and can be spun through the fingers in tight spots as easily as the best power-assisted American style wheel.

Volvo considers practicability a prime and canted the alloy spokes slightly downward from the horizontal to allow better vision of the instrument binnacle. Over to the right on the dashboard and only a glance away is the tachometer: It's a Smiths unit, swivel-based in a handsome plastic console of its own with redlining in the 6-7000 rpm sector. You can set it up to point through the wheel rim at you or carelessly brush it across to look into your passenger's eyes while you pull six G in top and then knock the needle back a cog with a shift to overdrive — if you dare. But it's near impossible to read at night.

Also on the top of the console is a new little tray — for carrying a pair of wrap-around glasses, a pair of fingerless driving mittens or . . .

The rest is familiar to Volvo lovers: the fabulous seats — reclining in this model — that hug you well enough without the standard seat belts and their fancy little centre-clip locks; the complete control centre — two speed wiper/washers,

Continued on page 191

There's little between them on price, and nothing to choose on performance. Now go fight somewhere else about it. This month's lucky dip: Volvo 123GT v Alfa-Romeo Giulia Super v Lancia 1.3 Rallye v Toyota 1600GT.

WELL, we've really done it this time! Our four car comparisons cause enough arguments at the best of times, but now we've really jumped in at the deep end. This month's giant comparison is guaranteed to cause fights that will last for years — Alfa-Romeo Giulia Super, Lancia Fulvia Rallye 1.3 coupe, Volvo 123GT, and Toyota Corona 1600 GT5.

These are four of the best-value high-performance cars on our market today. While they differ widely in concept, they are incredibly close in performance and price. The Alfa costs $3998, the Volvo $4160, the Toyota $4400 and the Lancia $4550; say, $550 spread over the four, which is larger than the normal four-car comparison spread but really doesn't matter much when you're spending this kind of money for a very personal car.

At the moment the Alfa sells the most, but the other three have only just reached the market, although the little Fulvia 2C coupe with the

FOUR
fire-eaters

1200 cc engine has been around for more than a year. In this area, it's interesting to see that the only other cars which offer any sort of a challenge are the Audi Super 90 ($3740), the BMW 1600 ($3988), the DS19 Citroen ($4480), the Falcon GT ($3890) and the Triumph 2000MD ($4150). For various reasons, mainly of price and character, none of these fits with the comparison except the BMW — but the NSW distributors don't seem very interested in having one tested.

The Alfa was supplied privately, although distributor Alec Mildren would not have hesitated had he been asked; the Lancia came from the distributors, Lambda Motors, the Volvo from British and Continental Cars and the Toyota from Australian Motor Industries.

Our usual disclaimers: All cars were stock, apart from radio and seat belts in the Alfa and 5½ in. rims and Michelin XAS tyres on the Volvo used for photography (the car used for figures was stock). All dimensional measurements were redone, apart from the makers' figures in the first section of the charts. Figures for horsepower, torque, compression ratio, fuel tank capacity, brake lining area and weight are from the manufacturers.

GENERAL: This was by far the most interesting giant comparison in the four years we have been doing them, mainly because each car represents an entirely different approach to being a high-performance thoroughbred.

FOUR fire-eaters

The Alfa is the only four-door car, for the 123GT is available only in the two-door Volvo form, but while the Toyota and the Volvo are uprated versions of bread-and-butter units the two Italian cars are original from the ground up. But the similarities — not only in performance figures — are really astonishing. Three — Alfa, Lancia, and Toyota — have ohc engines; two — Alfa and Toyota — have five-speed gearboxes, while the Volvo has an overdrive on top to equate five speeds; all have superb fully-reclining bucket seats; two (Alfa and Lancia) have discs all round.

The Volvo is patently the oldest design of them all; but old in this case should not be taken to mean out-dated. The B18 122S formula has

1. Alfa shows gentle roll, good stickability. 2. Lancia rolls slightly more, looks better balanced. 3. Volvo actually shows smallest roll angles, but more understeer. 4. Toyota has marked understeer.

been with us a while, but the 123GT is only a recent development. It is basically the two-door 122 body with the 115 bhp 1800S engine, but with radial covers, laminated screen, higher rated dampers, matched auxiliary lights, tachometer, alloy wheel, and the overdrive coupled to the 4.56 final drive from the wagon — the normal ratio is 4.1. What is equally remarkable is that the engine uses the standard P1800 bottom end even in its rally form of 145 bhp.

The Lancia grew up differently. It first saw the light of day as the Fulvia 2C coupe, a charming, rapier-

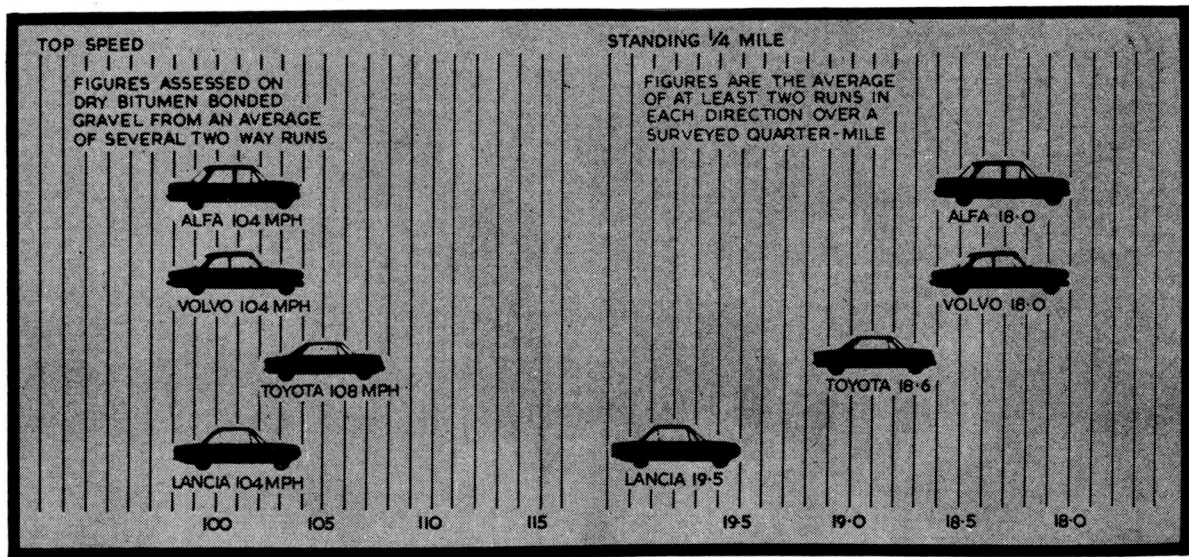

TOP SPEED

FIGURES ASSESSED ON DRY BITUMEN BONDED GRAVEL FROM AN AVERAGE OF SEVERAL TWO WAY RUNS

ALFA 104 MPH

VOLVO 104 MPH

TOYOTA 108 MPH

LANCIA 104 MPH

STANDING ¼ MILE

FIGURES ARE THE AVERAGE OF AT LEAST TWO RUNS IN EACH DIRECTION OVER A SURVEYED QUARTER-MILE

ALFA 18·0

VOLVO 18·0

TOYOTA 18·6

LANCIA 19·5

like flier that was typical Lancia but which lacked real torque from the 1216 cc 80 (DIN) bhp four. However, last year it gained another 82 ccs by dint of stroking from 67 to 69.7 mm, and while leaving the compression ratio on 9 to 1 the power in the Rallye 1.3 version went up to 87 at 6000 rpm and the torque from 76.7 at 4000 to 83.9 at 4500. There is, of course, the HF version with 101 bhp and 10.5 to 1 compression, if you like that sort of thing.

The 1600GT Toyota, on the other hand, sort of topsied. It started when engineers from Yamaha of Hamamatsu produced an experimen-tal twin-cam conversion on the Corona four, and then were con-tracted by Toyota to produce the dohc six for the 2000GT. The ex-perimental four then found its way into a works competition car, and before anybody could say knife it was a production model. It uses the seats from the 2000GT and the five-speed gearbox from the same source,

1. Alfa: Good arm and leg angles, poor support under thighs. 2. Lan-cia: Excellent wheel-pedal relation-ship, cushions too short. 3. Volvo: Wonderful seats and good position-ing spoilt by big wheel. 4. Toyota: Best driving position of the group.

although a four-slotter is available with the GT4 name-tag. The body is the two-door notchback Corona ccupe which we first saw in Austra-lia 18 months ago in 1600S form.

Probably the Alfa is now too com-mon and too much of an old friend to bear explanation. The four cylin-der dohc Giulia engine is by no means young, but there seems no reason to change it. In racing form it gives over 140 bhp, so the 98 bhp of the Giulia Super is small beer indeed. The car itself came from the Giulia TI, which was Alfa's bread-and-butter family car of about four years ago. But when people started

PERFORMANCE	Alfa Giulia	Lancia 2c Rallye	Volvo 123GT	Toyota GT5
Top speed	104 mph	104 mph	104 mph	108 mph
Standing $\frac{1}{4}$ mile	18.0 secs	19.5 secs	18.0 secs	18.6 secs
0-50 mph	8.6 secs	8.8 secs	7.7 secs	8.5 secs
30-50 mph	7.1 secs (4th)	11.0 secs	8.0 secs	NA (4th)
Fuel consumption	21.2 mpg	27.5 mpg	23.3 mpg	22.4 mpg
Maximum in gears:				
First	30 mph	30 mph	35 mph	37 mph
Second	50 mph	52 mph	52 mph	69 mph
Third	75 mph	80 mph	74 mph	92 mph
Fourth	95 mph	104 mph	104 mph	104 mph
Fifth	104 mph	—	95 mph (o/d)	108 mph
OVERALL DIMENSIONS				
Wheelbase	8 ft 3 in.	7 ft 7 in.	8 ft 6 in.	7 ft 11.3 in.
Track front	4 ft 3 in.	4 ft 3.1 in.	4 ft 3 in.	4 ft 2.8 in.
Track rear	4 ft 2 in.	4 ft 2.3 in.	4 ft 3 in.	4 ft 2 in.
Length	13 ft 7 in.	13 ft 0.5 in.	14 ft 7 in.	13 ft 5.8 in.
Width	5 ft 1.5 in.	5 ft 1.2 in.	5 ft 4 in.	5 ft 1.6 in.
Height	4 ft 8 in.	4 ft 3.2 in.	4 ft 10 in.	NA
Ground clearance	6.5 in.	4.7 in.	NA	7.1 in.
Fuel tank capacity	10 gals	8.5 gals	10 gals	9.9 gals
Weight	19.7 cwt	18.9 cwt	22.6 cwt	20 cwt

to race the TI Alfa brought out the TI Super with more poke and stiffer springing. But since the GTA — based on the Sprint GT — was homologated and became the racing four-seater the TI Super was converted to the Giulia Super as a slightly derated but better equipped road car.

ENGINES: Volvo's B18 four is the only pushrod in the group, but it has

Engines are all different, all beautiful. 1. Alfa; 2. Lancia; 3 Volvo; 4 Toyota.

proved itself to be so rugged and capable of producing infinite power that this is no disadvantage. It actually started as a 1400 cc mill developing 40 bhp, and this went to 80 bhp before it was found necessary to enlarge it to 1600 ccs and 85 bhp. Now it is 1778 ccs and can give 145 bhp; is there no limit to the resilience of this engine?

It is noisy over 4500 rpm, needs a slight touch of benzol to make it run well and stop running-on, is surprisingly economical at an esti-

Our usual wiper test shows up interesting points: 1. Alfa doesn't really clean much glass, has bad blind spot on driver's right; 2. Lancia isn't much better, with worse blind spot; 3. Volvo is best of all, but that's a shallow screen to start with; 4. Toyota is good for driver, not so hot for passenger-san.

mated average of around 26 mpg, and will spin to 6000 rpm. The noise is mainly induction roar for the two HS6 SU carburettors, but it still gets

SPECIFICATIONS

SPECIFICATIONS	Alfa Giulia	Lancia 2C Rallye	Volvo 123GT	Toyota GT5
Cubic capacity	1570 cc	1298 cc	1778 cc	1587 cc
Bore/stroke	78 by 82 mm	77 by 69.7 mm	84.14 by 80 mm	80.5 by 78 mm
Bhp at rpm	112 at 6500	87 at 6000	115 at 6000	110 at 6200
Torque lb/ft at rpm	NA	83.9 at 4500	112 at 4000	101.3 at 5000
Compression ratio	9.0 to 1	9.0 to 1	10 to 1	9 to 1
Carburettor/s	2 twin choke Weber	2 Solex C35 PHH	2 downdraft SU	2 sidedraft Solex PHH
Transmission	5 sp, all syn	4 sp, all syn	4 sp, all syn, with o/d	5 sp, all syn
Gear lever location	central floor	central floor	central floor	central floor
Suspension: front	coils, a/r bar	transverse leaf spring, a/r bar	coils, a/r bar	coils, a/r bar
rear	coils, a/r bar, Panhard rod, trailing arms, live axle	leaf spring, beam axle, torque rods, a/r bar	coils, torque rod, live axle	leaf springs, torque rods, live axle
Tyre size	155 by 15	145 by 14	165 by 15	6.45 by 14
Make (standard equipment)	Pirelli Cinturato	Michelin X or XAS	Pirelli Cinturato	Dunlop Grand Speed
Turns l to l	3.5	4.2	3.2	3.3
Turning circle	34 ft 10 in.	32 ft	31 ft	32 ft 6 in.
Brakes front/rear	disc/disc	disc/disc	disc/drum	disc/drum
Brakes diameter	11.2/8.6 in.	10.4/10.0 in.	10.8/9 in.	10.6/9.0 in.
Price	$3998	$4590	$4160	$4395

CALCULATED DATA

CALCULATED DATA	Alfa Giulia	Lancia 2C Rallye	Volvo 123GT	Toyota GT5
Final drive ratio	4.555 to 1	3.7 to 1	4.56 to 1	4.375 to 1
Mph/1000 rpm (top)	19.0 mph (5th)	16.9 (4th)	16 (4th), 21.3 (o/d)	21.5 (5th)
Piston speed at max bhp	2720 ft/min	3500 ft/min	3150 ft/min	3185 ft/min
Power to weight ratio	19.7 lb/bhp	24.2 lb/bhp	23.6 lb/ft	20.7 lb/ft

SPECIFIC DIMENSIONS

SPECIFIC DIMENSIONS	Alfa Giulia	Lancia 2C Rallye	Volvo 123GT	Toyota GT5
Boot lip height	1 ft 5 in.	2 ft 5 in.	2 ft 0½ in.	2 ft 5 in.
Depth of boot	2 ft 6 in.	1 ft 1½ in.	1 ft 10 in.	1 ft 7 in.
Mean interior width	3 ft 10 in.	3 ft 9 in.	3 ft 10 in.	3 ft 10 in.
Drivers' window				
Width (ex ¼ vent)	1 ft 4 in.	2 ft 2½ in.	1 ft 10 in.	2 ft 2½ in.
Front door aperture	2 ft 9 in.	3 ft 8 in.	3 ft 3 in.	3 ft 9 in.
Leg room rear	6 in.	1 in.	9 in.	3½ in.
Front seat travel	7¾ in.	6½ in.	10 in.	5½ in.
Steering wheel dia.	15½ in.	15½ in.	17 in.	15½ in.
Chest to wheel boss centre	1 ft 9 in.	1 ft 8 in.	1 ft 6 in.	1 ft 5½ in.
Rear vision mirror width	7½ in.	8 in.	8 in.	8¼ in..
Effective glovebox width	11 in.	1 ft 0½ in.	1 ft 8 in. (parcel shelf)	6 in.
Effective glovebox depth	4 in.	4 in.	7 in. (parcel shelf)	6 in.

EQUIPMENT

EQUIPMENT	Alfa Giulia	Lancia 2C Rallye	Volvo 123GT	Toyota GT5
Heater/demister	yes	yes	yes	yes
W/screen wipers	2 speed	2 speed	2 speed	2 speed
W/screen washers	yes	yes	yes	yes
Reverse lights	yes	yes	yes	yes
Boot light	yes	no	yes	no
Ashtrays	2 r, 1 f	1 f	2 r, 1 f	2 r, 1 f
Parcel shelf	yes	no	yes	yes
Grab handles	all round	ns front	ns front	n and os front
Headlamp flashers	yes	yes	yes	yes
Seat belts	no	no	2, 3 pts	2, lap
Window winder turns	4.8	2.5	3.2	4.0
Instruments	sp, tach, clock, water, oil, fuel, 10ths odo	sp, tach, fuel, oil, water, clock, 10ths odo	sp, tach, fuel, water, 10ths odo	sp, tach, fuel, oil, water, amps, 10ths odo
Handbrake location	centre facia	centre floor	right floor	centre facia
Door lock system	3 h, 1 k	1 b, 1 k	1 b, 1 k	1 b, 1 k
Boot lock system	b, k	b, k	latch, k	k
Keys	2	2	2	2
Lockable fuel cap	no	yes	no	yes
Interior bonnet lock	yes	yes	yes	yes
Choke system	manual	manual	manual	manual
Thru flow vent	no	no	no	yes
Under bonnet s/proof	yes	no	no	no
Armrests	2 f, 3 r	2 f	2 f, 3 r	3 f
Cigar lighter	1 f	1 f	1 f	1 f
Dress rims	no	yes	no	yes
Alternator	no	no	yes	yes
Bonnet stay	auto	manual	auto	auto
Ex rear mirror	no	1	no	2
Radio	no	no	no	yes
Recl seats	yes	yes	yes	yes
Rad ply tyres	yes	yes	yes	no (nylon)
Wood rim wheel	no	yes	no	imitation
Engine light	no	yes	yes	no
Mudflaps	no	no	yes	no

annoying in the indirects. Cruising in overdrive is, however, quite relaxed and quiet.

The Giulia engine is also noisy, but noisy in the Italian manner of cam gear thrash and exhaust crackle, rather than induction noise. It spins alarmingly freely, going quickly up to 6500 rpm. This is the same engine as used in the GTV, GTA, and GTZ; it differs from the Giulia TI in new valve timing and twin choke Webers.

The twin-cam conversion on the standard Corona crankcase has the inherent weakness of being only a three-bearing bottom end, but there is no sign of this in the smoothness and flexibility of this excellent four. The two twin-choke Mikuni-Solex 40PHH side-draught carburettors feed into the aluminium alloy head which carries the two camshafts driven by two-staged duplex roller chains. The chain tensioning and damping system is very elaborate, and helps make this one of the quietest overhead cam drives we have struck. In fact, this engine is quietest of the four.

The V4 was new for the Fulvia series of Lancias, with the 2C coupe originally using twin carburettors for more output than the little sedan. Now the 1.3 has given it torque as well as rpm, which should be very handy, for if the 2C had one fault it was the way steepish main road hills would haul off the top 15 mph of the cruising speed. The new engine also has an improved head and exhaust system — both new castings — plus two Solex C35PHH downdraughts. The head and crankcase are alloy, the block cast iron. The tachometer stops at 7000, but 6200 is the normal safe maximum. This engine has a freer, more turbine-like feel and sound than the others.

TRANSMISSIONS: The most appealing of the four is undoubtedly the Alfa's five-speed system, with a good-feeling chromed lever, sitting in a voluminous vinyl glove, that picks up the gears in a twinkling. The throws are longish, but "soft" and quite accurate. On the other hand, most of our drivers preferred the Lancia gearbox, for although it has only four speeds, operated by a lever working from almost on the toe-board under the facia, it is butter-smooth and lightning-fast and has the ability almost to beat your hand into the next slot. However, the clutch, although larger now, still shows signs of incipient slip. The Toyota gearbox is stiff and heavy (even after 6000 miles on the test car), with first hard to find on the move and reverse almost impossible to get any time. The movement from fifth back to fourth is the same as in the Alfa, but where in the Alfa one simply wipes the lever smartly backwards it has to be two conscious movements in the Japanese car. The Volvo is another thing again; the long lever works a very fine gearbox, but because of the lever length the feel is more notchy and less precise than in the others. On the other hand, the electric overdrive is first-class — it acts instantly, needs only a quick stab of the clutch to take up the shock. Direct top can be flicked in at almost 90 mph, and this makes the car astonishingly fast and beautifully controllable in all conditions in that 70-100 mph cruising band where these four cars will spend most of their days.

The overdrive, on a ratio of 0.756, is a little too tall for the available torque (it would probably be fine with a larger piston area) and in fact the car is faster in direct top than overdrive. The direct gearing is 16.0 mph per 1000 rpm in top, and overdrive 21.3; as you can see, the gap is large, but because overdrive is so tall and because it is so pleasant to use, the fuel consumption figure does not benefit accordingly, for you find yourself getting real joy out of accelerating hard in overdrive top.

The Lancia, which has a larger (7.874 in.) clutch than the 2C, pulls a 3.7 to 1 final drive, giving a gearing of 18.11 mph per 1000. It absolutely astonishes us with its ability to pull hard right through the range despite this very tall gearing for only 1300 ccs. The drag factor is very good, which helps, but it still is ridiculously flexible for its volume coupled with that gearing. The new head must work beautifully.

The Alfa has a ratio of 0.791 in fifth, which gears it at 19.8 mph, or 15.7 mph in direct top. The gearbox ratios are the best-spaced of the four. This is partly due also to a low final drive of 4.555 to 1, but the penalty for this of course is an ultra-low first gear. In the Toyota, fifth equates 21.5 mph per 1000, and fourth 18.0. This is excellent up the top end, but the ratios otherwise don't work all that well as first is too high for town work

PERFORMANCE: In terms of a comparison, we are now treading on dangerous ground. We could take refuge like cowards and tell you to check out the figures and make up your own mind; but these are all remarkably similar and in any case don't convey anything at all about how these cars cover the countryside. The Lancia, particularly, makes the stop-watches into liars.

However, one can't escape the fact that the Toyota is overall the fastest of the four, while the Alfa and the Volvo are neck-and-neck over a quarter-mile. The Lancia certainly does it with more ease and more precision than the others, while the Alfa seems to have the more spirit and fire in its job. Over give-and-take point-to-point conditions, given equal drivers, the Alfa would probably finish slightly ahead, as it would on a circuit. This is all caught up with ride, brakes, adhesion, and a dozen other things and is essentially a subjective assessment. For instance, at least one of our drivers, a real Lanciaphobile, swears that in the Rallye 1.3 he could beat the ears off any Alfa, while an Alfa-owning staff man is equally confident for his team.

HANDLING AND RIDE: With performance so equal on paper, this is the area where the cookie crumbles. Take the Alfa first: Its only real faults are faintly low-geared steering, a slight rubbery "walking" in the rear suspension — due doubtless to the amount of insulation in the very well located live rear axle — and a characteristic diagonal lurch when tossed hard across on lock. Otherwise, it strides across bad surfaces with no loss of adhesion or wheel tramp, is magnificent on loose and wet surfaces, and despite slight final oversteer, very easily controlled with power. Despite a spongy servo, it stops dramatically well, with little or no wheel-locking. Part of the trick of the game, of course, is that in an Alfa you can always have exactly the right gear and the right amount of power for any situation, which, in the final analysis, probably gives it that edge in point-to-point work. The steering is too low-geared, and winding on a lot of lock in a hairpin gets suddenly complicated if the tail lets go and you have to sort out the right amount of correction. There is some radial ply thump evident always.

The other Italian car is overall the best handler of the four. Despite front-wheel-drive, it has absolutely no feedback into the slightly too low-geared steering, no characteristic switch into oversteer when you shut the throttle in the middle of the corner — no bad habits whatsoever. It will not change line an inch regardless of how the surface alters through a corner. It is a consistent mild-to-neutral understeerer but can be made to oversteer in the wet or on gravel.

The main contribution to this superb handling is the steering — by far the best we have ever encountered on a test car. It is light, pin-sharp, and accurate in car-placing to the very last millimetre. The ride is also first-class, without transmitting any radial ply reflected effects. The brakes are as good as the Alfa's.

By comparison the Volvo feels cumbersome, with a big, vertical steering wheel and a lot more bulk to shift. But its handling is also well beyond reproach; almost neutral, with some feedback in the steering and some lost motion from the dead-ahead position. It is very hard to make the 123GT lose adhesion at the rear, and this car has the best ride of all over all types of surface. The feeling of "oneness" and solidity contributes greatly to the car's ability to cover big distances at very high averages. It just lacks that razor-sharp edge of spirit which the two Italian cars have. However, the brakes, if anything, are slightly better than the others, despite rear drums, mainly because the pedal is more progressive and transmits more information about what each pound of loading is doing to the deceleration.

The Toyota comes a poor fourth in this bracket, even though its handling is well and truly above average. The rear axle is not quite as well located as it should be, despite two trailing torque rods, and one can make an inside wheel lift and spin. The steering is heavy, particularly at town speeds, and the high-speed Japanese-made cross plies, despite fatter rims, don't quite match up. The ride is also on the firm side, although in a pleasantly accurate way. It understeers mildly and consistently, but is a predictable final oversteerer. Lighter, more sensitive steering would enable the enthusiastic driver to make more use of the considerable power available. Brakes have a peculiarly dead feel at first, but work well.

COMFORT AND CONTROL: The four vary a lot in their approach to looking after the driver. The best cockpit of all is the Toyota's. The driving position is faultless — a beautifully-positioned wood wheel with heavily raised and padded boss, seats that go right up behind the shoulders and out to behind the knees, good leg and arm angles. The seat reclining lever, partway up the outside edge of each squab, is a little awkward, but gives very fine adjustment.

It has a headlight flasher — rare among Japanese cars — and a central console which mounts the gear-lever also has a small odds-and-ends recess. The wheel spokes are painted dull black to cut glare, and clever recessing of the very clear and steady instruments, plus the use of satin-finish wipers (the only car of the four with these) means that there are no windscreen reflections at all. There is a blind spot in both rear pillars; however, the window winders are a little low-geared, although all glass is pleasantly tinted. All the interior trim is in black. The heater/demister works well and the washer has Toyota's usual powerful motor. A locker, with a padded lid, sits between the seats aft of the central console; This, a biggish glovebox and a deep well behind the rear squab supplement the fairly tight boot. The rear squab folds flat to form an extra luggage platform, but in any case the rear seat is not comfortable for two adults over any distance.

In the Lancia you get some unusual seats — heavily rolled around the edges with gaps in the squab centre piece for better circulation. The reclining buckets locate well against side forces, have a good range of adjustment and are very comfortable, but do not give enough support under the thighs. The two-spoked wood wheel is placed at good reach, but is a little low, so that most drivers will have to sit splay-legged. The armrest/door pulls are narrow and hard, and none of the switchgear is labelled, which is bad design, for there are several strange levers poking out

from under the dash which need labels. There is not really enough room for two adults in back on a long trip.

Full marks to the Lancia for its unique system of mounting the horn button in the usual place but using the centre of it to work the headlight flasher, so that you need only one hand to let the slowcoach know he's about to be engulfed by something very fast. The quarter vents are fixed and we found that the swivelling air vent system did not work as well on this car as on our last — or maybe the weather was hotter. There is a grab handle in front of the passenger and a reasonable glovebox. One brilliant touch is that the two keys are not only differently shaped but have different colored plastic ends. Carpeting is haircord, with a rubber insert under the driver's feet.

In the Alfa the driving position is also good, but the black plastic, alloy-spoked wheel is also set a touch low, forcing a splay-legged attitude. Speckled carpeting (with rubber inserts) and a reasonable standard of finish make things quite comfortable, although the rear vision mirror is too small and the door hardware fiddly. Instrumentation is clear white on black, but the same faults are there — lack of proper identification and insufficient seat support under the thighs. The seats are quite soft and well-shaped otherwise. You get big, shaped armrests which act as door pulls, courtesy lights on all doors (this is, of course, the only four door car in the group), and — like the Lancia — extra pockets under the dash on each side.

The Alfa does have one bad fault in that the pedals are not hung from the toeboard but work vertically from a floor mounting. This not only looks untidy, it also makes for some peculiar ankle angles at times. Apart from anything else it's damned old-fashioned and unnecessary.

The Volvo's fabulous seats, by far the best in the group with their orthopedic shaping and the way they grip from knee to small of back to shoulder, are spoiled by the use of an overlarge black plastic, alloy-spoked wheel. This is set with hardly any rake at all, and shorter drivers have to look under the top of the rim. This is the only car of the four to have pvc floor covering. There is a passenger's panic handle, but no glovebox, which is consistently annoying. There is a small odds-and-ends recess atop the facia next to the swivelling tachometer pod, plus map pockets in the doors, but you'd better just be glad that the Volvo has the largest boot of the four.

It is also undoubtedly the most comfortable inside for four passengers, who have more leg room than in the Alfa. It is also by far the safest car in a crash-engineering sense, and has the usual solid Volvo three-point belts which latch to a massive central hook. Instrumentation is old-fashioned and a bit garish, but all the switchgear is in easy reach and properly labelled. There is an excellent heater-demister, but no through-flow ventilation system. The standard of engineering is typically Volvo — which means first-class.

SUMMARY: If you think we're going to nominate the best out of this lot you deserve to be committed. That's a good way to get your head kicked in in a dark alley one night. If you go on styling it has to be the Lancia, for the Alfa is downright ugly, the Volvo old-fashioned and the Toyota a bit bland. If you want to race it at weekends, buy the Alfa. If you want it to last 10 years and still look good, take the Volvo. The Toyota is the only one of the four without anything specific to recommend it, but we'd just like to say here that it should get a new body by the Tokyo Show in October and then watch it sell.

Believe it or not, we're very glad we don't have $4600 to spend. It would worry us considerably . . . sleepless nights . . . grey hairs . . . it's nice to be poor.

\#

A MORE-GO

A dark-horse in the personalised transport market which really deserves a second look.

· AT-A-GLANCE:
Healthy new performance from 2-litre four . . . still noisy . . . very good leg- and headroom for four . . . sure-footed power brakes . . . accent on safety.

A SURVEY among Volvo owners has shown them to be one of the most dedicated single marque motorists. Porsche, of course, must take the honors with 95 percent of owners re-buying Porsche. Apparently solidarity, comfort, superb finish and safety must count more than swinging open the door to, say, the impressive plushness of a new Monaro. On the Australian market, the smallest Volvo sells for the equivalent of a GTS, even with V8 power.

When Ace Nolan of Swedish Motors suggested we take the lowliest Volvo — the 122S now powered by the B20, 2-litre engine, to test, we said, "Fine, Bill, we'd love to have a run in the S B20."

Now the latest Volvos haven't really quite turned us on. They are quality cars now selling in Australia for better than reasonable prices and Keran, Winkless, Andre Welinsky, Gerry Lister and co-piloting them in rallies has really shown their worth. Mind you those Repco-headed rally cars don't bear a great resemblance to an auto 144S off the showroom but they prove the basics are there. The 122S B20 quite rejuvenated our interest in the Swedish sedans. Out into the triffid tangle of Friday afternoon Mascot traffic, the 122S felt one of the fieriest stock Volvos we have had. It instantly took us back to the old Grey Lady — a 122S B18 test car Swedish Motors had back in the old days before things really began happening to Volvo in Australia. That was really a car, one of those gems where everything clicked and a bit of hard use only gilded the lily.

We instantly thought, well here's a vehicle well worth a long and close look. It justified this first assessment.

By the time we reached the office after putting the B20 through its

Latest 122S is externally unchanged from B18 series but new B20 engine gives new performance to betray the high-waisted, archaic lines.

The heart of the matter. Lusty 118 bhp four gives pleasing performance. Wide rod is leverage for power boosted, twin-circuit brakes.

VOLVO

wheels ROAD TEST

paces we knew it must be a good report.

There's performance to match the quality and maybe high waisted design went out in 1938 but there is a certain archaic charm. It's not unlike the Morgan *(tested this issue)* — safety design and control functions of 1969 blended with homely surrounds synonymous with purist motoring.

The B20 engine is carried throughout the Volvo range and has been the most significant capacity change for Volvos since the early 1960s. It came at the end of 1968 to give the

A new steering wheel, parking brake warning light (far right) and steering lock are additional interior touches in the B20 122S 2-door.

TECHNICAL DETAILS

MAKE:	Volvo	COLOR:	mid-blue
MODEL:	122S B20	MILEAGE. START:	1450
BODY TYPE:	2-door	FINISH:	1920
OPTIONS: Pirelli tyres $35		WEIGHT:	21 cwt
PRICE:	$3706		

FUEL CONSUMPTION:
Overall .... 28 mpg
Cruising .... 28-32 mpg

TEST CONDITIONS:
Weather, fine; surface, hot mix bitumen; load, two persons; fuel, premium

SPEEDOMETER ERROR:

Indicated mph:	30	40	50	60	70
Actual mph:	29.5	38.4	47.6	56.8	75,6

PERFORMANCE

Piston speed at max bhp .... 2500 ft/min
Top gear mph per 1000 rpm .... 17.7 mph
Engine rpm at maximum speed .... 5650 rpm
Engine rpm at cruising speed .... 3954 rpm
Lbs (laden) per gross bhp (power to weight) 20

MAXIMUM SPEEDS:
Fastest run .... 101 mph
Average of all runs .... 99.8 mph
Speedometer indication fastest run .... 106 mph
In gears: 1st, 37 mph; 2nd, 56 mph; 3rd, 85 mph; 4th, 99 mph.

ACCELERATION:
(through gears)
0-30 mph .... 4.0 secs
0-40 mph .... 5.2 secs
0-50 mph .... 8.5 secs
0-60 mph .... 10.2 secs
0-70 mph .... 15.5 secs
0-80 mph .... 19.8 secs
0-90 mph .... 28.0 secs

	3rd gear	4th gear
20-40 mph	5.3 secs	9.1 secs
30-50 mph	5.5 secs	7.5 secs
40-60 mph	5.4 secs	7.4 secs
50-70 mph	6.5 secs	8.2 secs

STANDING QUARTER MILE:
Fastest run .... 18.4 secs
Average of all runs .... 18.5 secs

SPECIFICATIONS

ENGINE:
Cylinders .... four in line
Bore and stroke .... 88.9 mm by 80 mm
Cubic capacity .... 1986 cc
Compression ratio .... 9.5 to 1
Valves .... ohv
Carburettors .... twin SUs
Fuel pump .... mechanical
Oil filter .... full flow, paper element
Power at rpm .... 118 bhp at 4800 rpm
Torque at rpm .... 123.2 lb/ft at 3500 rpm

TRANSMISSION:
Type .... 4-speed all syncromesh
Clutch .... sdp
Gear lever location .... central
Overall Ratios: 1st 3.13 to 1; 2nd, 1.99 to 1; 3rd, 1.36 to 1; 4th, 1.000 to 1; final drive, 3.25 to 1.

CHASSIS AND RUNNING GEAR:
Construction .... integral
Suspension, front .... coils, double wishbones
Suspension, rear .... live axle, trailing arm torque rods, panhard coils
Shock absorbers .... telescopic
Steering type .... cam roller
Turns 1 to 1 .... four
Turning circle .... 30 ft 4 in.
Brakes type .... four wheel disc power assisted dual circuit
Dimensions .... 8.6 in. dia

OVERALL DIMENSIONS:
Wheelbase .... 102 in.
Track, front .... 53 in.
Track, rear .... 53 in.
Length .... 14 ft 7 in.
Height .... 4 ft 11 in.
Width .... 5 ft 4 in.
Fuel tank capacity .... 13½ gals

TYRES:
Size .... 165 x 15
Make on test car .... Pirelli Cinturato

GROUND CLEARANCE:
Registered .... 6 in.

(Graph: Acceleration through gears with change points; X-axis "ELAPSED TIME IN SECONDS" 0-40; Y-axis "MPH" 0-90; markings: IST 37MPH, 2ND 56 MPH, 3RD 85MPH, STANDING ¼ MILE 18·5, TOP SPEED 99·8 M.P.H.)

GO-GO VOLVO

new heavier 144 extra zest. Now use this power in the old body that ran quite happily on 1.8 litres, thank you, and the result is most satisfying. The last series of 122Ss used a 115 bhp engine tuned as a rally conversion. Though the B20 has only 3 more bhp, a hefty 123 lb/ft of torque gives you solid wheelspinning power, in the two-door body.

Combine this with an excellent gearbox and you have a sports sedan that belies its outmoded looks. The gearbox is so good that it entirely compensates for an impossible heel-and-toe pedal position, strange for a car of this nature. We found when sidling up to lights first would slide in without hesitation at up to 25 mph. Integration with the clutch is precise and after short familiarisation, smooth progress is natural. There is some transmission whip on the overrun which is easily accounted for with early clutch throwout.

The Grey Lady was always one of our greatest arguments for a well-located live axle versus fully independent suspension. That car even under the most savage acceleration or braking would not put a wheel out of place. The new B20 122S has a similar axle location and if anything it is just too good. The axle is held so rigidly and the clutch bites so well, there is nowhere for the power to go but in wheelspin. Running our performance figures we found low rpm take-offs produced best times, bogging in wheelspin resulting from high rpm.

The tendency to wheelspin easily can be used to advantage in handling. Shod with Pirelli radials, the Volvo has high roadholding powers and an expected degree of understeer is almost non-existent. The 122S thrives on oversteer power sliding which is not only great fun but gives immense control. Even in wet conditions, the excellent rear axle location gives limited slip differential control, though of course it is a free axle. The wheels remain large at 15 inches

— a very good way to leap potholes. With well-bushed wishbone links and rear leaves, the ride is very good even over rough dirt road going. Volvos have proved their suspension solidarity and will take immense amounts of destruction testing before showing the first signs of strain in front sub-frame.

Slipping behind the wheel has one immediately grabbing for the seat and squab adjuster. When you realise the driver before was not a midget, the Volvo's archaism strikes harder. That is until you've run a few miles and decided, well, those Swedes aren't so stupid after all. Long arm driving is very well to impress your lady friend, but a big wheel, close upright seats and bent arms are far more relaxing and less fatiguing for long distances. Besides those orthopaedically-correct seats hold the torso so well there is no clinging on to the wheel for extra support needed for that drop of sporty driving. The seats can be adjusted for fore-aft movement, squab angle (a cam adjuster only not fully reclining), lumbar support and height. Because they are set high on individual platforms, the rear seat room is very good and unlike today's shoulder stiffening fastbacks, there is sufficient headroom to make it a genuine four seater.

At $3706 plus $35 for the optional Pirelli tyres, the 122S B20 falls into fairly fierce company in two-door sporty type cars — the Fiat 124S coupe, Toyota 1900SL, Capri and Monaro. Probably the genuine Volvo buyer would scarcely look at a Monaro or Capri as in the same class preferring the Volvo for its excellent finish and caring to ignore the outdated looks.

With the new B20 powerplant, there is performance to match any of the above coupes with handling and superb braking as well. The hidden extras are the safety design and Swedish ruggedness which will keep the 122S purring and untouched when others are on a second set of bearings.   #

VOLVO 123 GT

Continued from page 179

a choke that pulls out into your lap (you need it on cold mornings), a cigar lighter, heater demister/blower fan grab handles; and the extras — coat hooks, opening rear vents and map lights.

Our test car was painted in deep BRG which brought the initial glances which eventually saw the GT badges which led to questions which . . .

The paintwork was extended to the interior and matched with tan upholstery (black on the crash padding and below the door sill bottoms) and white perforated pvc trim for the headlining.

The GT comes as a two-door only, because Volvo figure the man who wants sporting performance won't mind sacrificing the extra weight to having his kids clamber in over the fold-down seat backs

A true wolf in sheep's clothing, it's virtually got the P1800S sports coupe's mechanicals. Engine and gearbox (complete with O/D top) come straight ex-P1800S and the rear axle ratio is added for obvious reasons: the standard 120 series sedan with type B18 engine doesn't have overdrive. The difference is 4.56 to 4.10, the latter being standard.

The P1800S engine is worth an additional 15 bhp on the normal 122S sedan: that gives it a rating of 115 bhp at 6000 rpm, whereas the 122S develops its peak at 5700. Torque is increased to 112 lb/ft at 4000 rpm from 108 lb/ft at 3500 rpm. Most of this comes from better breathing: The GT has a cleaner head and 10 to 1 compression ratio (122S is 8.7 to 1) plus a mild cam grind that you could almost tune on the exhaust note — it comes in around 3500 with a bellow, but otherwise still pulls smoothly and strongly.

Volvo also fitted competition shock absorbers to make sure that any increase in power was met with a proportional increase in handling ability. I suspect the springs got a gentle up-rating too, but this is a little more difficult to determine.

Volvo feels the braking department didn't need any attention apart from the addition of a power booster and what feels like competition linings, though there's no mention made of this in the literature on the car.

The clutch is a slightly soft point on this car though it is doubtful whether the average owner or even a particularly demanding one would ever discover any ultimate failure. It broke down on our acceleration runs, but recovered quickly from slip with a short cool-off period. Coincidentally the only troubles recently suffered by Volvos in endurance racing have been in the clutch, though locally it has been for different reasons.

Volvo attention to detail is meticulous but nowhere so apparent as under the bonnet where you could forgive a company for being a little sloppy and concentrating more on the functional than the sanitary side. But you only have to lift the counterbalanced lid and take a look in for the final evidence. Like the boot compartment it is automatically lit, and also like the boot it's clean. All exposed metal surfaces are enamelled like the exterior and there is little chrome — but lots of polished alloy: on the rocker boxes, carby bells, inlet manifold, brake booster and so on. The cooling system is sealed with a special overflow spill tank and all the electricals are housed in little alloy boxes in front along the guard, where you can get at them. There is also an alternator.

Price-wise the Volvo is a bit of a loner in this country. Its tamer sister, the 122S, sells for some $600 less and the distortion in extra value is due entirely to the import taxes these cars suffer. It's good for near-110 mph and will go down to low 17 seconds over the standing quarter mile with a few miles behind it. These days you have to pay out at least Ford Falcon GT-money for that sort of urge, so the Volvo still doesn't fall behind the pace. Remember, with that extra performance you still get good fuel economy and lots of sophistication. That counts.   #

the car with its high sills and small glass area, but he will find no fault in the handling and ride of this capable road machine.

Being heavy and firmly-sprung, it can take any type of road surface in its long stride, and it has a high margin of positive safety in its

122S-B-20

VOLVO

As the fans expected, this 2-litre version is the top performer among 4-cylinder Volvo models...

A CAR ROAD TEST

WHILE emphasis is shifting to the Volvo 144 and 164 models, both declared "manufactured in South Africa", the older 122S model remains popular — particularly with younger motorists — and continues in production here indefinitely as an assembled model.

The very functional 122S is a familiar car, stylish in a robust sort of way, smaller inside than the new models, but considerably lighter than the 144 and with a slightly smaller frontal area.

In the two-litre version introduced two years ago, it has gained considerably in torque and marginally in power output, with beneficial effects on overall performance.

BY HOW MUCH?

When Volvo's come under discussion — which is often — the point at issue is by how much the new 122S-B-20 outperforms the earlier B18-engined model, and that we can settle quickly by a quick comparison of this Test with that of the B18 published in June, 1967 (in Imperial measures, for convenience):

		B18	**B20**
0–50	..	9·2	**8·0**
0–60	..	13·1	**11·2**
0–70	..	18·4	**15·9**
¼ Mile	..	18·7	**18·0**

The B20 model has less than 3 per

cent increase in power output, but the torque increase — which has a pronounced effect in acceleration tests — is a healthy 9 per cent over the 1·8-litre model, and peaks at 3 500 rpm instead of the earlier 4 000 rpm.

Both cars are easily capable of 100-mph-plus, with not much difference between them in this department because of the very small power difference.

ECONOMY IMPROVES

The B20 model has very strong overtaking and gradient ability, and another benefit of the torque increase is that fuel economy actually becomes slightly better, in spite of the increase in bore size. For a heavy car, in fact, the km/l figures are remarkable.

Apart from the 6-cylinder 164, the 122S-B20 becomes the top Volvo performer — which is no more than the *aficionados* expected!

DATA AT 120

Min. noise level	83·5 dBA
0–120 through gears . .	18·3 sec
Economy at 120 . 8·5 km/l (11·8 litres/	100 km)
Braking from 120	4·7 sec
Reserve power at 120 .	0·076 grav
Max. gradient at 120 (top) .	1 in 13·2
Rpm at 120 (top)	4 285

HANDLING AND APPEAL

Quite naturally, the 122S lacks the modernity of the newer 144 models, but it remains a strong and appealing car, with pleasant and individualistic looks which are liked by owners.

A driver not accustomed to the 122S body feels rather immersed in

handling and the outstanding Volvo safety-braking system.

Most owners prefer radial-ply tyres for their extra adhesion and riding comfort.

SUMMARY

A great car in the Volvo tradition, this 122S-B20. The engine is smooth, very strong and virtually unburstable, the car is well-mannered, and its quality is uncontested.

It is likely to have a large and enthusiastic following for many years to come, among men who take their motoring seriously. ●

IMPERIAL DATA

Major performance features of this Road Test are summarised below in Imperial measures:

PERFORMANCE FACTORS:

Power/mass (lb/bhp)	23·8
Mph/1 000 rpm (top) . . .	17·45

ACCELERATION FROM REST (in seconds):

0–30	3·4
0–40	5·4
0–50	8·0
0–60	11·2
0–70	15·9
¼ Mile	18·0

MAXIMUM SPEED:

True speed . . .	101·1 mph

FUEL ECONOMY:

30 mph	49·2 mpg
45 mph	41·1 mpg
60 mph	33·7 mpg
70 mph	28·5 mpg

GEARED SPEEDS:

1st gear	33·4 mph
2nd gear	52·6 mph
3rd gear	76·8 mph
Top gear	104·5 mph

(Converted with reasonable accuracy from the Road Test results in metric measures.)

SPECIFICATIONS

ENGINE:
Cylinders	Four in line, B-20-B
Carburettors	Twin SU-HS6
Bore	88·9 mm (3·50 in.)
Stroke	80·0 mm (3·15 in.)
Cubic capacity	1 986 cm³ (121·5 cu. in.)
Compression ratio	9·3 to 1
Valve gear	Ohv, pushrods
Main bearings	Five
Aircleaners	Dry elements, foam-covered
Fuel rating	98-octane
Cooling	Water, 8·5 litres, expansion tank
Electrics	12-volt AC

ENGINE OUTPUT:
Max. power SAE	88·0 kW (118 bhp)
Max. power net	77·7 kW (100·4 bhp)
Peak rpm	5 800
Max. torque/rpm	166·2 N.m/3 500 (123 lb. ft.)

TRANSMISSION:
Forward speeds	Four
Synchromesh	All
Gearshift	Floor
Low gear	3·13 to 1
2nd gear	1·99 to 1
3rd gear	1·36 to 1
Top gear	Direct
Reverse gear	3·25 to 1
Final drive	4·10 to 1
Drive wheels	Rear
Tyre size	165S x 15

BRAKES:
Front	Discs
Rear	Discs
Boosting	Vacuum servo
Handbrake position	Inside driver door

STEERING:
Type	Cam and roller
Lock to lock	4·0 turns
Turning circle	9·5 m (31·1 ft.)

MEASUREMENTS:
Length overall	4·45 m (175·2 in.)
Width overall	1·62 m (63·8 in.)
Height overall	1·50 m (59·0 in.)
Wheelbase	2·60 m (102·4 in.)
Front track	1·31 m (51·6 in.)
Rear track	1·31 m (51·6 in.)
Ground clearance	0·21 m (8·3 in.)
Licensing weight	1 080 kg (2 381 lb.)

SUSPENSION:
Front	Independent
Type	Coils and stabiliser
Rear	Live axle
Type	Coils, torque and track rod

CAPACITIES:
Seating	4/5
Fuel tank	45 litres (10 gal)
Luggage trunk	0·5 m³ (17·5 cu. ft.)

TYRE PRESSURES:
Crossply:	Front	1·8 to 2·1 bars
	Rear	1·9 to 2·3 bars
Radial ply:	Front	1·9 to 2·2 bars
	Rear	2·1 to 2·5 bars

WARRANTY:
Six months.

BASIC PRICE:
Coast and Reef	R2 825

PROVIDED TEST CAR:
Lawson Motors Group, in association with Norton Motors, Cape Town.

TEST CONDITIONS:
Altitude	At sea level
Weather	Fine and mild
Fuel used	93-octane
Test car's odometer	4 000 km

INTERIOR NOISE LEVEL

ACCELERATION

FUEL CONSUMPTION

ENGINE SPEED

PERFORMANCE

MAKE AND MODEL:
Make	Volvo
Model	122S-B20B

PERFORMANCE FACTORS:
Power/mass (kg/kW)	13·9
Frontal area (m²)	2·43
Drag at 100 km/h (kg)	55·2
km/h/1 000 rpm (top)	28·0

(Calculated on licensing mass, gross frontal area, gearing and net power output.)

INTERIOR NOISE LEVELS:
	Min.	Wind	Road
Idling	48·0	—	—
60 km/h	70·0	72·0	80·0
80 km/h	74·5	77·0	83·0
100 km/h	78·5	81·5	86·0
Full throttle			See graph
Average dBA at 100			82·0

(Measured in decibels, "A" weighting, averaging runs both ways on a level road; "Minimum" with car closed; "Wind" with one window fully open; "Road" on a coarse gravel surface.)

ACCELERATION FROM REST:
0–50	3·6
0–60	5·1
0–70	6·3
0–80	7·8
0–90	9·9
0–100	11·7
0–110	15·0
0–120	18·3
400 m	18·0

OVERTAKING ACCELERATION:
	3rd	Top
40–60	4·1	6·2
60–80	4·2	5·4
80–100	4·6	5·8
100–120	5·7	6·7

(Measured in seconds, to true speeds, averaging runs both ways on a level road, car carrying test crew of two and standard test equipment.)

MAXIMUM SPEED:
True speed	162·3
Speedo reading	167

FUEL ECONOMY (litres/100 km in brackets):
60 km/h	16·0 km/l (6·2)
80 km/h	13·8 km/l (7·2)
100 km/h	11·4 km/l (8·8)
Full throttle	See graph

(Measured in kilometres per litre, averaging runs both ways on a level road.)

BRAKING TEST:
From 80 km/h:
First stop	2·8
Tenth stop	2·9
Average	2·9

(Measured in sec, with stops from true speeds at 30 sec intervals on a good bitumenised surface.)

GRADIENTS IN GEARS:
Low gear	1 in 2·9
2nd gear	1 in 4·0
3rd gear	1 in 5·7
Top gear	1 in 8·2

(Tabulated from Tapley g readings, car carrying test crew of two and standard test equipment.)

GEARED SPEEDS:
Low gear	53·8
2nd gear	84·6
3rd gear	123·7
Top gear	168·2

(Calculated to true speeds, at engine peak rpm — 6 000.)

PERFORMANCE

MAKE AND MODEL:
Make . . . . Volvo
Model 122S B20

INTERIOR NOISE LEVELS:

	Mech.
Idling	53
60 km/h	72
80 km/h	74
100 km/h	78
120 km/h	82

(Measured in decibels, "A" weighting, averaging runs both ways on a level road with the car closed.)

ACCELERATION FROM REST (in seconds)
0–60 km/h 4,3
0–80 km/h 6,7
0–100 km/h 10,1
0–120 km/h 14,6
400m sprint 16,5

OVERTAKING ACCELERATION:

	3rd	Top
40–60	 1,8	 2,9
60–80	. . . 3,4	 5,1
80–100	. . 3,6	 5,3
100–120	. . 4,6	 6,1

Measured in seconds, to true speeds, averaging runs both ways on a level road, car carrying test crew of two and standard test equipment.)

MAXIMUM SPEED:
True speed . .175,2 km/h (109,5 mph)
Calibration (in km/h):

Indicated:	60	80	100	120
True speed:	60	59,2	92,8	108,8

GRADIENTS IN GEARS:
Low gear . . . . .1 in 2,8
2nd gear . . . . .1 in 3,9
3rd gear . . . . .1 in 5,7
Top gear . . . . .1 in 8,2
(Tabulated from Tapley (x gravity readings, car carrying test crew of two and standard test equipment.)

GEARED SPEEDS (km/h)
Low gear . . . . , . . 51,8
2nd gear 81,6
3rd gear 119,3
Top gear 162,6
(Calculated at peak engine rpm – 5800)

TEST CONDITIONS:
Altitude . . . . at sea level
Weather . . . . Fine, warm and windy
Fuel used . . . . 98 octane

TEST CAR FROM: Alconi Developments of Johannesburg.

CONVERSION DETAILS

Alconi conversion for Volvo 122S

Stage I
Powerthrust Alconi exhaust . R 38,00
Modified intake manifold . . R 27,00
Fitting (including new carburettor needles and dynomometer tune-up . . . . R 50,00

Stage II
Parts as per Stage I . . . . . R 65,00
Plus Stage II polished ported and and modified cylinder head . R 45,00
Labour, fitting charges, gaskets, etc . . . . R 85,00
R195,00

Stage III
Optional cylinder head with larger valves R28,00

Over a dirt road, the car proved vice-free, even at high speeds. The slides were easily controllable.

The rugged, four-cylinder 122S motor looks like any other. But unobtrusive mods have lifted power output to a claimed 135 bhp (101,1 kW) SAE.

"Dyno-tuned" . . . the Alconi label.

SPECIFICATIONS

ENGINE
Cylinders . . . . Four-in-line
Carburettors . . . .Twin SU–HS 6
Bore . . . . 88,9 mm
Stroke . . . . 80,0 mm
Cubic capacity . . . . . 1986 cm³
Compression ratio . . . . .10 to 1
Valve gear . . . . . Ohv, pushrods
Main bearings . . . . Five
Air cleaner . . . . . Paper element
Fuel requirement 98-Octane

TRANSMISSION:
Forward speeds . . . . Four
Synchromesh . . . . .All
Gearshift position . . . . . .Floor
Low gear . . . . .3,13 to 1
2nd gear 1 . . . . .1,99 to 1
3rd gear . . . . .1,36 to 1
Top gear . . . . Direct.
Reverse gear . . . . . . .3,25 to 1
Final drive . . . . 4,1 to 1
Drive wheels . . . . .Rear.

WHEELS AND TYRES:
Road wheels . Standard pressed rims
Rim width 4J
Type of tyre . . . . . Pirelli radials
Tyre size . . . . 165 x 15

BRAKES:
Front . . . . Discs with Ferodo Formula 2430 (non-fade pads)
Rear . . . . Drums
Boosting . . . . .ATE booster
Handbrake position . . . . . Right of driver's seat on floor.

STEERING:
Type : Worm and roller
Lock to Lock : 4,0 turns
Turning circle : 9,5 metres (31 ft. 4 ins.

SUSPENSION:
Front : Independent
Type : Coils and stabiliser bar
Rear : Live axle
Type : Coils, torque and track rod.

MEASUREMENTS:
Length overall . . 4,45m (175,2 ins)
Width overall . . . .1,62m (63,8 ins)
Height overall . . . .1,5m (59,0 ins)
Wheelbase . . . . .2,6m (102,4 ins)
Front track . . . . .1,31m (51,6 ins)
Rear track . . . . .1,31m (51,6 ins)
Ground clearance . . .0,21m (8,3 ins)
Licensing weight . 1080 kg (2381 lb)

CAPACITIES
Seating 4/5
Fuel tank . . . . . 45 litres (10 gal)
Luggage trunk . . 0,5m (17,5 cu. ft)

techni**car** TESTS

A TUNED VOLVO

Simple mods bring out the latent power...

ONCE A Volvo man always a Volvo fan. Few cars have such a devoted band of supporters, which is probably why Volvo have managed to continue producing cars on what is a small scale, even by Continental standards. Volvos are strong cars, widely valued for their durability and for the fact that they rarely change their shape. They are more expensive than most cars of similar size, but the people who buy them expect to pay more for the quality they demand.

When Alconi of Johannesburg asked us to test their engine conversion for the legendary Volvo 122S, we were interested to see whether the car retained all its Volvo characteristics — including that impressive low speed torque. (Incidentally, the kit can be fitted to the 144S, which is powered by the same engine.)

The car which we tested was the one used as a test bed vehicle for the development of the conversion. The odometer showed 30 000 miles and despite the fact that most of this mileage had been completed with the

conversion installed, the engine had required no attention other than routine servicing. The owner was delighted with the conversion, which had not added any temperament to the engine's performance. In fact, the car is used as a family transport and is mostly driven by his wife.

The 1970 model car was brought to us by "Puddles" Adler, a director of Alconi, who had been closely involved with the development of the kit. Puddles told us that the conversion dated back two years. At that time, Alconi decided that there was scope for improving the engine efficiency of the 122S by eliminating deficiencies brought about by cost considerations in mass production.

With this in mind, Alconi went ahead with tests and found that a few finishing-off operations to engine components made a tremendous difference to performance. For example, a free-flow exhaust system developed by Powerthrust, in conjunction with Alconi's dynomometer, gave an extra 6 bhp! This is made up from heavier

gauge metal than usual and consequently, should have a longer life.

After removing the emission control device on the manifold, gas-flowing and polishing the exhaust and inlet manifolds boosted efficiency levels in both power and fuel consumption. Surprisingly enough, this operation, in conjunction with a carburettor needle swop to suit altitude conditions, had the effect of *decreasing* the carbon monoxide exhaust emissions by about 7 per cent — right through the rev range.

These promising results led Alconi to improve the cylinder head. They gas-flowed, polished, ported and relieved — and then balanced the combustion chamber capacities to within half a cubic centimeter.

This brought the power output up to around 97,5 kW (130 bhp) SAE. After optional 2mm larger valves from the Volvo P1800E (fuel-injected) engine had been installed, the power was up to 101,1 kW (135 bhp). The torque was also significantly stronger.

An impressive feature of the con-

Even after so many years the B20's lines look strong, simple and fast.

version is that it brings about an increase right throughout the rev range, both in power and torque. This is possible because Alconi purposely avoided changing cams — although they did conduct experiments with the Volvo "D" camshaft which showed that the bhp could be increased to 140, but only at the cost of losing power below 3 000 rpm and causing noisy tappet action.

Puddles told us that with more expensive and elaborate conversions involving changes of cam and carburettors, further improvements would only be marginal — and would almost certainly involve some loss of tractability.

We drove the car for a considerable distance and we were impressed by the complete absence of running-on, something which often plagues the "10-to-one-compression" Volvos. Moreover, the car showed no tendencies to pre-ignition or pinking on 98-octane petrol. Any symptoms of this kind had been eliminated by polishing and re-designing the combustion chambers.

The highest true top speed recorded by us was 175,2 km/h (109,5 mph) but this was not quite a true reflection of the car's capabilities as the engine developed a mild miss at about 6 300 rpm which may have been due to faulty fuel delivery or an ignition leak. On our first run we managed to reach the magic figure of 110 mph but on the reverse run, the miss became worse and subsequently the car refused to exceed 109 mph in either direction. We therefore feel that the purchasers of this kit could reasonably expect a true speed of 110 mph if their car was in perfect tune.

The restricting factor would be gearing rather than power, for at the car's maximum speed, the engine was doing approximately 6 300 rpm — some 500 rpm above its peak output level. Alconi could have corrected this by using higher gearing to bring the engine "back on to the cam" at top speed using either larger diameter wheels or a different diff ratio, or they could have changed the cam. However, as already mentioned, Alconi were unwilling to sacrifice low-end flexibility. They also felt that a higher top speed would have been of little practical value in view of the blanket speed limit of 120 km/h.

In developing the kit, Alconi's basic objective was to give the B20 engine an all-round increase in power, aiming it at the man who would like his Volvo to be brisker without having to drive it like a cammy car.

The converted car remains a good road vehicle and its power increase has

been obtained with a minimum of fitting effort.

The total price of the kit as fitted to the B20 we tested is R223 and it can be bought in three optional stages as follows: Stage One: Powerthrust Alconi exhaust R38; modified intake manifold R27; and fitting (including new carburettor needles and dynomometer tune-up) R50. Stage Two: parts as per stage one, R65; polished, ported and modified cylinder head (on exchange), R45; and fitting charges, gas-

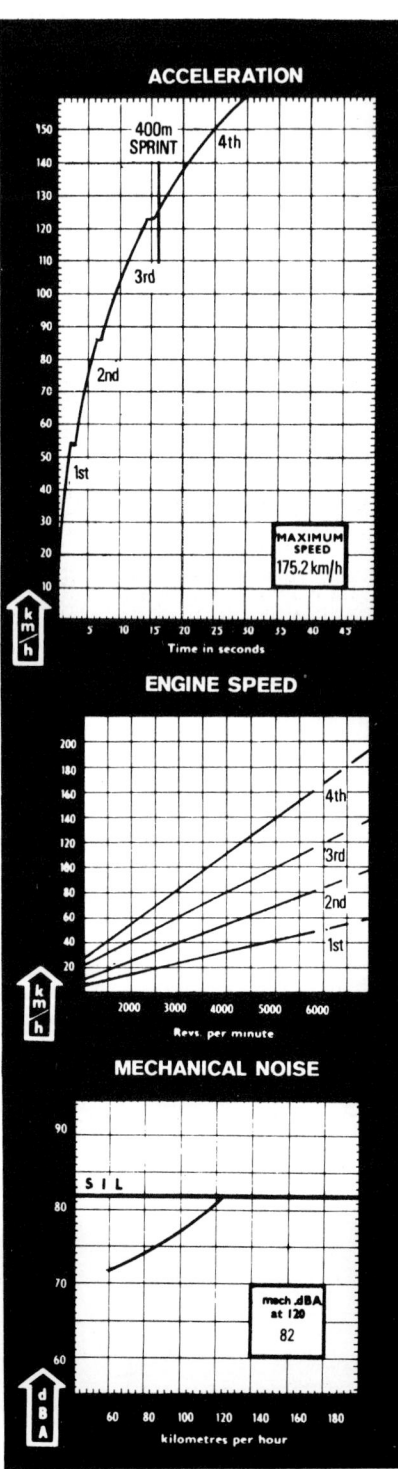

kets etc. R85 — or a total of R195. Stage Three: the modified cylinder head with 2 mm larger valves — R28 extra.

As can be seen, an enthusiast can save himself R85 by fitting the Stage Two kit and making the necessary adjustments himself. Alternatively, a smaller sum could be saved by fitting the kit at home and then having the assembly tuned on a dynomometer.

The car we tested had Koni shock absorbers and was shod with Pirelli 165-15 tyres on the standard 4j,15 rims. The Pirellis were inflated to 1,65 bars (24psi) in front and 1,79 bars (26 psi) at the rear. The ride thus achieved was firm and very stable. To counter any risk of brake fade, the test car had Ferodo Formula 2430 pads fitted to its front discs.

The proof of the power is on the drag-strip: just how much does the Alconi kit improve performance? The difference is quite impressive when our test results are compared with those of the standard Volvo B20, as tested by our associate CAR in its November 1970 issue.

Top speed has improved from 161,7km/h (101,1mph) to 175,2km/h (109,5mph) and the acceleration times show a similar increase.

	Standard B20	Alconi B20
0..60km/h	5,1	4,3
0..80km/h	7,8	6,7
0..100km/h	11,7	10,1
0..120km/h	18,3	14,6
400m sprint	18	16,5

Acceleration from 0 to 60 mph improved from 11,2 to 8,6 seconds! This improvement is not proportional to those shown above because the car was privately owned by a friend of Puddles' and we were anxious to protect the machinery during most acceleration runs. However, during the 0 to 60 mph runs, we really let her rip because Alconi believed that this is the figure for which most people look when making comparisons.

It may come as a surprise to many Volvo owners that a worthwhile increase can be gained from such relatively simple modifications to the engine — one which makes it seem almost as if a larger engine has been installed.

The kit should be of special interest to owners of automatic Volvos who would like to modify their cars but are hesitant to do so for fear that they might lose some bottom-end torque — and so adversely affect the operation of their transmission. With this kit, the problem does not apply. A similar conversion is available for the Volvo 164. ●

VOLVO 122-S & P 1800

Used sporting cars for people who think

BY THOS L. BRYANT

YOU PROBABLY DON'T know the names Assar Gabrielsson or Gustaf Larsson and there's really no reason why you should. But, they were the founders of AB Volvo in 1924 and thus the originators of the Swedish car industry. Gabrielsson was a management expert at SKF, a Swedish ball bearing manufacturer, and Larsson was a young engineer with a considerable interest in automobiles. After three years of hard work, they produced their first car in April 1927, a touring car with a 28-bhp 4-cyl engine. Production that first year totaled 297 cars but over the past 50 years, production has risen nearly a thousandfold. Some of the cars from the Gothenburg plant in recent years have been especially noteworthy to the enthusiast and thus form the basis for this "Used Car Classic."

122-S

THE FIRST Volvos to appear in America arrived in the mid-1950s and were 444 models. Everyone went around commenting on how cute they were and how they resembled, from the rear, a 1946 Ford. Well, the 444 was in production in 1944, so

perhaps this was another case of our ethnocentricity getting in the way of the facts. However, the 444 and later the 544, which appeared in 1958, proved to be popular among enthusiasts and the Swedish car manufacturer was off and running in the U.S.

In 1956, Volvo showed the 120 Series to the press in Sweden and reaction was quite favorable to the new design. In Europe the car would become known as the Amazon, but in the U.S. the moniker was 122-S. It took two years for the 122-S to make the journey to the American market and in the September 1959 issue of R&T there appeared a laudatory road test report on the car: "And a refreshing new car it is, too: pleasant looking, easy (and fun) to drive, economical and durable in the extreme. It is also refreshing to find a company that actually does something to make its product safe for the occupants, and does it without asinine statements that the public won't buy safety."

The R&T report went on to say, "...the newest import is a handsome car in a reserved way, with no evident ostentation or gaudiness." Also, "...a close examination of the car, along with many miles behind the wheel, brought favorable comments from every tester and rider. Design, construction and general quality are obviously excellent, and there is a pervasive feeling of durability."

The early 122-S was powered by a 1586-cc engine which had been used in the 444 since 1956. In the 122-S, the engine, designated the B-16, developed 85 bhp at 5500 rpm and 87 lb-ft torque at 3500 rpm. In Sweden, the B-16 engine could be purchased with a single carburetor and detuned output of 60 bhp, but the export model had twin SU carburetors. All of the motoring press was enthusiastic about the performance capability of the 122-S, saying the 4-cyl engine was extremely flexible and one of the most free-revving rocker-arm engines around; pointing out, too, that the car's performance was one of its outstanding attractions, especially considering that it was a family 4-seater. The R&T test in 1959 showed a top speed of 92.0 mph and a 0-60 mph time of 16.2 sec. Certainly not breathtaking performance, but

admirable for a 4-door sedan with a 4-cyl engine.

The 122-S was built as a unit body/frame car for safety and durability. Other safety features included optional shoulder belts and such standard items as a padded instrument panel, dished steering wheel and a collapsible steering column. The package tray on the passenger's side was also designed to collapse under impact and the sun visors were thickly padded—all of this long before governmental regulations came into being.

The suspension design of the 122-S was carried over from the 444 and 544: coil springs and tube shocks all around with A-arms and an anti-roll bar in front, while at the rear there were trailing arms and a Panhard rod for lateral location of the live axle. This combination worked very well and we noted in our original test report that "corners can be taken with gusto, though with considerable body roll and squealing of tires . . ." We also noted that the ZF steering was precise and transmitted good road feel to the driver, although it seemed slower than its 3.2 turns lock to lock would indicate.

For 1962, Volvo added a 2-door sedan to the 122-S line and upped the engine displacement to 1780 cc. Power rose from 85 bhp to 90 at 5000 rpm and the torque went from 87 lb-ft to 105 at 4000 rpm. The increased displacement of the B-18 engine was accomplished by an increase in the bore (from 3.13 to 3.31) while the stroke remained the same as in the B-16 engine. At the same time, disc brakes were now standard on the front instead of drums all around as on the earlier models, and a 12-volt electrical system replaced the 6-volt setup. The new B-18 engine was reported by Volvo to have greater fuel economy as a result of increasing the number of crankshaft bearings from three to five,

full machining of all combustion chambers and two additional intake ports resulting in direct induction to each cylinder. Our tests, however, didn't support the factory claims. We reported that the 122-S with the B-16 engine would return 24–27 mpg (R&T, September 1959) while the B-18 engine delivered 21–26 mpg (R&T, May 1962). The larger engine did show an improvement in performance, however, as the acceleration time from 0–60 mph dropped from 16.2 sec to 14.5.

The new engine was also used in the P 1800 which had been introduced just shortly before the 122-S 2-door, so let's take a look at that model.

P 1800

THE P 1800 was designed in 1959 and introduced in late 1961 as a 1962 model. It shared the same suspension and engine with the 122-S but it had 10 more horsepower (100 @ 5500 rpm vs 90 @ 5000) as a result of different carburetion. The body was designed by Frua of Italy and the first couple years' production was built in England with the drivetrain components being shipped over from Sweden. Late in 1964 production was transferred to the Volvo factory in Sweden.

R&T tested the P 1800 in February 1962 and made the following comments about its performance: "All cars of this type that come our way for test get a thorough wringing-out on twisty roads and there the Volvo, despite its weight and soft ride, gave a fine performance. There is a dreadful amount of lean while cornering, but the driver can't feel it inside the car and it doesn't seem to affect the handling. Bends, fast or slow, can be taken with *élan*—just a touch of steadying understeer being present at all times."

We also noted that the P 1800 was not meant for sprinting and that it would take a drubbing in acceleration from less expensive cars, but went on to say, "In doing that for which it was intended, fast steady cruising, the P 1800 is superb and it gave us the impression it would run forever at near maximum speed."

With the move to production of the 2 + 2 coupe in Sweden the car became known as the P 1800S. The basic changes were a revised interior including new seats, less fancy wheel covers and a boost in the bhp figure to 115 at 6000 rpm. Despite the increased horsepower, performance was virtually unchanged (0–60 mph in 13.9 sec for the 1800S vs 13.6 sec for the original P 1800) and it was more or less a case of Volvo continuing with what they considered a basically good car that needed only refinement. By the time of the R&T road test of the 1800S (August 1966) our staff had become less than enthralled with the styling of the car: ". . . staff opinions on the 1800S styling were generally unenthusiastic, with low marks going to the chromium sweepspear and the semi-finned rear fenders, both cliches of a bygone American era."

In 1969, the 1800 went through another evolutionary step, with the engine jumping from 1800 cc to a full 2 liters with fuel injection. The car was designated the 1800E and there were other important changes as well, including 4-wheel disc brakes, Michelin radial tires, aluminum alloy wheels and a strengthened gearbox. The eggcrate grille was replaced by a simpler and more attractive one composed of horizontal bars, but the basic styling was the same.

The 2-liter, fuel-injected engine gave the 1800E greatly improved performance, lowering the 0–60 mph time to just over 10 sec. The 2-liter engine was also used in the sedans of the day, but only the 1800 received the fuel injection and thus had 12 more bhp and 7 more lb-ft torque than the carbureted Volvo engine at no increase in engine speed. Our comments in the road test of the 1800E (February 1970) went like this: "The engine, noted for its durability rather than refinement—it's neither mechanically smooth nor quiet in the coupe—has good low-speed torque as well as the ability to pull nicely all the way to its 6500-rpm redline, and in overdrive the car will now do an honest 115 mph. The 0–60 mph and ¼-mile times are quite respectable too, putting the 1800E into the same class with such cars as the Alfa 1750, BMW 2500 or Mercedes 280SL. Furthermore, the engine runs cleanly without any trace of emission-control leanness symptoms and uses very little more fuel than the earlier test car."

PHOTOS BY GORDON CHITTENDEN

Introduction of 1800E in 1969 brought with it a nicely upgraded dash layout.

Volvo 1800S.

Volvo 1800E.

Our 1970 test shows we were quite favorably impressed with the new gearbox: "The hefty lever on the new gearbox gives one an impression of unbreakability that is borne out by the gearbox itself; we manhandled the box unmercifully in the acceleration tests and found it capable of taking the fastest slam shifts without a crunch."

In summarizing the 1800E in 1970, we noted that while the car was up to date in performance, handling and braking, the styling, accommodations and use of available space had fallen off the pace of cars at that time, and called on Volvo to design a new model.

Buying A Used Volvo: What to Look For

ALTHOUGH WE are covering rather distinct models in this "Used Car Classic," we are going to make some generalizations in this area of the story and, of course, point out individual quirks or characteristics where necessary. The engines, from the B-16 to the B-18 to the 2-liter B-20, are workhorses. R&T printed an Owner Survey covering the 1800, 122-S and early 144 Volvos in the March 1969 issue and not a single car in that report had required an engine overhaul. We were impressed by that and said in the survey, "... we're going to project 110,000 miles as an average life between overhauls in a Volvo." When searching for a good, used Volvo then, chances are the engine will not require major work, but it should be checked over anyway, and preferably by someone who knows Volvos pretty well.

Some of the early 122-S models came with a 2-speed automatic transmission that was, frankly, terrible. In keeping with the tenor of this report, we would omit those cars with an automatic gearbox from consideration in purchasing a used Volvo because of the diminished performance. The 4-speed manual gearboxes found in all of the models have proved nearly indestructible. Having said that, however, we should point out that on nearly all the early 1800 coupes, the overdrive unit shared a common oil reservoir with the gearbox. The Laycock de Normanville overdrive used 30W engine oil for lubrication and many owners never bothered to check the oil level. The result was that they fried the overdrive. So, perhaps the first question to ask in examining an 1800 is, "Does the overdrive work?"

Unlike many of our earlier "Used Car Classic" subjects, Volvos have not been prone to body rust problems. This is not to say that it never occurs, but that it's not especially common. However, we should mention the early 1800 coupes had an annoying tendency to leak water through the cowl vent in front of the windshield. Volvo is quick to point out that this was a problem on those cars that were built in England. Often, this is a result of the deterioration of the rubber molding around the vent hatch or it may be that the drain tubes that allow excess water to run out of the vent opening are clogged.

Our 1969 Owner Survey indicated that many 1800 coupe owners found the instruments rather unreliable, that cooling system problems could occur in all models, that 122-S cars tended to have window winding problems and 10 percent of those responding to the survey reported oil leaks from the differential and gearbox. The survey also showed that 12 percent of the cars had clutch difficulties.

By and large, it would be safe to say that the Volvo models covered in this "Used Car Classic" probably have fewer major problems than most any other car we've reported on in the past. The prospective buyer must exercise caution, of course, to be sure the car he or she wants is in good mechanical condition, but with the Volvos the chances seem to be better than average. ⊕

TOUGH SWEDE IS A SAFE BET

Strong, reliable and long-lasting, Volvo's sturdy Amazon gets the once-over from **Kim Henson**

L AUNCHED IN SWEDEN IN 1957, AND REACHING BRITISH shores in 1962, the distinctive Amazons have gained a thoroughly deserved reputation for being well-built, solid and enduringly reliable. Indeed, they are still eminently practical for daily use. Incidentally, although 120 Series cars are universally known as 'Amazons', officially this term should only be used in Sweden, since the name applied to a pre-war German motorcycle; its use by other makers is restricted.

All models are traditional front-engine, rear-drive machines, with a choice of power outputs and body styles. Saloons were produced in two- and four-door forms, all providing generous leg and head room in both front and rear compartments, and all having ample and sensibly shaped luggage boots, with lids opening from bumper level. Volvo also built a particularly useful estate version, endowed with a commendably large load compartment and a

Sturdy Scandinavian: Amazons first seen in the UK in 1957

horizontally divided tailgate. This incorporates a rear window which can be left open when required (to carry especially long loads, for example).

The cars were initially powered by 1.6-litre fours, but by the time imports to Britain commenced in 1962, more powerful 1.8-litre motors were installed, the engine capacity (and output) increasing again in 1968 when 2-litre units were fitted. For each capacity, both single- and twin-carburettor versions were produced. Also offered was a sporty, two-door version of the 1.8-litre variant (with a twin-carb engine from the P1800 sports car, plus uprated running gear). Designated 123GT, this model was only available during 1967/8 and so is now quite rare – and sought-after.

The GT offers excellent performance for a saloon of its time, with 60mph appearing from rest in just over ten seconds and a top speed well in excess of 100mph. It also handles well, courtesy of its uprated suspension. However, even if you can't find a GT, don't dismiss the lower-powered cars. None of the Amazons disappoint on the road; all are capable of cruising easily at 70mph, and the 2-litre versions (in particular) offer power and torque in quantity. A single-carb B20-powered model will scoot to 60mph in 12 seconds or so, and is ultimately capable of over 100mph.

These Volvos are easy to conduct in town as well as on the open road, with plenty of low-speed pulling power making light of heavy traffic and a particularly tight turning circle aiding parking manoeuvres.

Overdrive was a useful option on the saloons (and standard-fit on the GT), and it is worth finding a car so equipped, especially if you frequently undertake long journeys, to enjoy the benefits of more relaxed high-speed cruising, improved fuel consumption and, ultimately, longer engine life. If the car you buy hasn't got overdrive, don't worry, it can easily be fitted retrospectively. Amazon specialist South Service can tackle this job for you for an outlay of approximately £750, including a reconditioned gearbox and a rebuilt overdrive unit. The work immediately adds around £500 to the value of the vehicle, so is a good investment in all respects.

Many other improvements offered by specialists such as South Service make the cars even more suitable for daily use. These include uprated springs and shock absorbers, dynamo to alternator conversions, quartz headlamp installations (very easy), heated rear screens (about £200 fitted) and performance uprating. Twin carburettors, performance manifolds and exhaust systems, higher lift camshafts (and so on) are all available. The brakes are already very effective, usually only requiring the installation of competition quality pads/linings to upgrade them.

Fuel consumption is reasonable, considering the substantial weight of these cars, with realistic expectations from 1.8-litre models being 25mpg or so in town (the 2-litre B20-engined cars use a little more), and better than 30mpg on long trips.

What you should pay

£500-£1000 *Restorable, essentially sound examples*

£2500 *Tidy cars requiring cosmetic attention; add £1000 for 123GT*

£4000 *Will buy you a really good (but not concours) saloon or estate*

£5000+ *Excellent (not perfect) 123GT models – if you can find one*

History (UK) of the Amazon

1962 (Jan) *Amazons introduced to UK (launched Sweden 1957). 121: single carb, 75bhp, 1.8-litre B18A engine. 122: twin carb, 90bhp, B18D; disc front brakes*
1962 (Sept) *221 estate. Saloons gain reverse lights*
1963 (Sept) *Load compartment carpet for estate*
1964 (Oct) *New grilles feature three vertical bars; interior updates. 121 now has disc front brakes. New 131 (two-door)*
1965 (Sept) *Automatic transmission option. Output of 122 now 95bhp (higher compression ratio and new camshaft)*

Earliest known UK survivor is this 122S; twin carbs, 1.6-litres

Characterful grille on early car

1966 (May) *132 arrives (two-door version of 122)*
1966 (Oct) *Power increases; 122 now 100bhp; 121, 85bhp*
1967 (Apr) *222 introduced (twin-carburettor estate)*
1967 (May) *Debut of rally-bred 123GT, with 115bhp engine, overdrive, uprated suspension,* servo-assisted brakes, wide wheels, fog and spot lamps, full instrumentation, inc tacho
1967 (Aug) *All variants feature anti-burst door locks, collapsible steering column and servo-assisted brakes*
1968 (Sept) *B20 versions powered by 1986cc engine (as fitted to 140 Series); single carb, 90bhp; twin carb, 118bhp. Steering column lock, dual-circuit brakes standard*
1969 (Sept) *Head restraints installed as standard*
1970 (Aug) *Imports end*

FOR *Rugged build, less rust-prone than most contemporaries, long-lasting drivetrains, ahead-of-their-time safety features*
AGAINST *Some parts prices expensive and rising, best-of-bunch GTs are scarce*

FRONT WINGS
Most vulnerable section is around and above headlamp; check here first, then the rear corners

FRONT INNER WINGS/PILLARS
Severe rust along inner wing tops and around pillars usually hidden by outer wings

WHEELARCH LIPS
Rear wheelarch edges are often badly perforated; repair sections are the answer

PHOTOGRAPHY BY COLIN BURNHAM

Checklist

Engine Given regular oil and filter changes, motors are pretty bulletproof. If the engine is not noisy, and is not using oil, it's probably okay. Start the engine (if possible, from cold) and listen for knocking from the bottom end.

The valve stems do not have conventional oil seals, so high oil consumption can be due to oil escaping down the valve guides, rather than cylinder/piston wear. If the top of the engine is covered with an oily coating, indicating fuming, it is due for a rebuild.

Watch for oil leakage from the felt seal for the rear main bearing; kits are available for fitting more efficient, modern seals.

Transmission The gearbox should be quiet in all gears. If it is less noisy in top gear than in the other ratios, the bearings are worn.

Driveline vibration indicates wear in the propshaft couplings. On cars with overdrive, check that it engages when required.

Running gear Front suspension lower swivel joints may be badly worn and, in really bad cases, can fall apart, regardless of MoT.

Raise and support the front of the vehicle. Holding the wheel, attempt to move the hub units up and down – any more than just perceptible movement indicates major wear.

Check the bushes supporting the upper wishbones (those for the lower wishbones survive longer). On pre-1966 cars, check all grease points for evidence of regular attention (later cars had sealed joints).

Also check the steering idler bush, the lower seal in the steering box, the rubber and fabric coupling in the column and, on pre-1966 models, for weakening of the pressed steel arms supporting the rear axle.

The need to use a strong puller to remove the rear brake drums can discourage regular servicing. Amazons take 165x15 radials, which are now becoming more difficult to find.

Bodywork Always inspect the front inner wings and adjacent box sections along their upper edges from under the bonnet and from beneath each front wheelarch. Remedial work here is time-consuming. The same is true of badly weakened front door pillars/bulkheads; inspect closely from under the wheelarches.

With the exception of the front outriggers,

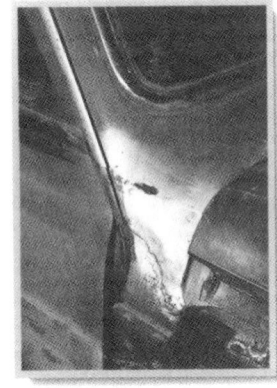

SCREEN PILLARS
Holes at the base of a screen support are bad news, indicating severe rusting within the pillar

ENGINE
If an Amazon engine has an oily coating, it is 'tired' and will soon need an overhaul

Amazon underbodies seldom suffer from serious rusting. The floors, too, usually survive well, although a leaking windscreen seal can cause havoc with the front floorpans, the windscreen pillars and the lower edge of the screen frame. There is no way to establish how bad the problem is unless the screen is removed. The rear window can leak too; look/feel for dampness of the upholstery around its sealing rubber.

Amazon engines rugged; this is a single-carb, 1.8-litre 121

The front wings are prone to rusting around the headlamps and in the upper and lower rear corners. Outer sills are rust-prone at their rear ends but the inners usually remain intact.

Open the boot and inspect the seams between the boot floor and the wheelarches, the spare wheel well and the lower extremities of the double-skinned bootlid.

Inspect the joint between the fuel tank's filler neck and the tank itself. The presence of a leak here can be confirmed by smell, and by running a finger around the seam.

The bottom edges of the rear wings and the doors are rot-prone, as are reflector mountings and the bumper ends. Some brightwork is expensive: to replace all sections of both bumpers can cost £1000.

The rear bodywork of estate models has its own potential rust traps, including the ledges beneath the rear side window rubbers, the tailgate and the rear bumper mountings. However, the fuel tanks usually survive better than on saloons.

Interiors Headlinings, steering wheels, door trims and floor coverings (usually rubber matting) all survive well. However, seat stitching (as opposed to the covering material itself) can separate, especially if the support webbing in the base of the seat has perished/stretched, allowing the seat to drop.

SEATS
Seat stitching can fail, often because webbing in seat base has deteriorated; this is easily replaced

INTERIORS
Trim and interior parts survive well; horn contacts can corrode internally

Why I own one
Graham Mills, 1966 121 four-door saloon

ADVERTISING COPY WRITER GRAHAM Mills has always liked the distinctive styling of the Amazons, and was looking for one when his 121 (single-carb, 1.8-litre saloon) became available in 1986. It had been on display in a Volvo showroom and, when the garage closed, came up for sale. Alas, the earlier history of Graham's car is not known.

The Volvo is Graham's only vehicle, and until recently he used it daily to drive to and from work in London. These days the car is used and enjoyed at weekends, and its outings include long treks into the country.

'It's great to drive,' says Graham. 'It keeps up well with traffic, and is comfortable for long distances.' Initially he found the car 'a bit wallowy' on corners, but he soon got used to this, and says that driving an Amazon is easy if you bear this characteristic in mind and drive accordingly. Other aspects appreciated by Graham are the 'excellent' visibility, the car's solidity, and the feeling that other road users keep clear of Amazons, presumably because of their perceived strength.

Graham has had no difficulty in keeping his car on the road. He uses South Service in London for spares, service and repairs, and praises the firm's ability to be able to supply any parts required.

Graham has been well pleased with his car, and thoroughly recommends Amazons as attractive, affordable, practical and reliable classics which drive well and which survive better than most.

Owner's logbook

Purchase price £1500
Miles when bought 80,000
Miles driven since 60,000
Insurance £180
Problems In 11 years, Graham's car has required only one clutch, one driveshaft, the repair and respray of one door (due to rust), and a little welding to the underbody and inner wing areas. The only other trouble has been a sticking valve
Costs None recently

Northern *exposure*

Sweden's warrior
princess is far from
svelte, but the
Volvo Amazon's
enduring character
captivates
Mick Walsh

I'm worried. Since turning 40 I've started reading classified ads from the back. I've stopped that hopeless quest to find a bargain Ace and now first scan V for Volvo. For the last 20 years I've barely given these Swedish classics a second glance and now I'm mysteriously attracted to its rugged charms. I haven't yet traded leather jackets for Harris tweed, acquired a labrador, started a slipper-only rule at home or bought a pipe… but I do crave a Volvo Amazon.

I've told sniggering friends that I'm planning a sortie into historic rallying but, to be honest, it isn't dramatic images of Tom Trana powersliding an Amazon on the '65 Acropolis or, more recently, Frank Fennell's impressive results with his loyal 120 that have triggered this Volvo yearning. No, I'm smitten by its chunky aesthetics and simple uncluttered form. That high waistline and short roof snubbed styling trends even when the first B16 was debuted at the Örebro show in September 1956. The Amazon almost invented retro styling, its plain looks harking back to a fading era of American styling. Looking like a scaled down, slightly frumpy version of the classic Chrysler 300 series – but without the flashy chrome trimmings – the Amazon could have come from Virgil Exner's studio. Some dismiss that distinctive nostril grille, pronounced V-form bonnet mould and safe, square jaw profile as more East than West, more large-scale Moskvitch than shrunken Yank but stylist Jan Wilsgaard's creation for me is as classic as a Victorinox Army knife or a Rolex Oyster watch.

I've caught myself day dreaming of Anita Ekberg lookalikes in brightly coloured ski jumpers driving two-tone Amazons to deserted sand dunes and at the video store I now ask for *Wild Strawberries* and not *The Wild One*. Something strange is happening.

Determined to make my first encounter with an Amazon memorable, I plotted a trip back to its homeland. I tracked down a '60s ice racing champ, Kent Olson, whose talents with a Volvo on the frozen tracks earned him a works Alfa drive. Olson even had a sorted Amazon readied for some fun but a leaking ozone layer and global warming has resulted in the mildest Swedish winter in decades. We had the car but no ice. Maybe next year.

First impressions are important so I planned a local adventure, a trip to the bleak wastelands of Dungeness, an unfashionable area that attracts arty types much like the Amazon. Photographer Jane Bown captured this area in a set of moody monochromes and film-maker Derek Jarman built his own shrine here at Prospect Cottage by the track to the nuclear power station. Here, as a student, an Amazon first caught my attention parked alongside a sleek DS and trendy VW Karmann Ghia in the garden of Corbusier-style beach house near Winchelsea. What better place to discover if a Volvo has soul beneath its rugged shell and no risk mechanics.

Former *C&SC* art editor John Blundell agreed to loan his loyal 1966 122S for my solo sortie to the south coast. Blundell has owned NKX 2D for 16 years. The Amazon followed a Riley One Point Five and Ford Pilot in 1972 and still Blundell determinedly refuses to own a modern. In recent times a Jag MkII has usurped the Amazon from a garage to life outside at the kerb but from its smart appearance you'd never guess it had clocked up

150,000 miles and still runs with mostly original components: "I wanted something tough with bags of character that could drive three climbing friends and equipment to Wales every weekend. It has the lock of a taxi and I always feel safe driving it. With that high door line and narrow windows, its like being in an Alvis armoured car," reports Blundell. "Amazons are fantastically strong as proven one winter on an icy road near Oxford. I lost it, clouted the kerb and ended up stationary on top of a roundabout. Despite the huge bang, the car drove on with no problems. Everything is extremely accessible and they are a joy to work on. What other classic can you walk down to a modern dealer and still buy spares off the shelf? If I ever sell it, I know I'll just end up buying another." Most years Blundell heads for the Western highlands and the Amazon always wins over the Jag for the long trip: "Overdrive is essential for motorways and the wind noise gets tedious but for those rough tracks to isolated campsites in the Hebrides or Skye, the Amazon's ground clearance is a big plus."

I collected John's trusty 122S late on a Sunday night and, highlighted by street lights, she looked as sharp as Volvo's original brochure artwork. On the first turn of the key, the engine thrummed into life: "It's never failed me," reports Blundell who mocks the opening scene in *Four Weddings and a Funeral* when Hugh Grant's 122S refuses to start. Bathed in a eerie green glow from the speedo, I soon find myself hustling the Amazon home around the Richmond one-way system, late-night clubbers applauding as rubber squealed and the body leaned into one tight roundabout. During the short run home, its creamy worm and peg steering, gutsy low-down torque and the direct action of the tall gearlever all impressed. For that brief fling in the dark, this middle-aged

I've caught myself day - dreaming of Anita Ekberg lookalikes

Volvo didn't short-change in character.

I was up before the sun for my drive to Dungeness. Insulated with thermals and loaded with favourite driving tapes – Chet Baker, Ry Cooder and Lyle Lovett – the Amazon rumbled off into sleepy Surrey suburbia. I cursed the hopeless wing and tinted rear-view mirrors as I battled with impatient truckers on the M25 but, in overdrive top, the Amazon is happy at a relaxed 70mph. Push on harder and you'll tire quicker than the car from wind noise and an intrusive engine.

By the time we reached Royal Tunbridge Wells, the horizon had turned into a dramatic Turner-esque skyscape and, for the first time, I experienced the interior in natural light. The dash looks like a cheap 1960s transistor radio with its split finish of ribbed black plastic and shiny painted metal. The ribbon speedometer with weedy numerals is pure Dagenham dustbin design and the heater controls could have come from a cheap toaster. The quality Jaeger clock mounted on the ashtrays looks totally out of place here. There's no glovebox but an undertray and deep, pleated door pockets offer useful storage. Passengers are forewarned of body roll by a chunky plastic grab handle. The floorpan is trimmed with a massive one-piece rubber mat. It's supremely functional and as

Functional interior, with rubber mat and strip speedo. Prototypes featured kidney-shaped cluster and more sporty round instruments

Engine bay is clear and practical: ideal for the home mechanic. Twin SU carburettors of 122S offered welcome extra power

basic as a Beetle but the way the controls work and feel is inspiring. The broad plastic wheel rim feels good between the fingers, the tall gearstick has a splendidly precise action and the overdrive stalk on the short steering column flicks up and down ratios with military precision. Every action reassures about the quality of Volvo engineering.

Local lorries block up the A21 and the Amazon's all iron 1.8-litre ohv four feels grumpy when you call upon all its 95 horses to get by. It's performance (0-60mph in 13 secs, 50-70mph in 8.8 secs and noisy maximum 100mph) is by no means sleepy but its gritty response makes you feel the engine is pulling more than the 21½cwt it weighs. Like all simple pushrod fours, the Volvo unit can be transformed with tweaks to the top end such as a hot cam (D- or K-type) and larger inlet valves. Race and rally boys push their engines to 7000rpm but in standard form nothing happens over 4500rpm and it's smoothest well below that. But burbling along in overdrive top at the legal limit, with the revs nudging 3500rpm, the Amazon feels as if it will roll on to Morocco and back without a thought. The engine inspires trust but it never sparkles.

The rewarding A268 rolls over the Rother Levels before dropping across to the lofty township of Rye, and here I soon learn to respect the limits of Amazon roadholding. On a fast, clear righthander we go in too fast and after near-terminal understeer and vintage-style sawing of the big wheel just get through. The low-geared steering didn't inspire but the situation is nothing a set of De Carbon shocks, tough bushing and toe-in couldn't sort.

Later, on the coastal backroad that runs around the drainage ditches started by the Saxons to reclaim the marshes, not even the bored sheep stare as the Amazon yaws around the many great bends. The body rolls but the it hugs on to the apex in a determined manner and it's easy to understand why the Amazon performed so well on track and stage in anger. The independent front end (wishbones, coil springs and anti-roll bar) and live rear axle offer a good compromise between ride and handling which, to an American buyer in the early '60s must have been a revelation. No wonder they marketed it there as the 'family sports car'. The good ship Amazon may have looked dated from its launch but would outdrive any pony-car on a mountain pass. The rear end hops around when the surface gets rough and, on faster bends, roll oversteer can be easily induced but the messages come early. Only the heavy brakes fail to inspire and one spirited dash resulted in fade.

Locals seemed to approve of the Amazon as it hummed by and I caught several looking up and smiling with approval. Resilient machines are appreciated here and during a breakfast in New Romney I was heartened to see the local light railway bringing hordes of children to school from along the coast. The sight of noisy youngsters baling out of the miniature carriages that once ferried Laurel and Hardy was straight off a Pathe newsreel. I motor back to Dungeness and soak up the isolation of this unique place. The Amazon looks strangely at home among the lonely, weatherbeaten shacks of the luggers and shrimpers. It takes a tough soul to survive here and this Swede fits in perfectly. It may never move your heart like an Alfa Giulietta saloon, or dazzle like a Citroën DS but you can't help but admire its strong and loyal charms. If your lifestyle demands you depend on a classic, there is no other choice. If I ever move somewhere as desolate but surprisingly beautiful as Romney Marsh, I wouldn't think twice. ◆

in brightly coloured ski jumpers driving two-tone Amazons

Early Amazon featured stylish two-tone paint for American market. US ad campaign ('If you'd like a good used VW see your Volvo dealer') was very agressive

Beach glamour, Swedish style. Amazon Estate P220, initially for home market only, featured redesigned suspension to lower the floor, and split tailgate

Volvo Amazon

Volvo built their reputation on this car - tough and reliable. And despite its looks, this no heavyweight slouch

Solid, dependable and long-lasting, this is the car that Volvo built their enviable reputation on. They may be foreign, but fear not, they are utterly conventional.

The first examples were imported to Britain in 1959, but for our purposes the post-1962 cars are a better bet. These had significant improvements like front disc brakes, 12-volt electrics and an 1800cc engine in place of the earlier 1600. The base 121/122 model is perfectly acceptable, but the best combination of value and performance is found in either the 122S model as featured here or the two-door 131/132S available from 1965 up until 1969 - two years after the four-door had been phased out.

Twin SU carbs on the 122S help boost power by 10bhp and there is the always desirable option of overdrive on the four-speed gearbox. Post-'65 cars had servos, too, which give a more confidence-inspiring feel to the brakes, which, though perfectly adequate, you previously had to work hard for.

Doug Payne has owned his 122S for over five years. 'It was bought for my wife,' he told us, 'but I prefer driving this to our Citroën so I tend to use it most.

'I like the fact it's so basic, there's nothing to go wrong. It's always started first touch and the only thing that's ever gone wrong was a puncture. You can hardly blame the car for that. If it did go wrong you don't need much knowledge to fix one. There are no clever electronics to worry about - even the radio is the original, and it still works.'

I took it out for a spin and was immediately struck by how comfortingly familiar it felt, despite never having driven one. There was nothing to learn, none of the awkwardness you normally feel at first in a strange car, apart from the brakes.

Doug agrees: 'Its only fault is the lack of a servo, they fitted them as standard from the next year. I was going to fit one, but got used to the brakes before I got round to it.'

Although Amazons look big and heavy, they're not. Their 2,400lb is the same as a new Escort. 'It's lively enough,' Doug feels, 'but you've got to use the gearbox to get the best out of it. It cruises best at 65. On a run, it'll do 28-30mpg which can't be bad.'

'I'm starting to get the urge to own something different. If someone came along with £2500 I'd wave the car goodbye. But I'm not too worried if it stays.' Doug can be contacted on 01733 348160.

Club

Volvo Enthusiasts Club,
4 Goonbell, St Agnes, Cornwall TR5 0PH.

Volvo Owners Club,
18 Macaulay Ave, Portsmouth, Hants. PO6 4NY.

'There was nothing to learn, none of the awkwardness you normally feel at first in a strange car '

SPECIFICATION (122S-B18)	
Engine	1780cc 4-cyl
Brakes	Disc/drum
Top speed	93mph
0-60mph	14.5secs
Fuel cons.	25-30mpg
C1-2 price	£1750-3400